Polymer-Solvent Molecular Compounds

To my wife Chantal, and my daughters Delphine and Aurélie

To the memory of my parents

Cover pictures Left: spherulites from syndiotactic polystyrene/*trans*-decalin compound. **Right:** fibrils from syndiotatic polystyrene/naphthalene compounds after naphthalene sublimation.

Polymer-Solvent Molecular Compounds

Jean-Michel Guenet

ELSEVIER

AMSTERDAM • BOSTON • HEIDELBERG • LONDON
NEW YORK • OXFORD • PARIS • SAN DIEGO
SAN FRANCISCO • SINGAPORE • SYDNEY • TOKYO

Elsevier
Linacre House, Jordan Hill, Oxford OX2 8DP, UK
Radarweg 29, PO Box 211, 1000 AE Amsterdam, The Netherlands

First edition 2008

Copyright © 2008 Elsevier Ltd. All rights reserved

No part of this publication may be reproduced, stored in a retrieval system
or transmitted in any form or by any means electronic, mechanical, photocopying,
recording or otherwise without the prior written permission of the publisher

Permissions may be sought directly from Elsevier's Science & Technology Rights
Department in Oxford, UK: phone (+44) (0) 1865 843830; fax (+44) (0) 1865 853333;
email: permissions@elsevier.com. Alternatively you can submit your request online by
visiting the Elsevier web site at http://elsevier.com/locate/permissions, and selecting
Obtaining permission to use Elsevier material

Notice
No responsibility is assumed by the publisher for any injury and/or damage to persons
or property as a matter of products liability, negligence or otherwise, or from any use
or operation of any methods, products, instructions or ideas contained in the material
herein. Because of rapid advances in the medical sciences, in particular, independent
verification of diagnoses and drug dosages should be made

British Library Cataloguing in Publication Data
A catalogue record for this book is available from the British Library

Library of Congress Cataloging-in-Publication Data
A catalog record for this book is available from the Library of Congress

ISBN–13: 978-0-08-045144-2

For information on all Elsevier publications
visit our web site at books.elsevier.com

Printed and bound in Slovenia

08 09 10   10 9 8 7 6 5 4 3 2 1

Working together to grow
libraries in developing countries

www.elsevier.com | www.bookaid.org | www.sabre.org

ELSEVIER   BOOK AID International   Sabre Foundation

Contents

Foreword	ix
About the Author	xi
Introduction	xii
PART I: Investigation tools	**1**
1 Phase diagrams	**3**
Gibbs phase rules	3
Types of transformation	6
The solvent crystallization method	10
Metastability	10
Effect of pressure	10
Spinodal decomposition	11
Incongruently melting compounds	12
Size and nucleation effects	13
Flory's theory	14
2 Scattering and diffraction	**16**
Theoretical	16
Scattering by cylindrical objects	19
Scattering and diffraction by helices	20
Effect of contrast	24
Scattering by polydispersed systems	25
Scattering by semi-rigid systems	27
Scattering by randomly aggregated systems	29
Scattering by a regularly folded chain	29
Experimental aspect	30
Neutron scattering and diffraction	30
X-ray scattering and diffraction	32
Light scattering	33
3 NMR Spectroscopy	**36**

4	**Some theoretical approaches**	**41**
	A fractal view of thermoreversible gels	41
	Fibril fractal dimension vs chain persistence length	43
	The critical gelation concentration C_{gel}	44
	Fibril stability	45
	Rheological behaviour: modulus vs concentration	47
	On the definition of thermoreversible gels	49
	Chain conformation	51
	Scale-free networks	53

PART II: Biopolymer complexes — 55

5	**Cellulose**	**57**
	Cellulose I and II	57
	Complexes of cellulose	60
	Complexes with soda. Cellulose I–cellulose II transformation	60
	Complexes with liquid ammonia	64
	Complexes with amines	64
	Complexes with inorganic acids	68
	Cellulose III	68

6	**Agarose**	**70**
	The sol state	72
	The gel state	75
	Thermodynamics	75
	Gel nanostructure	79
	About the sol–gel transition	81
	Gel rheology	82
	Gels from modified agarose	83

7	**Amylose**	**87**
	A-amylose	88
	B-amylose	90
	V-amylose	91

8	**Chitin**	**96**

9	**Deoxyribonucleic acid (DNA)**	**99**

PART III: Synthetic polymer complexes — 105

10	**Atactic poly[vinyl chloride]**	**107**
	The gel state	110
	Formation and melting behaviour	110
	Gel morphology	113
	Gel nanostructure	114
	Nanostructure of gels from chemically modified PVCs	121
	The pregel state	126
	Molecular structure of PVC aggregates	126

	Gel rheology	131
	PVC gels	131
	Gels from chemically modified PVCs	136
11	**Polyaniline**	**137**
	Properties of solutions	137
	Solid state molecular structure	141
	Gels	145
12	**Poly[vinylidene fluoride]**	**150**
13	**Liquid-crystalline polymers**	**160**
	PART IV: Intercalates and clathrates	**163**
14	**Poly ethylene oxide**	**165**
	Clathrates and inclusion compounds	166
	Intercalates	169
	Intercalates through molecular recognition	169
	PEO/hydroxybenzene compounds	172
15	**Isotactic polystyrene**	**180**
	Thermoreversible gels	180
	Thermodynamics: temperature–concentration phase diagrams	181
	Morphology and molecular structure	186
	Rheological properties	193
	Assemblies of spherulites	195
	Thermodynamics	196
	Morphology and molecular structure	198
	Solvent-induced structures	201
	Hybrid systems: polymer and self-assembled structures. metallogels	203
16	**Syndiotactic polystyrene**	**210**
	Solvent-induced compounds	214
	Solvent sorption and crystallization	214
	Thermal stability	218
	Morphology	220
	Crystalline structures	222
	Solution-cast compounds	226
	Single crystals	226
	Spherulitic textures. Phase diagrams	228
	Compounds from high-melting-point solvents	232
	Thermoreversible gels	242
	sPS/benzene	242
	sPS/toluene	246
	sPS/chloroform	247
	sPS/tetrahydronaphthalene (tetraline)	248
	Effect of formation path on melting behaviour	250
	Multiporous networks	251

17	**Atactic polystyrene**	**255**
	Thermodynamics	255
	Molecular structure	260
	Mechanical properties	263
18	**Syndiotactic poly[*p*-methyl styrene]**	**265**
	Non-solvated forms	265
	Clathrates	266
19	**Stereoregular poly[methyl methacrylate]s**	**271**
	Evidence for compound formation	273
	Temperature–concentration phase diagrams	273
	Spectroscopic methods	276
	Gel molecular structure	278
	Gel rheology	283
Conclusion		**284**
References		**286**
Index		307

Foreword

Crystallizable polymers represent a large proportion of the polymers used for manufacturing all types of products, and, as a result, have received the continued attention of scientists these past 60 years. The study of molecular compounds from crystallizable polymers, particularly from synthetic polymers, has been the growing subject of interest for the last 20 years or so. This is the case probably because many systems have been only recently discovered, as is the case of poly[oxyethylene] intercalates, or because the polymers have been newly synthesized, as is the case with syndiotactic polystyrene and syndiotactic poly[*para*-methyl styrene]. The present book is intended for bridging the gap and giving a description of the systems that have been extensively investigated. This book is divided into four parts. Part I is devoted to the presentation of important investigation techniques and some theoretical approaches. I have presented here what I felt was necessary for the understanding of this book. As will be discovered, temperature–phase diagrams are abundantly described, so a short section on how to map out and to decipher phase diagrams seemed to me necessary. Similarly, scattering and diffraction techniques are commonly used in the investigation of these systems and therefore a section on scattering and diffraction by rods, rod-like structures, and helices has been included, as the equations dealing with such systems seldom appear altogether in papers or books. A short section on a special aspect of NMR not frequently used but powerful for deciding whether compound formation takes place or not has also been written with the help of Professor J. Spěváček. Finally, a section is given on some theoretical approaches, which pertains essentially to the fibrillar gels. Parts II–IV describe the different polymers and biopolymers that are known to produce molecular compounds and that have been, and still are, extensively studied. Part II is therefore devoted to biopolymers, as these were certainly the first polymers known in relation to the production of molecular compounds, chiefly with water. These compounds very often already exist in the living systems from which the biopolymers have been extracted. In this case compound formation arises primarily from hydrogen bonding of the solvent molecules, in most cases water molecules, with the biopolymer. Part III deals with synthetic polymers where compound formation is either due to hydrogen bonding or to electrostatic interactions. Finally, Part IV describes intercalate and clathrate systems in which compound formation is mainly due to a molecular recognition process.

This book has been written over a period of several years, both in Strasbourg at Institut Charles Sadron, but also in my holiday house in Salviac in southern France during summer vacations. So, I have clearly to thank my wife Chantal and my two daughters, Aurélie and Delphine, for coping with a husband and father stuck to his computer for hours, and being absent-minded for the rest of the time. My thanks also go to my colleagues who gave me valuable advice and/or read some chapters of this book to provide me with necessary criticisms and corrections. In particular, it is my pleasure to thank Dr Annette Thierry for reading the text on clathrates and intercalates, Dr Cyrille Rochas for perusing the section that deals with biopolymers, and Dr Alberto Saiani and Professor Yves Grohens for reading the section on PMMA gels. As aforementioned, Professor J. Spěváček helped me write the short section on NMR. I also benefited from discussions on poly[oxyethylene] intercalates with Professor Marcel Dosiere, who invited me as a guest-professor to Mons-Hainaut University, Belgium, on several periods. Discussions with Professor Hideyuki Itagaki on polystyrene systems during a stay as guest-professor in his laboratory at Shizuoka University, Japan, were more than helpful. I am also grateful to Professor Arun Kumar Nandi from IACS, Calcutta, who introduced me to the realm of poly[vinylidene fluoride]. Professor Larry Belfiore was also kind enough to read some of the chapters of this book and suggest appropriate corrections. I also wish to express my gratitude to my former PhD students and post-doctoral fellows Mohamed Ramzi, Christophe Daniel, Alberto Saiani, Sandrine Poux, Daniel Lopez, Biswajit Ray, Chinnuswamy Viswanathan, and, in particular, Sudip Malik, with whom we carried out investigations on syndiotactic polystyrene compounds, and who was of considerable help to me. Other colleagues were kind enough to provide me with the literature on different topics: Dr Henri Chanzy on cellulose and amylose, Dr Pascal Damman on poly[oxyethylene] intercalates, Professor Olli Ikkala on polyaniline, Professor Yashin Cohen on liquid-crystalline polymers, Professor Alain Domard on chitine, and Professor Vittoria Vittoria, Professor Gaetano Guerra, Professor Vittorio Petraccone, Dr Oreste Tarallo and Dr Christophe Daniel on syndiotactic polystyrene and syndiotactic poly[*para*-methyl styrene]. Note that, as far as it was feasible, figures appearing in this book have been redrawn from the original data for the sake of standardization. The references to the papers from which these data have been drawn are given when appropriate in the figure caption.

Finally, although I have tried to be as objective as possible, I would like to emphasize that this book is not just a review but a personal view on the domain.

About the Author

Professor Jean-Michel Guenet is Directeur de Recherche CNRS at Institut Charles Sadron, Strasbourg, France, a CNRS-owned laboratory associated with the Université Louis Pasteur. He was born in 1951 in Blanc-Mesnil, which is a suburb of Paris, France. He graduated in 1974 from Paris XIII University as a material science engineer and obtained a PhD degree in 1980 at Louis Pasteur University in Strasbourg. He was a post-doctoral fellow for one year at Bristol University, UK, with Professor A. Keller on a grant from the Royal Society. He was a visiting scientist at NIST with Professor G.B. McKenna, Gaithersburg, USA, in 1985, a visiting professor at Université de Mons-Hainaut, Belgium, with Professor M. Dosière from 1995 to 2004, and an invited professor at Shizuoka University, Japan, with Professor H. Itagaki in 2002. Apart from being the author of about 150 papers, he has also written a monograph on thermoreversible gels published by Academic Press in 1992 (*Thermoreversible Gelation of Polymers and Biopolymers*). In 1990, he was awarded the Dillon Medal of the American Physical Society for his work on polymer gels. He was the first non-American to receive this award. He is also the founder of a series of conferences entitled *Polymer-solvent Complexes and Intercalates*.

On ne doit pas exiger de cette classe d'hommes (les chercheurs) qu'ils professent et qu'ils enseignent, mais qu'ils inventent et qu'ils publient: car les découvertes sont rares; elles sont le fruit d'un long travail, de pénibles méditations; elles ne se commandent pas, et ne sont pas susceptibles d'être assujetties aux heures périodiques d'un cours public.

Antoine-Laurent de Lavoisier, *Réflexions sur l'instruction publique*

Introduction

The commonly accepted definition for a *compound* is a *chemical substance consisting of two or more elements chemically bonded with a well-defined ratio*. This ratio is termed the *stoichiometry* (from the Greek στοιχειον or *stoicheion*, which means 'element' and μετρειν or *metrein*, which means 'measure'). In many cases, the bonds are so strong (covalent bonds for instance) that the compounds exist in the three states, solid, liquid and gaseous (e.g. C_6H_6). The compounds of interest here are those that exist in the solid state but vanish in the liquid state, or even before melting through a *solid–solid transformation* process. This is so simply because the bonds between atoms or molecules are weaker and of the order of magnitude of kT. These systems are therefore not chemical compounds and are usually designated as *molecular compounds*. As will be discovered throughout this book, molecular compounds can also be designated as *crystallo-solvates, complexes, clathrates, intercalates,* or *inclusion compounds,* the use of which depends upon the system and/or the field of research.

Molecular compounds have been known for decades for atomic systems, such as some metallic alloys ($CuTi_2$, $AuSn$, $AuIn$, etc.), for minerals, such as salt hydrates ($NaCl.2H_2O$ etc.) and clays, and even for organic molecules such as urea inclusion compounds. There exist two categories of compound that display two different types of phase behaviour which are best described by the temperature–concentration (or composition) phase diagram (Figure 1): the *congruently melting* type, and the *incongruently melting* type.

Congruently melting compounds behave as a pure substance in that they have a well-defined melting temperature at the stoichiometric composition. For instance the alloy Cr_7C_3 melts spontaneously at $T = 1745°C$ without prior solid transformation (Anderson 1987).

Incongruently melting compounds transform into another phase (solid phase or another compound of different stoichiometry) before the occurrence of the macroscopic melting

Introduction

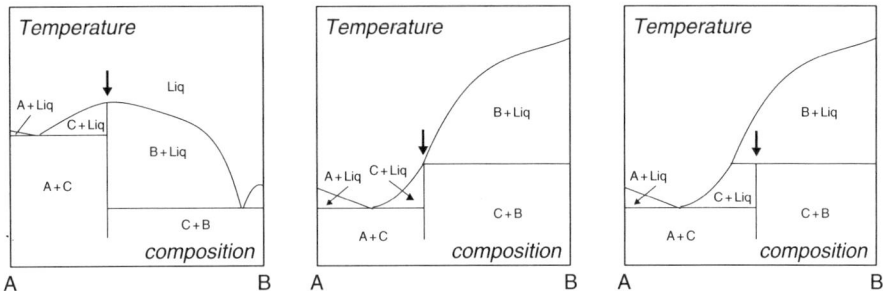

Fig. 1. Schematic temperature–composition phase diagrams for binary systems. **Left:** congruently melting compound; **right:** incongruently melting compound; **middle:** singular-melting compound. The phases in the different domains are labelled; C = compound, A and B = the two components of the binary system. The arrows indicate compound stoichiometry.

of the sample through a *solid–solid phase transition*. For instance NaCl.2H$_2$O transforms into pure NaCl crystals by releasing water molecules at $T = 0°C$. Similarly, the alloy Cr$_3$C$_2$ transforms into graphite and a liquid (Anderson 1987). Note that *peritectic transformation* is very often used for designating the incongruently melting process although it is not, strictly speaking, correct. As a rule, a peritectic system is closer to the case of a solid solution. Incongruently melting compounds are only a special case of peritectic.

The intermediate case between the congruently melting and the incongruently melting compound, designated as *singular-melting compound*, actually behaves like a congruently melting compound, in that no transformation of the type *compound→solid phase + liquid* occurs, but direct melting. This is the case for alloys such as AuSb$_2$ (Chevalier 1989).

Molecular compounds possess a molecular arrangement in the solid state where each position of the crystalline lattice is occupied by a given atom or molecule (see Figure 2). The crystal unit cell of a molecular compound is not that of any of its constituent but one of its own. As a result, the diffraction pattern of a molecular compound differs from that of any of its components, and gives therefore an additional hint, together with the T–C

Fig. 2. Schematic representation by means of a 2D lattice of the molecular arrangement of a pure substance (**left**), of the solid solution of a binary mixture (**middle**) and of a binary molecular compound (**right**); here the stoichiometry is 1 grey molecule (atom)/4 black molecules (atoms).

phase diagram, of compound formation. This is clearly at variance with a *solid solution* where the atoms or molecules are dispersed randomly on positions of the crystalline lattice of the main component. In this case, increasing the minor component eventually leads to a phase separation. In the case of a molecular compound, gradual addition of the minor component increases the fraction of compound until the stoichiometric composition is reached. On reaching the liquid state through heating, all the components are randomly dispersed, and therefore what was previously a molecular compound with specific properties has simply become a homogeneous, liquid mixture.

In the case of macromolecules, aqueous molecular compounds were already known at the turn of the twentieth century for biopolymers such as cellulose and amylose. Some compounds of these polysaccharides with organic molecules, such as amines, were also studied. Since the synthesis of synthetic crystallizable polymers such as polyethylene, and, later, stereoregular polymers such as isotactic polypropylene, much attention has been paid to their crystallization behaviour, as well as to their molecular structure. In particular, the chain trajectory within the heterogeneous semi-crystalline environment was a matter of bitter debate between those scientists promoting regular chain-folding in the lamellar crystals, and those advocating random re-entry. The controversy was just fading away as growing evidence for regular folding were collected, when polymer–solvent intercalates were discovered for poly[oxyethylene] (PEO) (Point and Coutelier 1985; Point et al. 1986a; Point and Demaret 1987) and suggested for isotactic polystyrene (iPS) thermoreversible gels (Guenet 1986; Guenet and McKenna 1988). In the former case, this discovery was largely fortuitous. It turned out that the supposed growth of PEO single crystals in dilute solution of this polymer in a mixture of isomers actually produced PEO intercalates with one of the isomers of the solvent (see Intercalates in Chapter 14). The large difference in melting point between pure PEO crystals ($\sim 63°C$) and PEO intercalates ($\sim 90°C$) settled the issue, which might otherwise have gone unnoticed. Since then, other molecular compounds have been discovered with the newly synthesized syndiotactic polystyrene (Immirzi et al. 1988; Vittoria et al. 1988), but also with poly[vinylidene fluoride], poly[vinyl chloride], and many others that will be detailed in this book.

In the early days of polymer science, most known compounds were formed between biopolymers and water through hydrogen bonds. As a result, it was generally admitted that strong interactions, such as hydrogen bonds, were required. Yet, simple organic solvents could also do the trick thanks to a molecular recognition process between the microstructure created by the helical form of the polymer and the shape of the solvent (intercalates). Schematically speaking, one could therefore distinguish between two classes of compounds: *enthalpic* compounds formed through strong interactions, either hydrogen bonds or electrostatic interactions, and *entropic* compounds for which the driving force is molecular recognition and only involves weaker interactions of the van der Waals type. To be sure there exist hybrid systems where both mechanisms come into play. This is the case with some PEO intercalate compounds where hydrogen bonds are also involved.

The morphologies of these polymer–solvent molecular compounds can be of the *spherulitic* type or of the *fibrillar* type for the same polymer depending on the solvent used. The latter situation usually promotes the formation of thermoreversible gels. As will be detailed in this book, the interaction at the molecular level between the polymer and the solvent molecules is mainly responsible for these morphologies by altering the chain persistence length.

Despite the wealth of investigations, there still exists a common, sometimes deeply rooted, prejudice in that crystallization and, in particular, thermoreversible gelation would be favoured in poor solvents on account of a phase separation process. This comes from a view inherited from Flory's theory on polymer solutions for which *liquid–liquid phase separation* occurs on cooling below the θ-temperature, namely on crossing the *binodal line* (Flory 1953). This line delimits a so-called miscibility gap, within which solutions of crystallizable polymers should supposedly be quenched to produce gelation. In early studies (Feke and Prins 1974), it was even considered that thermoreversible gelation resulted from *spinodal decomposition*, one of the two possible mechanisms of the liquid–liquid phase separation process (see Chapter 1). Yet, crystallization and gel formation are more likely to proceed through *liquid–solid phase separation* on cooling below the *liquidus* line, which does not imply that the solvent is necessarily poor. On the contrary, compounds can form in good solvents, 'good' in the sense of Flory's theory, namely well above the θ-temperature of the polymer–solvent couple. Indeed, a good solvent is capable of promoting aggregation of the type *polymer–solvent–polymer*, through molecular recognition for instance, particularly if the organized state (crystal in most cases) is more stable than the disordered state (the liquid). This is the case for many systems, as will be discovered throughout this book.

Before presenting polymer–solvent molecular compounds, the main investigation tools widely used for investigating and characterizing these systems are briefly described in the four chapters of part I.

The butterflies spread their sails on the sea of light. Lilies and jasmines surge up on the crest of the waves of light, The light is shattered into gold on every clouds, and it scatters gems in profusion.

Rabindranath TAGORE in *"Gitanjali"* (1912)

PART I

Investigation tools

The easiest and most straightforward way of investigating molecular compounds is initially to map out the temperature–concentration phase diagram. This diagram provides invaluable information about compound formation, the different phases and domains, and, in many cases, the stoichiometry(ies). This should be the first step of any study. Once the various phases and domains have been identified, then and only then should further investigations be carried out. In particular, molecular structure and morphology can be investigated by use of scattering and diffraction techniques that will help establish the nature of the crystal unit cell and the type of helical structures involved. There will be a feedback which will help to confirm the outcomes from the phase diagram and resolve any ambiguities. The chapters in Part I are therefore dedicated to the presentation of the various techniques and theories – not necessarily detailed in standard textbooks – that may be useful for achieving these investigations.

CHAPTER 1

Phase diagrams

Thermal analysis is a powerful tool for establishing temperature–concentration phase diagrams (T–C phase diagram), although its outcomes often require it to be complemented with optical microscopy observations and/or spectroscopic investigations (infrared spectroscopy, X-ray diffraction). Thermal analysis can be carried out by means of classic calorimetry or differential scanning calorimetry. These techniques allow the observation of thermal events such as first-order transitions (melting, solid transformation, gelation), second-order transitions (lambda transition) or pseudo second-order transitions (glass transition).

Thermal analysis will provide two parameters: the temperature, T, at which the transition occurs and the associated latent heat, ΔH, for first-order transitions, or the associated jump of heat capacity, ΔC_p, for second-order transitions. The knowledge of T and ΔH, or T and ΔC_p, allows one to construct the temperature–concentration phase diagram. This section deals with a short description of how phase diagrams must be established and how they should be interpreted. Basically, these phase diagrams are the same as those used for characterizing metal alloys and for which a subsequent literature is available.

Gibbs phase rules

In 1876 Gibbs worked out in one stroke the theory about phase equilibria, which paved the way to phase diagrams. The most important variable derived from this theory is the *variance*, v_G, which stands for the number of parameters (including pressure and temperature) that can be varied without modification of the state of a system. The expression of the variance is obtained by considering that for N components distributed in φ phases in thermal equilibrium, the chemical potential of one component is the same in all the phases:

$$\mu_A^\alpha = \mu_A^\beta = \mu_A^\gamma = \ldots$$

$$\mu_B^\alpha = \mu_B^\beta = \mu_B^\gamma = \ldots$$

$$\mu_C^\alpha = \mu_C^\beta = \mu_C^\gamma = \ldots$$

$$\ldots\ldots\ldots\ldots\ldots$$

There are, accordingly, $(\varphi - 1)$ independent equations for each component, and a total of $N(\varphi - 1)$ relations. In addition, there exist $\varphi(N - 1)$ independent variables related to concentrations of the components in all the phases. Including pressure and temperature, the number of independent variables amounts to $2 + \varphi(N - 1)$.

The variance is then simply the number of independent variables minus the number of relations between them:

$$v_G = N - \varphi + 2 \tag{1.1}$$

or at constant pressure, which is often the case with polymer–solvent systems:

$$v_G = N - \varphi + 1 \tag{1.2}$$

The variance is nothing but the dimension of the locus where phases coexist. Consider a two-component system at constant pressure. If $\varphi = 1$ (one phase) then $v_G = 2$, which means that a phase exists on a locus which is a **surface** (a two-dimension object). If $\varphi = 2$ (two phases), then $v_G = 1$, which implies that the locus is now a **line** (a one-dimension object). Finally, if $\varphi = 3$ (three phases) then $v_G = 0$ and the locus reduces to a **point** (a zero-dimension object). A two-component system can therefore only produce a maximum of three phases at constant pressure.

The dimensional aspect of the variance allows one to account quite simply for the occurrence of ***temperature-non-variant transformations***. These transformations always take place at the same temperature independent of the concentration. An illustration of this statement is given in Figure 1.1, which shows two possible types of phase diagram (one wrong and one correct) representing a eutectic system. In the upper diagram the transformation from $\alpha + \beta$ into $\beta + liquid$ occurs on a line with no special specifications.

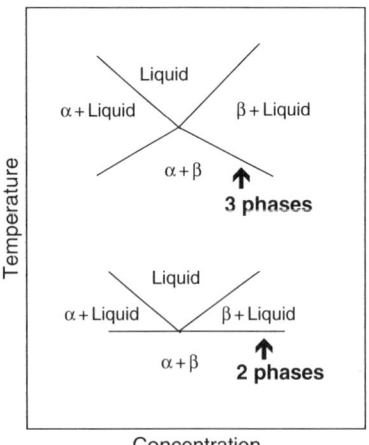

Fig. 1.1. Two schematic phase diagrams for a eutectic melting, one wrong (**upper**), and one correct (**lower**). See text for details.

As a result, three phases coexist on this line: $\alpha + \beta +$ liquid in conflict with the value of the variance, which allows only two phases to coexist on a line.

The correct way to draw the phase diagram is shown in the lower diagram of Figure 1.1. Here, the transformation from $\alpha + \beta$ into $\beta + liquid$ occurs on a line but at constant temperature. As a result, on this line only two phases coexist, in the present case $\beta + eutectic\ liquid$ (this means liquid of eutectic composition), which complies with the value of the variance.

Another important outcome from Gibbs' theory on phase diagrams is **Tamman's plot**, which is obtained through the application of the **lever rule**. **The enthalpy (or latent heat of fusion) associated with a first-order transformation must vary linearly as a function of concentration**. Regrettably, this plot is all too often merely overlooked, although it provides valuable information when appropriately used (e.g. see Carbonnel et al. 1970; Rosso et al. 1973). For instance, as will be seen below, the stoichiometry of compounds can be easily determined. It is worth emphasizing that any significant departure from a linear variation of the latent heat with concentration should be examined carefully as it may convey something important (variation of crystallinity with concentration, intermingling of two endotherms that are not well-resolved, etc.).

So far we have considered systems at equilibrium and one may wonder to what extent the Gibbs phase rules can be applied to non-equilibrium systems. The notion of equilibrium essentially means that the phases formed are stable, and display no further evolution with time. This will obviously depend upon the mobility within the system: a glass of crystalline polymer seems to be *stable*, although actually being *metastable*, simply because motions are frozen, which impairs crystallization. Again, thermal analysis allows one to distinguish between a *stable* and a *metastable* phase. ***If, on heating, a phase transforms through the occurrence of an endotherm, then the phase was stable***. Conversely, observation of an exotherm while heating therefore implies that the phase was metastable (note that the reverse statement holds true on cooling: the formation of stable phases produces exotherms). This obviously should be tested as a function of the scanning rate.

As opposed to what is commonly believed, stable phases are not necessarily formed while very slowly cooling a homogeneous solution. Domains where different types of stable phases coexist can be attained on cooling very rapidly. The capability of a system to reach such domains depends strongly upon the formation kinetics of the phases occurring at high temperature. A well-known example is the competition between *crystallization* and *liquid–liquid phase separation*: if the crystallization rate is slow enough, the system can reach the miscibility gap, thus allowing liquid–liquid phase separation to take place (e.g. see Zarzicky 1970). As will be seen in this book, the same can occur between *crystallization* and *gelation*. Also, as will be shown in the case of sPS and/or iPS compounds, the temperature–concentration phase diagram is independent of the path followed for reaching a T,C coordinate: cooling solutions of concentration C_o to a given temperature T_o is equivalent to decreasing the concentration from $C = 1$ to $C = C_o$ at constant temperature T_o provided that the systems were totally amorphous to start with (Ray et al. 2002; El Hasri et al. 2004).

As has been discussed both theoretically and experimentally by Guenet for this scenario (Guenet 1996b; Guenet 2003), non-variant transformations can be observed for systems prepared under drastic cooling conditions provided that the quenching temperature is the same (see also Koeningsveld et al. 1990). Correspondingly, if Gibbs phase rules apply, the Tamman plot gives information about the different phases and their relative proportions.

As a rule, it is equally important to establish both the formation phase diagram and the melting phase diagram (which is sometimes referred to as a *state diagram*). Their shapes should be similar, and the formation enthalpies should be equal to the transformation enthalpies.

Types of transformation

The various types of transformations that can be observed in polymer solutions forming polymer–solvent compounds are brought together in the hypothetical phase diagram shown in Figure 1.2. Starting from the low-concentration polymer side, these are: *liquid–liquid phase separation* giving rise to a *monotectic transition* (M), *congruently melting*

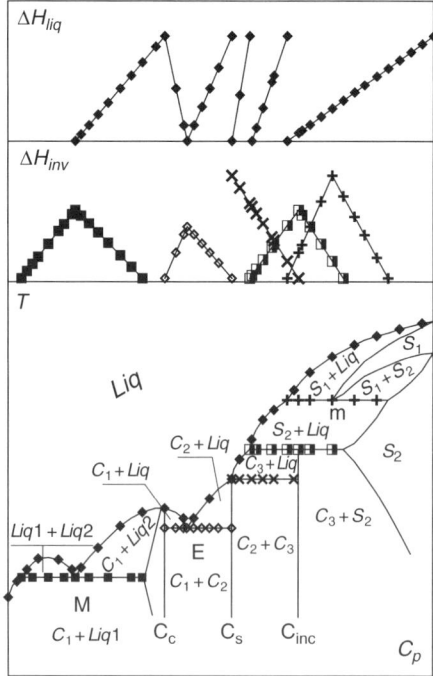

Fig. 1.2. Schematic T–C phase diagram. The thermal event and its associated enthalpy are shown with the same symbol. ΔH_{liq} = enthalpy associated with the liquidus lines; ΔH_{inv} = enthalpy connected to the non-variant transformations. For further details see text.

compound (C_c), *eutectic melting* between compound C_1 and compound C_2 (E), *singularly melting compound* (C_s), *incongruently melting compound* (C_{inc}), and *metatectic transformation* (m).

We shall now describe the events occurring for each case, together with the effect on the associated enthalpy. For the sake of clarity, the transition temperatures and the associated enthalpies are shown in Figure 1.2 using the same symbol.

A *monotectic transition* occurs when a liquidus line meets with a binodal line below which liquid–liquid phase separation takes place (this is why it is sometimes referred to as the miscibility gap). At the monotectic temperature and the monotectic concentration the reaction is:

$$C_1 + Liq1 \longrightarrow C_1 + Liq2$$

The enthalpy associated with the monotectic transition increases first, becomes maximum at the monotectic concentration, and then decreases to zero. The concentration at which this enthalpy becomes zero corresponds to the concentration of the solid solution formed between the solvent and compound C_1 (a solid solution is a random dispersion of one component into the other).

For polymers, the binodal line is located at low concentration as the maximum of the binodal for monodisperse polymers; i.e. the critical temperature T_c, is written (Flory 1953):

$$\frac{1}{T_c} = \frac{1}{\theta}\left[1 + \frac{1}{\varphi_1}\left(\frac{1}{\sqrt{x}} + \frac{1}{2x}\right)\right] \quad (1.3)$$

where φ_1 is the polymer volume fraction, x is proportional to the polymer molecular weight, and $T_c = \theta$ for infinite molecular weight. The maximum of the binodal cannot therefore be seen nor can its effects be observed at concentrations as high as 50% (w/w) as opposed to what is stated by Roels et al. 1994.

The way in which liquid–liquid phase separation proceeds will be discussed later.

The observation of a *congruently melting compound* when mapping out the phase diagram implies the existence of a mutual organization between the polymer and the solvent. This type of compound behaves as a pure substance as the reaction is:

$$C_1 \longrightarrow Liq$$

Here, the enthalpy associated with the liquidus line (terminal melting) is maximum at the stoichiometric concentration C_γ (Figure 1.3). In principle, the **thermodynamic** stoichiometry of the compound, namely the number of solvent molecules/monomer, S_φ, can be calculated from C_γ through the simple equation:

$$S_\varphi = \frac{1 - C_\gamma}{C_\gamma} \times \frac{m_p}{m_s} \quad (1.4)$$

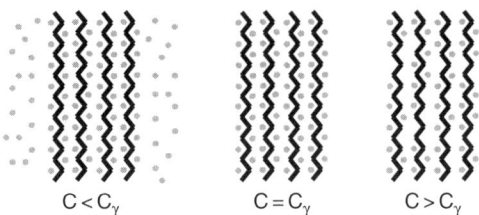

Fig. 1.3. Schematic representation of a compound system for a concentration below the stoichiometric concentration where there is an excess of solvent molecules (**left**), at the stoichiometric concentration (**middle**), and above the stoichiometric concentration (**right**) where there is a deficit of solvent molecules.

where m_p and m_s are the molar mass of the monomer and of the solvent, respectively. In the case of polymers, one has to consider two other aspects for calculation of the stoichiometric composition: (1) the degree of crystallinity is not 100%; (2) the amorphous domains may be more solvated than the compound.

Note 1.1 The semi-crystalline phase

When dealing with crystalline polymers, the crystalline lamellae are intimately interconnected with amorphous domains, so the fate of the former depends on the fate of the latter. They can therefore be considered as one phase, namely the semi-crystalline phase. Under these conditions, a solid phase will consist of crystalline polymer lamellae alternating with amorphous domains containing (solid solution) or not containing (solid) solvent molecules. The solvated amorphous phase not connected in one way or another with the semi-crystalline phase is then designated as the liquid phase. Note that the crystalline phase can also be a compound.

Taking X_c as the degree of crystallinity and S_c the crystal stoichiometry, namely the number of solvent molecules per monomer in the compound phase (namely the real value of the compound stoichiometry), and S_a as the number of solvent molecules per monomer in the amorphous phase, one obtains:

$$S_c = \frac{1 - C_\gamma[1 + (1 - X_c)S_a(m_s/m_p)]}{X_c C_\gamma} \times \frac{m_p}{m_s} \tag{1.5}$$

If $X_c = 1$ or $S_c = S_a$, one then retrieves equation (1.4). Now, if the amorphous phase is more solvated than the compound, the 'thermodynamic' stoichiometry derived from equation (1.4) is likely to be larger than the actual value calculated from equation (1.5). This may pertain to sPS compounds for which S_φ and S_c are often at variance (see Chapter 16).

The events occurring at a *singularly melting compound* are basically the same as those observed for a congruently melting compound. In fact, this type of compound is simply intermediate between congruently melting and incongruently melting compounds. The enthalpy associated with the melting of C_2 is therefore maximum at the stoichiometric concentration, and then decreases to zero to the stoichiometric concentration of C_3 in the present case. The enthalpy related to the liquidus line starts from zero and goes linearly up to the concentration at which the incongruently melting line meets the liquidus.

An *incongruently melting compound* transforms into a liquid and a solid solution (or another compound) before melting:

$$C_2 \longrightarrow S_2 + Liq$$

There is both partial melting of the system and release of solvent from the crystalline lattice of the compound (Figure 1.4).

The enthalpy connected with the incongruently melting line displays a sawtooth variation with a maximum corresponding to the stoichiometric concentration. The enthalpy associated with the liquidus line varies in the same fashion as that shown for the singularly melting compound.

Finally, *metatectic melting* can be observed when a polymer possesses two stable crystalline forms in the solid state. At the metatectic point the reaction is:

$$S_2 + Liq \longrightarrow S_1$$

The enthalpy connected with the metatectic transition again exhibits a sawtooth variation. The enthalpy associated with the liquidus line starts from zero and is maximum at $C_p = 1$.

Eutectic melting can also occur between two compounds. Usually, it consists of interspersed domains of one domain within the other. When one type of domain starts melting it acts as a 'solvent' to the other type – hence the drop of global melting temperature at the eutectic point. For the sake of simplicity, namely having all types of transformation on one figure, the eutectic case has been illustrated in Figure 1.2 by considering an eutectic occurring between two compounds C_1 and C_2 of different stoichiometric compositions. At the eutectic concentration the following reaction takes place:

$$C_1 + C_2 \longrightarrow Liq$$

The enthalpy connected with the eutectic melting displays a sawtooth variation peaking at the eutectic point (experimental examples can be found in Smith and Pennings 1974; Wittmann and St John Manley 1977).

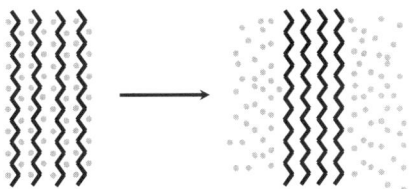

Fig. 1.4. Schematic representation of the incongruent melting of a compound: the compound (**left**) transforms into a solid phase with solvent release (**right**).

The solvent crystallization method

In many cases a given phase coexists with pure or nearly pure solvent. The composition of this phase can be accordingly determined by use of the solvent crystallization method (for details see Klein and Guenet 1989). In the case of polymer–solvent compound one has therefore direct access to the stoichiometry.

The principle of the method consists in lowering the temperature of the two-phase system so as to achieve crystallization of the free solvent. The melting enthalpy of the free solvent is then measured as a function of the polymer weight fraction. The experimental enthalpy ΔH for a given polymer weight fraction x_p is:

$$\Delta H = \Delta H_o \times x_{fs} \qquad (1.6)$$

in which ΔH_o is the melting enthalpy of the solvent and x_{fs} the weight fraction of pure solvent in the sample. This parameter is expressed as follows:

$$x_{fs} = 1 - x_p(1 + \bar{\alpha} m_s/m_p) \qquad (1.7)$$

where m_s and m_p are the molar mass of the solvent and of the monomer, respectively, and $\bar{\alpha}$ the number of solvent molecules per monomer (namely the 'thermodynamic' stoichiometry S_φ when dealing with a compound). Gibbs lever rule implies that ΔH varies linearly as a function of x_p. By determining the polymer weight fraction x_p^o at which ΔH vanishes, the value of $\bar{\alpha}$ is straightforwardly derived:

$$\bar{\alpha} = \frac{1 - x_p^o}{x_p^o} \times \frac{m_p}{m_s} \qquad (1.8)$$

Metastability

The existence of metastable phases is a common phenomenon with polymers on account of restricted mobility, and correspondingly slow transformation kinetics. Three cases deserve discussion as they are within the scope of this book: (1) the effect of the existence of a stable phase at high pressure; (2) the liquid–liquid phase separation mechanism; and (3) the melting behaviour of compounds.

Effect of pressure

Figure 1.5 shows a hypothetical pressure–temperature phase diagram (P–T phase diagram), revealing the existence at high pressure of another crystalline phase S_2 in addition to that S_1 observed at normal pressure. There is a metastability extension of the melting line where $S_2 \rightarrow$ *liquid* within the domain of existence of S_1 (dotted line). (Note that these metastability extensions also exist in T–C phase diagrams.) This means that on the left side of this line the phase S_2 can exist in a metastable form, which should transform into the metastable form S_1 to achieve equilibrium. What may experimentally occur is the simultaneous growth of S_1 and S_2 at different rates. It can even be that, at

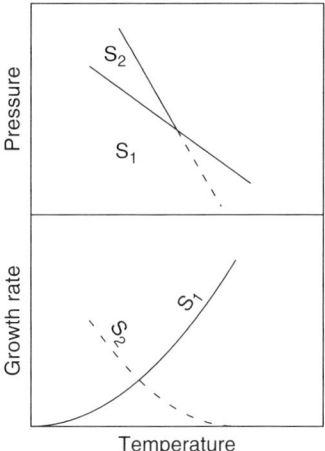

Fig. 1.5. Upper: hypothetical pressure–temperature phase diagram for a solid with two crystalline phases. **Lower:** the possible growth rate of each phase as a function of temperature.

low temperature, the metastable phase S_2 grows more rapidly than S_1 (see for instance Figures 12.4 and 14.20). Now, if mobility is sufficiently low, S_2 may display a very long lifetime, giving the wrong impression that it is a stable phase.

Under these conditions one may observe three situations: (1) at high temperature the presence of S_1 only; (2) at intermediate temperature a mixture of S_1 and S_2; (3) and at low temperature the occurrence of S_2 only. The state of the system essentially will then be determined by the different kinetics rates (*kinetically-controlled* system).

In principle, as has been stated earlier in this section, S_2 should transform on heating while giving off an exotherm, which indicates its metastability status.

Spinodal decomposition

Similar kinetic effects come into play in the case of a liquid–liquid phase separation. Let us first examine the case where the equilibrium phase diagram is of the type drawn in Figure 1.6. Two situations are worth contemplating: (1) the system is cooled slowly; and (2) the system is rapidly quenched.

In the first case, the system will simply first phase separate into two liquids that will eventually produce two crystalline phases on crossing the monotectic temperature T_M. Liquid–liquid phase separation will proceed via a *nucleation and growth* mechanism.

If the polymer solution is quenched very rapidly, many situations can take place that produce different results. If the system is *quenched above T_M* in such a way as to reach the spinodal region (domain below the spinodal curve), phase separation will proceed through a *diffusion process* (no nucleation step) because below the spinodal the system is under a highly *unstable state*. Such a diffusion mechanism is said to generate a network

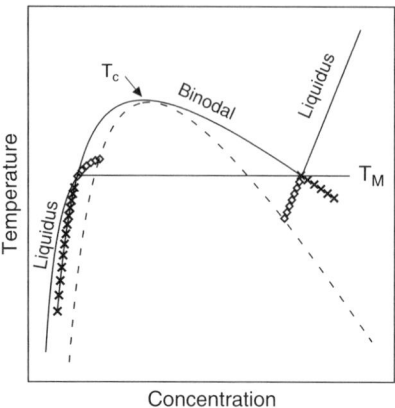

Fig. 1.6. Representation of the spinodal curve (dotted line) and the metastability extensions ($(\times)$ = metastability extension of the bimodal; $(\Diamond)$ = metastability extension of the liquidus) in the case of a monotectic transition. T_M is the monotectic temperature, T_c the critical temperature.

structure at the early stage of the decomposition, which vanishes rapidly through the usual Ostwald ripening (Cahn and Hilliard 1959).

If the solution is quenched very rapidly but this time well below T_M as well as within the spinodal region, liquid–liquid phase separation must occur first followed by crystallization, as has been demonstrated by Cahn (1961). Crystallization is therefore liable to pin down the network structure created at the early stage of the spinodal decomposition. This is why this mechanism once became popular in accounting for the thermoreversible gelation of polymers and biopolymers (for a thorough discussion of this point see, for example, Guenet 1992).

Finally, if the solution is quenched below T_M but within the domain bounded by the binodal and the spinodal curves, the system stands in a so-called *metastable state*, and in all cases liquid–liquid phase separation through *nucleation and growth* will occur first, followed by crystallization.

In the case of systems for which the equilibrium liquidus line (i.e. at very slow cooling rates) does not normally cross but stands above the binodal, the same situation as that described above may be observed by using very high cooling rates (see for example Guenet 1996b). The same principles apply.

Incongruently melting compounds

Metastability extension also occurs in the case of an incongruently melting compound, as is shown in Figure 1.7.

The metastable extension of the liquidus into the domain above the incongruently melting temperature gives rise to a metastable congruent melting. If $T_{metastable} - T_{stable}$ is low

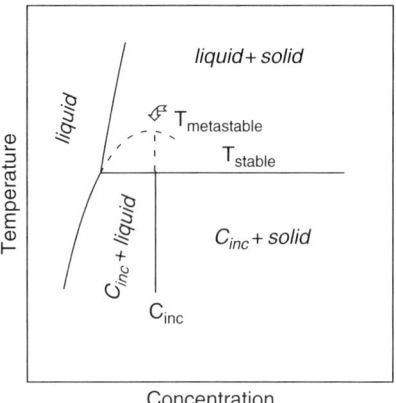

Fig. 1.7. The metastability extension of the liquidus into the domain above the incongruent melting highlights the existence of a metastable congruent melting.

enough, the incongruent melting may be simply bypassed by using appropriate heating rates. As a result, one may observe an endotherm due to the congruent melting of the compound. Note that the melting process still involves an endothermic event because the molecular compound is a stable phase. *One should not confuse a metastable phase and the metastable melting of a stable phase.* In the latter case, an endotherm is observed, yet the magnitude of the associated latent heat is strongly heating-rate dependent. This has been for instance observed in iPS thermoreversible gels (Guenet et al. 1985). Sometimes, again depending upon the heating rate, the metastability of the melting event is revealed by the exotherm of recrystallization of the metastable liquid because the system may still be located below the melting liquidus of the solid phase. If the metastable congruent melting point is located well beneath the terminal liquidus, then the liquid produced at $T_{metastable}$ may be in a situation of high undercooling and may recrystallize very rapidly so as to form the final solid phase.

Size and nucleation effects

The finite size of a phase may have consequences on the phase diagram. For instance, if it varies with the composition of the system, non-variant transformation will no longer appear since melting processes are strongly size-dependent. This can be shown by considering the free energy ΔF_o as derived by Gibbs for objects of finite size:

$$\Delta F_o = S\sigma - V\Delta f \tag{1.9}$$

where S and V are the surface and the volume of the object, respectively, σ the surface free energy and Δf the free energy per unit volume. It is more convenient to introduce the melting point depression ΔT due to the finite size:

$$\Delta T = T_m^o - T_m \tag{1.10}$$

where T_m^o and T_m are the melting points for an infinite and a finite size, respectively. Introducing the melting enthalpy Δh_f, a quantity that is size-independent, and by further assuming that it does not depend upon temperature, then Δf can be written:

$$\Delta f = \Delta h_f \left(1 - \frac{T_m}{T_m^o}\right) \quad (1.11)$$

At equilibrium $\Delta F_o \leq 0$, so one obtains the relation:

$$T_m = T_m^o \left[1 - (S\sigma/V\Delta h_f)\right] \quad (1.12)$$

Because of the surface free energy, the term ΔF_o only becomes negative once a critical size is reached. Below this size the particle is unstable and vanishes as $\Delta F_o > 0$. The critical size $\rho_c = V_c/S_c$ is then written:

$$\rho_c = \frac{\sigma T_m^o}{\Delta h_f \Delta T} \quad (1.13)$$

Usually, to reach the critical size a subsequent undercooling with respect to the melting temperature is required. Under these conditions nucleation is said to be *homogeneous*. Impurities may act as critical nuclei, in which case the nucleation is designated as *heterogeneous*. In the latter case, no significant undercooling is needed.

Owing to the dependence of the melting temperature on crystal size, the shape of a T–C phase diagram may sometimes be deceptive. For instance, the observation of a maximum in the melting temperature as a function of composition may arise from the existence of a compound but may as well be due to a size effect. Only the variation of the melting enthalpy allows one to choose between the two cases: if the melting enthalpy varies linearly and also exhibits a maximum, this points undoubtedly towards a compound. Conversely, if the degree of organization varies with composition, the melting enthalpy variation will depart from linearity, and only the melting temperature evolution is reliable provided no size effect comes into play. **Clearly, T–C phase diagrams are only meaningful when cross-examined in the light of Tamman's diagrams.**

Flory's theory

Although the theory developed by Flory to account for melting point depression is not directly applicable to the case of compounds it deserves, nevertheless, to be presented (Flory 1953). The theory considers an amorphous polymer solution in equilibrium with polymer crystals. If no variation of heat capacity occurs at the melting the general relation holds:

$$\frac{\Delta \mu}{T_m} = \Delta H_m \left[\frac{1}{T_m^o} - \frac{1}{T_m}\right] \quad (1.14)$$

where $\Delta \mu$ is the variation of chemical potential, T_m the melting point in presence of solvent, T_m^o the melting point of the pure polymer and ΔH_m the melting enthalpy. Flory and Huggins respectively (Flory 1942; Huggins 1942) have expressed the chemical potential of a moderately concentrated solution:

$$\Delta \mu_p = RT \left(V_p/V_s\right) \left[(\varphi_p - 1) + \chi_1 (1 - \varphi_p)^2\right] \quad (1.15)$$

where R is the gas constant, φ_p the polymer volume fraction, V_p and V_s the molar volumes of the polymer and of the solvent, respectively, and χ_1 the Flory interaction parameter. Equating (1.14) and (1.15) finally yields:

$$\frac{1}{T_m} - \frac{1}{T_m^o} = \frac{RV_p}{\Delta H_m V_s} \times \left[(1 - \varphi_p) - \chi_1 (1 - \varphi_p)^2\right] \quad (1.16)$$

For *ternary systems*, the melting point variation as a function of solvent composition at constant polymer volume fraction is written:

$$\frac{1}{T_m^{1,2}} - \frac{1}{T_m^1} = -\frac{RV_p}{\Delta H_m V_1} \times G_{\varphi_p}(\varphi_i, \chi_{ij}) \quad (1.17)$$

where $G_{\varphi_p}(\varphi_i, \chi_{ij})$ is written:

$$-\varphi_1 - \frac{\varphi_2}{x_2} - \chi_{12}\varphi_1\varphi_2 + \left[\chi_{1p}\varphi_1 + \chi_{2p}\frac{\varphi_2}{x_2} + 1\right](1 - \varphi_p) - \chi_{1p}(1 - \varphi_p) \quad (1.18)$$

In equations (1.17) and (1.18) the symbols used are: $T_m^{1,2}$ and T_m^1, the melting temperature in the binary solvent and in solvent 1, respectively, at a given polymer volume fraction; V_1 the molar volume of solvent 1; χ_{ij} the different interaction parameters; x_2 the ratio of the molar volume of solvent 1 to solvent 2; and φ_i the different volume fractions related through:

$$\varphi_1 + \varphi_2 + \varphi_p = 1 \quad (1.19)$$

These theories do not take into account the case of polymer–solvent compounds. Until a more appropriate theory is derived, Klein and Guenet have suggested that the above theory can be used provided some simple modifications are performed (Klein and Guenet 1989). They assume that, after compound melting, the solution consists of solvent molecules and of *solvated chains*, namely chains that carry the same number of solvent molecules as in the 'crystalline' state. Under these conditions, it is only necessary to introduce a new interaction parameter Λ_1, which stands for the interaction parameter between the solvent and the *solvated chains*, and to rescale the polymer volume fraction by the stoichiometric volume fraction φ_γ.

$$\frac{1}{T_\gamma} - \frac{1}{T_\gamma^o} = \frac{RV_\gamma}{\Delta H_{m\gamma} V_s} \times \left[(1 - \varphi_p/\varphi_\gamma) - \Lambda_1 (1 - \varphi_p/\varphi_\gamma)^2\right] \quad (1.20)$$

where T_γ^o and T_γ are the melting temperatures of the system at the stoichiometric concentration (i.e. the pure compound) and at a volume fraction φ_p, respectively, and V_γ and $\Delta H_{m\gamma}$ the molar volume and the melting enthalpy of the compound. Equation (1.20) is obviously only valid for $\varphi_p < \varphi_\gamma$.

CHAPTER 2

Scattering and diffraction

Theoretical

The general expression for the intensity scattered by a binary system composed of molecules a and b is written:

$$S(q) = \overline{A}^2(q)S_a(q) + \overline{B}^2(q)S_b(q) + 2\overline{A}(q)\overline{B}(q)S_{ab}(q) \qquad (2.1)$$

where $\overline{A}^2(q)$ and $\overline{B}^2(q)$ are the coherent scattering amplitudes of molecules a and b, and $S_a(q)$, $S_b(q)$ and $S_{ab}(q)$ are the scattering factors which characterize the spatial correlations between molecules a, molecules b and molecules a and b. q is the momentum transfer, which reads:

$$q = \frac{4\pi}{\lambda}\sin\theta/2$$

where λ is the radiation wavelength and θ the scattering angle.

Note 2.1 Momentum transfer

Crystallographers use s instead of q, and 2θ instead of θ, so s reads:

$$s = \frac{2}{\lambda}\sin\theta$$

The expression of such a scattering factor is:

$$S_a(q) = \sum_i \sum_j <\exp iqr_{ij}> \qquad (2.2)$$

where r_{ij} is the distance between two molecules a. Similar expressions hold for $S_b(q)$ and $S_{ab}(q)$. The scattering amplitude, which characterizes the correlations within one molecule, is written:

$$\overline{A}(q) = \sum_k a_k < \exp iq\rho_k > \qquad (2.3)$$

where a_k is the scattering factor of atom k (or scattering length for neutron) and ρ_k its distance from the centre of mass of the molecule.

Now, we shall consider two main q-ranges: the *small-angle scattering* range and the *diffraction* range. The *small-angle scattering range* is such that $A(q)$ reduces to a constant, which implies $q\rho_k < 1$. One obtains:

$$A(q) = A = \sum_k a_k \qquad (2.4)$$

This is achieved when the experiment is performed at low q-values, i.e. at low resolution, typically less than $q \approx 3\,\text{nm}^{-1}$. Under these conditions **the system can be regarded as a continuum** where one type of molecule is embedded in the other type. This is the *incompressibility hypothesis*, which states that any point of the system contains a molecule a or b. $S(q)$ reduces then to:

$$S(q) = [< A > - < B >]^2 S_a(q) = [< A > - < B >]^2 S_b(q) \qquad (2.5)$$

The cross-term $S_{ab}(q)$ has vanished; only correlations between molecules a, on the one hand, and molecules b, on the other hand, are relevant. The term $(<A> - <B>)^2$ is the contrast factor.

Conversely, in the *diffraction range* the resolution is much higher because larger q-values are involved, so the system must be regarded as it is, i.e. *discontinuous*. Under these conditions relation (2.1) still holds, which, in particular, means that the cross-term $S_{ab}(q)$ must be taken into account.

Here, it is worth comparing the diffracted intensity between *solvated crystals*, on the one hand, and *non-solvated crystals*, on the other hand. In the case of *solvated crystals*, changing the labelling type of the solvent will necessarily alter the diffraction pattern, and particularly the relative ratio between diffraction peaks. This is so thanks to the existence of the cross-term arising from the mutual organization between solvent and polymer.

In the case of *non-solvated crystals*, the cross-term $S_{ab}(q)$ can be dropped provided the crystals are large enough. The intensity is reduced for those crystals embedded in a liquid:

$$S(q) = < A(q) >^2 S_c(q) + < B(q) >^2 S_l(q) \qquad (2.6)$$

with obvious meaning as to the subscripts. One realizes that changing the labelling type of the solvent will have no effect on the diffraction pattern of the polymer. This can be achieved by means of neutron scattering. Neutron diffraction experiments can therefore reveal the existence of polymer–solvent molecular compounds, since both the hydrogenous and the deuterated molecules of a solvent are, in many cases, available.

In this chapter we shall consider in particular the scattering and diffraction by objects possessing cylindrical symmetry such as helices, fibres and the like. These types of object are seldom treated in recent textbooks, yet they pertain to many molecular structures and morphologies encountered in thermoreversible gels, such as fibrils or helices.

The intensity scattered by any system in the scattering range under the 'dilute regime' condition is simply expressed as:

$$I(q)_{C=0} = KCMP(q) \tag{2.7}$$

in which K is a constant (which includes the calibration constant of the apparatus used for the measurements and the contrast factor of the scattering species ($(<A> - <B>)^2$), C and M are respectively the concentration and the molecular weight of the scattering species and $P(q)$ is the form factor. The form factor contains all the interference effects arising from the shape of the particle under study. Its determination is therefore of prime importance.

Note that 'dilute regime' means in its broader sense that intermolecular scattering can be neglected.

In most cases calculation of the scattering function for all values of q is an insuperable task. As experiments are usually carried out in a limited range of scattering vectors, it is convenient to perform the calculation by making approximations. Fibrillar systems are very instructive on this point (Figure 2.1). The differing lengths of the system define different scattering domains.

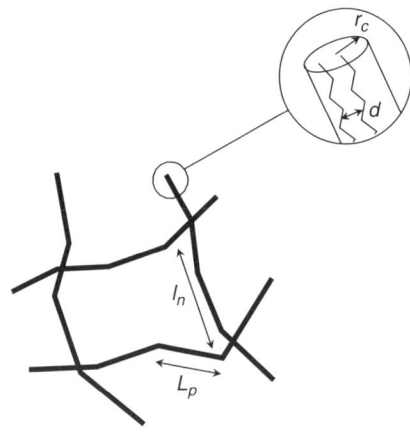

Fig. 2.1. Sketch of a fibrillar system. The blow-up shows details within a fibril. l_n = gel mesh size; l_p = fibril 'persistence length'; r_c = fibril cross-section radius; d = smallest distance between chains within one fibril.

Scattering and diffraction

The scattering range is defined by considering the product of q with the distance of interest. For $ql_n < 1$ the effect of the mesh size comes into play, possibly showing some diffraction maximum. For $ql_n > 1$ together with $ql_p < 1$, the scattering function will be sensitive to the longitudinal fractal dimension (see Chapter 4). For $ql_p > 1$ together with $qr_c \ll 1$ the system will scatter like an array of diluted cylinders. For $qr_c \approx 1$, the cross-section will affect the scattering curve. Finally for $qd > 1$, single chain scattering can be observed.

Scattering by cylindrical objects

Fournet has derived a general expression for the form factor of a uniform cylinder of length L and radius r_H (Fournet 1951):

$$P(q) = \int_0^{\pi/2} \frac{\sin^2(qL\cos\theta)}{q^2 L^2 \cos^2\theta} \frac{4 J_1^2(qr_H \sin\theta)}{q^2 r_H^2 \sin^2\theta} \sin\theta \, d\theta \tag{2.8}$$

where θ stands for the angle between the scattering vector q and the cylinder long axis.

In most of the cases discussed in this book (fibres or helices), L is much larger than r_H ($L \gg r_H$), and the condition $qL \gg 1$ is generally satisfied. Under these conditions the term:

$$\frac{\sin^2(qL\cos\theta)}{q^2 L^2 \cos^2\theta} \tag{2.9}$$

rapidly equals zero except for $qL\cos\theta \approx 0$, i.e. for $\theta \approx \pi/2$ (Pringle and Schmitt 1971). This means that only those cylinders perpendicular to the scattering vector will contribute to the cross-section scattering. $P(q)$ is then rewritten:

$$P(q) = \frac{4 J_1^2(qr_H)}{q^2 r_H^2} \int_0^{\pi/2} \frac{\sin^2(qL\cos\theta)}{q^2 L^2 \cos^2\theta} \sin\theta \, d\theta \tag{2.10}$$

After performing the integration, equation (2.10) is written:

$$P(q) = \frac{\pi}{qL} \frac{4 J_1^2(qr_H)}{q^2 r_H^2} \tag{2.11}$$

This relation can be generalized to all types of cross-sections:

$$P(q) = \frac{\pi}{qL} \varphi(q\sigma) \tag{2.12}$$

where $\varphi(q\sigma)$ is the scattering by a cross-section of characteristic size σ. $\varphi(q\sigma)$ is expressed as:

$$\varphi(q\sigma) = \int_\sigma \rho(r) J_o(qr) 2\pi r \, dr \bigg/ \int_\sigma \rho(r) 2\pi r \, dr \tag{2.13}$$

where $2\pi\rho(r)\,dr$ is the distribution function of the scattering sites.

For a hollow cylinder of outer radius r_H and inner radius γr_H the following relation is thus derived:

$$P(q) = \frac{\pi}{qL} \left\{ \frac{2}{(1-\gamma^2) q r_H} (J_1(qr_H) - \gamma J_1(q\gamma r_H)) \right\}^2 \quad (2.14)$$

which for $\gamma \approx 1$ (thin-walled hollow cylinder) reduces to:

$$P(q) = \frac{\pi}{qL} J_o^2(qr_H) \quad (2.15)$$

As the intensity is proportional to $MP(q)$, the intensity always contains the ratio $M/L = \mu_L$, which is the *mass per unit length* of the cylinder.

Scattering and diffraction by helices

The same principle applies to helices. Here, we shall first consider uniform helices for which Schmitt and later Pringle and Schmitt have derived the detailed form factor (Schmitt 1970; Pringle and Schmitt 1971). In actuality, the scattering by uniform helices oriented perpendicular to the incident beam (i.e. perpendicular to the scattering vector) was derived by Cochran and colleagues in the early fifties (Cochran et al. 1952). The way Pringle and Schmitt re-derived this form factor is, however, more instructive as the close similarity between cylinders and helices can be thus highlighted. These authors consider the cross-section shown in Figure 2.2, which is valid for both single ($\varphi = 0$) and double helices.

The cross-section scattering $\varphi(q\sigma)$ is then written:

$$\sum_o^\infty \varepsilon_n \cos^2(n\varphi/2) \frac{\sin^2(n\omega/2)}{(n\omega/2)^2} [g_n(qr_H, \gamma)]^2 \quad (2.16)$$

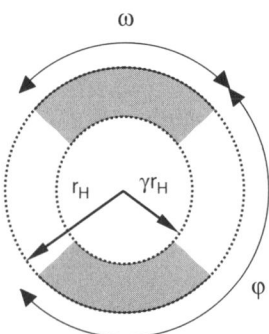

Fig. 2.2. Grey areas highlight the helix cross-section considered by Pringle and Schmitt. The dotted lines show the equivalent hollow cylinder.

where $\varepsilon_o = 1$ and $\varepsilon_n = 2$ for $n \geq 1$, and where $g_n(qr_H, \gamma)$ is written:

$$g_n(qr_H, \gamma) = \frac{2}{r_H^2(1-\gamma^2)} \int_{\gamma r_H}^{r_H} rJ_n\left(qr\sqrt{1-a_n^2}\right) dr \qquad (2.17)$$

in which the terms a_n are defined as follows:

$$a_n = 2\pi n/qP \quad \text{for } q > 2\pi n/P$$
$$a_n = 1 \quad \text{for } q \leq 2\pi n/P$$

where P is the pitch of the helix, which corresponds to the elevation along the helix long-axis after one turn of the helix. Note that for perfectly symmetrical double helices, i.e. $\varphi = \pi$, relation (2.16) is always zero for odd n.

Interestingly, for the term corresponding to $n = 0$ (i.e. when $0 < qP < 2\pi$) the scattered intensity reduces to relation (2.14). This means that, at low resolution, helices scatter as solid cylinders, and that the scattering is only dependent upon the helix radius, not upon its pitch. Little information can therefore be gained in the small-angle scattering range.

If one now considers infinitely thin helices (i.e. $\omega = 0$ and $\gamma \approx 1$), it can be shown that $g_n(qr_H, \gamma)$ reads:

$$g_n(qr_H, \gamma) = J_n\left(qr_H\sqrt{1-a_n^2}\right) \qquad (2.18)$$

This relation is valid for randomly oriented helices. Usually, oriented samples are preferred and used for determining a helical structure. As a result, the helices are parallel to the direction of orientation and perpendicular to the incident beam. One can then define two scattering vectors: q_ζ (perpendicular to the helix axis) and q_2 (parallel to the helix axis), which are commonly designated as $2\pi R$ and $2\pi \zeta$. The following relation holds:

$$4\pi^2 R^2 = q^2 - 4\pi^2 \zeta^2 \qquad (2.19)$$

In the reciprocal space ζ amounts to n/P, where P is the pitch of the helix and n an integer. Eventually, relation (2.18) is written:

$$g_n(qr_H, \gamma) = J_n(2\pi R r_H) \qquad (2.20)$$

and the scattered intensity:

$$I(R) \propto J_n^2(2\pi R r_H) \qquad (2.21)$$

This again shows that, for randomly oriented systems, only those helices perpendicular to the incident beam will contribute to the scattering (and diffraction at larger scattering vectors). This is the reason why oriented samples are preferred in the diffraction range. The summation in equation (2.16) implies the existence for oriented samples of diffraction layers, the positions of which are numbered using n. Equation (2.21) was first derived by Cochran et al. 1952.

Equations (2.16) and (2.21) hold for $n > 0$ so long as the helical structure can be regarded as continuous, i.e. so long as $q\xi_s < 1$, where ξ_s is the size of the basic constituent of the helix (the monomer unit for instance). At higher resolution, the helix is clearly discontinuous, made up of atoms of different types and various scattering cross-sections. Cochran and colleagues suggested regarding a discontinuous helix as the product of a continuous helix of pitch P and a set of planes perpendicular to the z-axis and spaced h apart (h, designated as the *axial rise*, is the distance between the projections onto the z-axis of identical scatterers lying on two consecutive units). The ratio P/h is then the number of units per helix turn. Several helix turns may be necessary for the helix to repeat on itself, a fact which is expressed by means of two integers u and v. A helix is then defined as u units occurring in v turns ($P/h = u/v$). As a result, the pitch still corresponds to the elevation after one turn of the helix but is no longer the repeat distance c. The distance over which the helix repeats itself is for rational helices $c = vP$, where v is the integer defined above. Note that irrational helices also exist for which there is no true repeat distance (the α helix for instance).

The Fourier transform of the above product is zero everywhere except for:

$$\zeta = \left(\frac{m}{h}\right) + \left(\frac{n}{P}\right) \quad (2.22)$$

in which m and n are integers. By introducing the layer line index l ($l = vP\zeta$), equation (2.22) becomes:

$$l = um + vn \quad (2.23)$$

In contrast with continuous helices, for a given l several values of n are possible in view of relation (2.23). As a result, for a given layer line, the intensity will be the sum of the intensities corresponding to the different possible values of n, that is:

$$I(R, l/c) \propto \sum_n J_n^2(2\pi R r_H) \quad (2.24)$$

Note that for continuous helices l and n are the same. Here, for *randomly oriented helices*, a second summation over l should be performed on relation (2.24) in a similar fashion to that done by Pringle and Schmitt for continuous helices (relation (2.16)).

From the diffraction pattern of oriented systems the ratio u/v (number of units per turn) can usually be deduced by determining the first intense layer line and the layer line containing a meridional spot (i.e. on the z-axis). The first intense layer line occurs when relation (2.23) is satisfied for $n = 1$ while the first meridional spot is seen when $n = 0$ and $m = 1$. The ratio $l(1st\ meridional\ spot)/l(1st\ intense\ layer)$ yields u/v. A theoretical example is given in Figure 2.3 for a 5_2 helical structure by using relation (2.24).

A practical example is given in Figure 2.4, in which is displayed the diffraction by a PEO/resorcinol complex (4_1 helix). As the diffraction pattern is recorded on a flat plate, the layer lines are located on hyperbolas.

In the diffraction range relation (2.24) is more precisely written for oriented helices:

$$I(R, l/c) \propto \sum_n G_{n,l} G_{n,l}^* \quad (2.25)$$

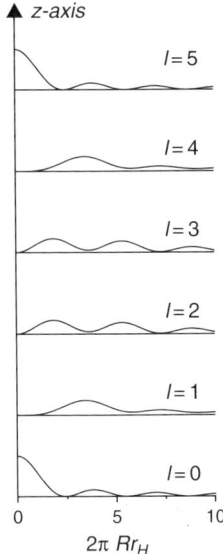

Fig. 2.3. Diffracted intensities calculated from relation 2.24 for a discontinuous 5_2 helix. The layer lines are indicated from $l=0$ to $l=5$. The first intense layer occurs for $l=2$ (layer containing largest number of maxima) while the first meridional spot is seen for $l=5$.

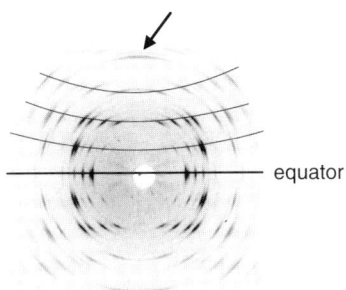

Fig. 2.4. Fibre diffraction pattern for PEO–resorcinol complex. Due to the geometry of the experimental set-up the layer lines appear as hyperbolas. The first meridional reflection at $l=4$ is shown by arrow. (Courtesy of M. Dosière.)

where:

$$G_{n,l} = \sum_j a_j J_n(2\pi R r_{Hj}) \exp\left[i\left(-n\varphi_j + 2\pi l z_j/c\right)\right] \qquad (2.26)$$

where r_{Hj}, φ_j and z_j are the polar coordinates of atom j and a_j its scattering power (or scattering length).

Note that the front factor π/q is absent for oriented helices. Inspection of relation (2.10) shows that for $\theta = \pi/2$ this term simply equals unity.

For those readers interested in obtaining more information on the diffraction by helices the book by Tadokoro is a leading reference (Tadokoro 1979).

Effect of contrast

Consider the case of a helix where the outer part consists of a shell of solvent molecules in strong interaction (case 1) or the case of a double helix with solvent occluded between the two strands (case 2). Under these conditions the contrast throughout the sample is no longer homogeneous. Let us determine the cross-section form factor $\varphi(qr_H)$ for case 1 in the low-resolution range:

$$\varphi(qr_H) = \left[\int_0^{\gamma r_H} A_{in} J_o(qr) 2\pi r dr + \int_{\gamma r_H}^{r_H} A_{out} J_o(qr) 2\pi r dr \Big/ A \right]^2 \quad (2.27)$$

where A reads:

$$A = \int_0^{\gamma r_H} A_{in} 2\pi r dr + \int_{\gamma r_H}^{r_H} A_{out} 2\pi r dr \quad (2.28)$$

where A_{in} and A_{out} are the scattering power of the inner part and the outer part, respectively. Introducing:

$$A_m = \gamma^2 A_{in} + (1-\gamma^2) A_{out} = A/\pi r_H^2 \quad (2.29)$$

$\varphi(qr_H)$ is written after performing the integations:

$$\varphi(qr_H) = \left[\frac{2A_{in}\gamma}{A_m qr_H} J_1(q\gamma r_H) + \frac{2A_{out}}{A_m qr_H} (J_1(qr_H) - \gamma J_1(q\gamma r_H)) \right]^2 \quad (2.30)$$

In the q-range where $qr_H < 1$ relation (2.30) reduces to:

$$\varphi(qr_H) = 1 - \frac{q^2}{4} \frac{r_H^2}{A_m} \{A_{in}\gamma^4 + A_{out}(1-\gamma^4)\} + \ldots \quad (2.31)$$

This equation highlights that only an apparent cross-section radius can be determined in this q-range unless $A_{in} = A_{out}$. By toying with the scattering amplitude of the solvent the scattering pattern can be drastically altered. A typical example of such a situation is encountered in the case of polyelectrolyte/surfactant complexes where the polymer is deuterated and the surfactant is protonated (Ray et al. 2003). Using deuterated solvent or hydrogenous solvent produces totally different scattering patterns, as shown in Figure 2.5.

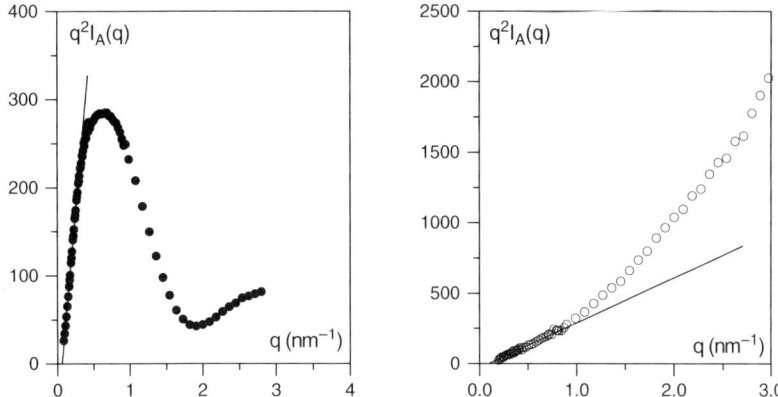

Fig. 2.5. Scattering patterns for $iPS_D/CTAB_H$ in deuterated n-butanol (**left**) and in hydrogenous n-butanol (**right**). The straight lines stand for the linear variation at small q-values. The upward departure from linearity in the case of hydrogenous n-butanol indicates the occurrence of an apparent 'negative radius'. (Data from Ray et al. 2003.)

Scattering by polydispersed systems

Oster and Riley (1952) have derived the form factor of a collection of n cylinders:

$$P(q) = \frac{\pi \mu_L}{qn^2} \varphi(q\sigma) \sum_{j=1}^{n}\sum_{k=1}^{n} J_o(qr_{jk}) \tag{2.32}$$

Again strong correlation between cylinders exist only when they are parallel to one another. Guenet (1994) has shown that these correlations vanish for a pair of cylinders whose long axes cross at a fixed angle.

Relation (2.32) is useful when aggregates are made up with a discrete number of cylinder-like structures. In the case of fibre, which can contain thousands of helical structures, Guenet (1994) has derived another approach by analogy with chemical aggregates (Daoud et al. 1984). The cross-section radii of the fibres are said to be characterized by a distribution function of the type $w(r) \sim r^{-\lambda}$, where $0 < \lambda < 3$ and with two cut-off radii, r_{min} and r_{max} (see Scale-free networks in Chapter 4). Two regimes are then considered: the *transitional regime*, which is reached for $qr_{max} > 1$ but otherwise with no special condition on qr, and the well-known *Porod regime*, which is reached for $qr_{min} > 1$. For reasons that will be obvious in what follows, the product $q^4 I(q)$ rather than $I(q)$ will be more useful. In the *transitional regime* the intensity is then written:

$$\frac{q^4 I_A(q)}{C} = 4\pi^2 \rho \left[A(\lambda) q^\lambda - \frac{1}{\lambda r_{max}^\lambda} \right] \bigg/ \int_{r_{min}}^{r_{max}} w(r)\, dr \tag{2.33}$$

where ρ is the density and $A(\lambda)$ is expressed as follows:

$$A(\lambda) = \frac{\Gamma(\lambda)\,\Gamma(\frac{3-\lambda}{2})}{2^{\lambda}\Gamma(\frac{\lambda+1}{2})\,\Gamma(\frac{\lambda+3}{2})\,\Gamma(\frac{\lambda+1}{2})} \qquad (2.34)$$

which eventually yields for $\lambda = 1$:

$$\frac{q^4 I_A(q)}{C} = \left[2\pi^2 \rho q - \frac{4\pi\rho}{r_{max}}\right] \times \frac{1}{Log(r_{max}/r_{min})} \qquad (2.35)$$

Interestingly, in all cases the extrapolated intercept q_o with the q-axis yields r_{max} by means of the following relation:

$$\frac{1}{r_{max}} = [\lambda\pi A(\lambda)]^{1/\lambda}\,q_o \qquad (2.36)$$

which, for $\lambda = 1$ reduces to:

$$r_{max} = \frac{2}{\pi q_o} \qquad (2.37)$$

In the *Porod regime* the intensity reduces to the well-known form:

$$\frac{q^4 I_A(q)}{C} = 4\pi\rho \int_{r_{min}}^{r_{max}} \left[1 + \frac{3}{8q^2 r^2} + \ldots\right] w(r)\,dr \bigg/ \int_{r_{min}}^{r_{max}} w(r)\,dr \qquad (2.38)$$

Equation (2.38) eventually yields for $\lambda = 1$:

$$\frac{q^4 I_A(q)}{C} = 4\pi\rho \left[\frac{1}{r_n} + \left(8q^2 r_{min}^3 Log\frac{r_{max}}{r_{min}}\right)^{-1}\right] \qquad (2.39)$$

where r_n is the mean number-averaged cross-section radius of gyration.

Note that the *Porod regime* is usually reached from above owing to additional $1/q^6$ terms which give rise to a maximum in the $q^4 I(q)$ vs q representation. These $1/q^6$ terms are enhanced when different curvatures, i.e. crossings between cylinders, are present. The magnitude of the maximum is therefore related to the number of crossings between cylinders (i.e. between fibres).

The scattering vector, q^* at which the *transitional regime* and the *Porod regime* intersect is shown to be written:

$$\frac{1}{r_{min}} = [\lambda\pi A(\lambda)]^{1/\lambda}\,q^* \qquad (2.40)$$

which, for $\lambda = 1$ reduces to:

$$r_{min} = \frac{2}{\pi q^*} \qquad (2.41)$$

The determination of q_o and of q^* yields r_{min} and r_{max} without the need for absolute calibration. As a result, a system of known r_{min}, r_{max} and density could be used as a calibration standard. Note that, independent of λ, one ends up with the simple relation:

$$\frac{r_{min}}{r_{max}} = \frac{q_o}{q^*} \tag{2.42}$$

Scattering by semi-rigid systems

Thus far, rigid and straight cylinders have been dealt with. While a structure can be locally straight it may be globally worm-like. A typical example is the worm-like chain: for distance smaller than a so-called persistence length l_p it scatters as a rod while for larger distances the chain statistics is Brownian (Heine et al. 1961; Luzzati and Benoit 1961). The calculation of the form factor of such a structure with a finite length L is not straightforward and to date only semi-analytical expressions have been derived (Yoshisaki and Yamakawa 1980). The asymptotes of the form factor can be derived rigorously only in the limit of very long chains (Des Cloiseaux 1973). Two regimes are expected:

$$P(q) \propto \frac{6}{q^2 l_p^2} \quad \text{for } q l_p < 1 \tag{2.43}$$

$$P(q) \propto \frac{\pi}{q l_p} + \frac{2}{3 q^2 l_p^2} \quad \text{for } q l_p > 1 \tag{2.44}$$

In a $q^2 I(q)$ vs q representation, one should therefore observe two asymptotes: a flat asymptote at low-q and a straight line at high-q. The intersect between these two regimes yields q^* which reads:

$$q^* = \frac{16}{3 \pi l_p} \tag{2.45}$$

In practice, however, owing to the finite size of the chains, the flat asymptote is not straightforwardly observed unless L/l_p is very large. The best way to determine l_p remains to fit the data with the pseudo-analytical expression obtained by Yoshisaki and Yamakawa (1980). Brûlet et al. (1996) have proposed the use of three different functions depending on the q-range:

1. For $q l_p \leq 2$

$$P_1(q) = \frac{2}{X}(\exp -X - 1 + X) + \frac{2 l_p}{15 L}\left[4 + \frac{7}{X} - \left(11 + \frac{7}{X}\right)\exp -X\right] \tag{2.46}$$

   where $X = q^2 l_p L/3$.

2. For $2 < q l_p < 4$

$$q^2 l_p L P_2(q) = 6 + 0.547(q l_p)^2 - 0.01569(q l_p)^3 - 0.002816(q l_p)^4 \tag{2.47}$$

3. And relation (2.44) for $q l_p \geq 2$.

$P_1(q)$ had been derived by Sharp and Bloomfield (1968), $P_2(q)$ from Yoshisaki and Yamakawa (1980).

Note that the chain, while being locally rod-like, may not take on globally Brownian statistics. As a result, a more generalized expression can be written for $ql_p < 1$ by considering the so-called fractal dimension D_f:

$$P(q) \propto \frac{1}{(ql_p)^{D_f}} \qquad (2.48)$$

In the case of objects with finite cross-section r_σ, the condition $l_p > r_\sigma$ usually holds. As a result, provided that $ql_p > 1$, all the above expressions can be used without modification.

Other models of semi-rigid chains have been developed. Hermans and Hermans (1958) have considered the case of N freely jointed rods. They have derived the following expression.

$$P(q) = \frac{1}{N}\left[2\Lambda(\beta) - \frac{4}{\beta^2}\sin^2\frac{\beta}{2}\right] + \frac{2}{N^2}\Lambda^2(\beta)\left[\frac{N(1-\nu) - (1-\nu^N)}{(1-\nu)^2}\right] \qquad (2.49)$$

in which $\beta = ql_r$, l_r being the length of the rods, $\nu = \sin\beta/\beta$, and with:

$$\Lambda(\beta) = \frac{1}{\beta}\int \frac{\sin t}{t}dt = \frac{1}{\beta}Si(\beta) \qquad (2.50)$$

A still more elaborate model has been considered by Muroga (1988). It consists of N rods of length A alternating with Gaussian sequences containing n elements of length a. Introducing the fraction of rods $f = NA/L$ (L = chain contour length), and $\omega = qna^2/6$, Muroga has obtained the following expression:

$$P(q) = \frac{f^2}{N}\left[2\Lambda(\beta) - \frac{4}{\beta^2}\sin^2\beta^2/2\right] + \frac{2f^2}{N}\Lambda^2(\beta)\left[\frac{\exp(-\omega)}{1-\nu\exp(-\omega)}\right]$$

$$-\frac{2f^2}{N^2}\Lambda^2(\beta)\left[\frac{\exp(-\omega)}{1-\nu\exp(-\omega)}\right]\left[\frac{1-(\nu\exp(-\omega))^N}{1-\nu\exp(-\omega)}\right]$$

$$+(1-f)^2\left[\frac{2}{N\omega} - 2\left(\frac{1-\exp(-\omega)}{N\omega}\right)^2\left(\frac{\nu(1-(\nu\exp(-\omega))^N)}{(1-\nu\exp(-\omega))^2}\right)\right. \qquad (2.51)$$

$$\left.-2\frac{(1-\exp(-\omega))(1-\nu)}{N\omega^2(1-\nu\exp(-\omega))}\right]$$

$$+\frac{2(f-f^2)\Lambda(\beta)(1-\exp(-\omega))}{N\omega(1-\exp(-\omega))}\left[2 - \frac{(1+\exp(-\omega))(1-(\nu\exp(-\omega))^N)}{N(1-\exp(-\omega))}\right]$$

Scattering by randomly aggregated systems

Guenet and Picot have derived the form factor for randomly aggregated Gaussian chains:

$$I(q) = \frac{P_o(q)}{N_c^2} \times \left[\frac{2P_o^{N_c+1}(q) - N_c P_o^2(q) - 2P_o(q) + N_c}{[1 - P_o(q)]^2} \right] \quad (2.52)$$

where $P_o(q)$ is the chain form factor and N_c the number of chains in one aggregate (Guenet and Picot 1981). This calculation can still be used with rod-like systems by using the form factor of a rod for $P_o(q)$.

Scattering by a regularly folded chain

A chain regularly folded along the growing face of a crystal may be regarded as a slab of length L_c, width l_c and thickness δ_c (Figure 2.6). This assumption is valid for $L_c \gg d_c$, where d_c is the average distance between adjacent stems, and for $\delta_c \approx d_c$ where δ_c is the stem diameter. The latter condition implies adjacent or near-adjacent folding.

For $L_c \gg l_c$, the analytical expression derived for infinitely-thin sheets ($\delta_c \approx 0$) by Porod can be used (Porod 1948, 1952):

$$P(q) = \frac{2}{qL_c} \left\{ \frac{\pi'}{ql_c} \left(\int_o^{ql_c} \frac{J_1(x)}{x} dx \right) + \frac{1}{qL_c} \left(\frac{\sin ql_c/2}{ql_c/2} \right)^2 - \frac{\sin qL_c}{(qL_c)^2} \right\} \quad (2.53)$$

in which π' is a complex function that rapidly reaches the value of π beyond the Guinier regime. Two main regimes are expected for $qL_c >$ with $ql_c < 1$, on the one hand, and for $qL_c >$ and $ql_c > 1$, on the other. For $qL_c >$ with $ql_c < 1$, the last two terms can be dropped and the Bessel function can be developed which eventually gives:

$$P(q) \approx \frac{\pi}{qL_c} \exp -q^2 l_c^2/24 \quad (2.54)$$

For $qL_c >$ and $ql_c > 1$ the integration can be performed from 0 to ∞ which yields:

$$P(q) \approx \frac{2\pi}{q^2 l_c L_c} \quad (2.55)$$

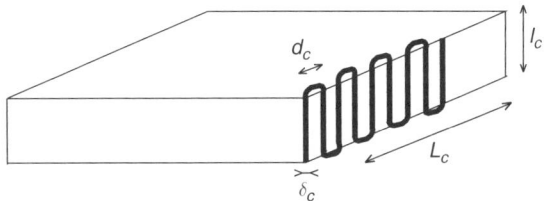

Fig. 2.6. Chain regularly folded along the growth face of a single crystal (adjacent folding).

To take into account the effect of slab thickness another approach can be used. It consists in considering a long rod-like object (length L_c) possessing a rectangular cross-section of length l_c and width δ_c. The form factor of such a structure has been derived by Mittelbach and Porod (1961) on the condition that $L_c \gg l_c$ and $l_c > \delta_c$:

$$P(q) \approx \frac{\pi}{qL_c} \frac{2}{\pi} \int_0^{\pi/2} \left[\frac{\sin ql_c/2 \cos\theta}{ql_c/2 \cos\theta} \times \frac{\sin q\delta_c/2 \sin\theta}{q\delta_c/2 \sin\theta} \right]^2 \sin\theta \, d\theta \qquad (2.56)$$

This gives for $qL_c >$ and $ql_c > 1$ with $q\delta_c < 1$:

$$P(q) \approx \frac{2\pi}{q^2 l_c L_c} \times \exp{-q^2 \delta_c^2/12} \qquad (2.57)$$

Note that the intensity is proportional to $M/L_c l_c$, a quantity which is a **_mass per unit area_** (μ_S).

This is a simple model that may be relevant for solution-grown crystals or crystals grown from the bulk near the 'equilibrium' melting temperature (e.g. see Guenet 1980; Guenet and Picot 1983). Under other crystallization conditions, regular folding tends to be interrupted by the competition between two chains crystallizing on the same growth plane or chains entering different lamellae (e.g. see Sadler and Keller 1977; Spells and Sadler 1984). Spells and co-workers have tackled this issue by considering the probability of adjacent folding, and developed computer-simulated models (Sadler and Spells 1984; Spells and Sadler 1984; Sonntag et al. 1995).

Experimental aspect

This section is chiefly intended to giving a short outline of the main scattering techniques, i.e. neutron, X-ray and light scattering, highlighting their advantages as well as their shortcomings. For further reading there exists a wealth of specialized treatises (e.g. see: neutron, Higgins and Benoit 1994; X-ray, Guinier 1956; light, Huglin 1972).

The choice of the most appropriate scattering technique essentially relies upon two parameters: the *accessible q-range* and the *contrast factor*. Depending on the q-range and on the charateristic length(s) of the system under study, different domains of the scattering curve will be explored. The rule of thumb states that the distance investigated is about $1/q$.

Neutron scattering and diffraction

Small-angle neutron scattering requires *thermal neutrons*, i.e. neutrons that have been slowed down while crossing heavy water. Their wavelengths are ranging typically between 0.3 and 2 nm. The current neutron scattering set-up allows access to the following q-range:

$$0.01 < q \ (\text{nm}^{-1}) < 3$$

which allows one to explore distances ranging from about 100 nm down to 0.3 nm. This corresponds to molecular dimensions encountered in many polymeric systems (e.g. chain radius of gyration, persistence length, fibre cross-section).

The main advantage of neutrons is, however, the high contrast that can be brought about by using deuterium-labelled species in lieu of their hydrogenous counterparts. It turns out to be so because the scattering lengths of hydrogen ($a_H = -0.375 \times 10^{-12}$ cm) and of deuterium ($a_D = 0.675 \times 10^{-12}$ cm) are utterly dissimilar, especially so far as their sign is concerned (see Table 2.1). This has considerable effects, as the scattering amplitude of a molecule containing n atoms is written:

$$A = \sum_{i=1}^{i=n} a_n \qquad (2.58)$$

There exists a major drawback: hydrogen also gives off a high incoherent signal. This incoherent signal must be measured separately and then subtracted from the experimental spectrum to obtain the coherent scattering. This procedure becomes exceedingly delicate when the ratio signal to noise is low (typically lower than 1.1). Incorrect subtraction may yield an utterly irrelevant coherent scattering curve.

Note that a deuterated matrix can be used instead, in which case the labelled species are hydrogenous. This inverse labelling is used for increasing the signal to noise ratio so as to deal with the problem arising from the incoherent scattering of the hydrogenous species. Indeed, the incoherent scattering from deuterated species is usually negligible.

The coherent scattering amplitude for a molecule is calculated by means of relation (2.4). The contrast factor in the case of an object of scattering amplitude A and molar volume v_o immersed in a medium of scattering amplitude B and of molar volume v_m is written:

$$K = \frac{N_A}{m_o^2} \left(A - \frac{v_o}{v_m} B \right)^2 \qquad (2.59)$$

where N_A is Avogadro's number and m_o the molecular weight of the basic constituent of the object (a monomer unit for instance).

Provided deuteration does not alter too much the thermodynamic properties, the resulting contrast allows one to 'single out' an object among its matrix of identical objects, and therefore to have access to its scattering form factor. For instance, the chain conformation can be studied in concentrated solutions (Daoud et al. 1975) or gels (Guenet 1987). The intermolecular terms can even be cancelled by using an appropriate mixture of deuterated and hydrogenous solvent that matches exactly the coherent scattering of the hydrogenous chains matrix. For instance, in the case of ternary systems, the scattered intensity is written:

$$I(q) \propto \left(a - \frac{v_A}{v_S} s \right)^2 S_A(q) + \left(b - \frac{v_B}{v_S} s \right)^2 S_B(q) + 2 \left(a - \frac{v_A}{v_S} s \right) \left(b - \frac{v_B}{v_S} s \right) S_{AB}(q) \qquad (2.60)$$

Table 2.1. Coherent scattering lengths a_c and incoherent cross-section σ_{inc} ($\sigma_{inc} = 4\pi a_{inc}^2$) for the commonest atoms (Bacon 1975).

atom	$a_c (10^{-12}$ cm)	σ_i (barn)
H	−0.375	80
D	0.670	2.2
C	0.662	0.01
N	0.940	0.4
O	0.575	0.04
F	0.574	0.2
S	0.285	0.01
Cl	0.958	2.8
Br	0.685	0.4

where a, b and s are the scattering amplitudes of A, B, and S, respectively, and $S_i(q)$ the form factor of A, B, and S respectively. If $a = v_A s/v_S$ then the scattered intensity is related to the structure of B only, and vice versa if $b = v_B s/v_S$ (e.g. see Lòpez and Guenet 1999).

The same contrast effect is used in neutron diffraction experiments; the way in which the contrast is calculated differs as the incompressibility hypothesis no longer applies. Also, *fast neutrons* can be used possessing wavelengths lower than 0.1 nm, which gives access to very large transfer momenta. Note that for a two-component system, four structure factors can be determined by making use of the four labelling possibilities. This means, particularly in the case of polymer–solvent compounds, that the relative ratio of intensities corresponding to different crystallographic planes will depend upon the labelling. For instance, depending on whether hydrogenous solvent or deuterated solvent is used, the diffraction pattern will be significantly modified (appearance or disappearance of some reflections) as opposed to what would occur in the absence of a compound. Neutron diffraction can be used to demonstrate the existence of a compound in the case of systems poorly organized (few reflections) where the usual symmetry rules are barely applicable.

Note that neutron scattering can also be used for giving circumstantial evidence of the occurrence of a compound. Indeed, the composition of a mixture of deuterated and hydrogenous solvent required for matching the coherent scattering of a polymer system can be calculated. If the experimentally determined composition differs conspicuously from that calculated, compound formation is likely to occur (e.g. see chapter 6 for instance).

X-ray scattering and diffraction

Neutrons are massive particles that interact with the nuclei of atoms, whereas X-rays are constituted of photons that are scattered by the electronic cloud surrounding atomic nuclei. Small-angle scattering can be performed with X-rays. Considerable development of this technique has taken place in the past few years with the creation of synchrotron sources that produce intense flux and better geometrically defined beams. The typical accessible range of transfer momenta is:

$$0.1 < q \text{ (nm}^{-1}) < 2.5$$

For small-angle scattering, the electronic structure factor can be regarded as unity, and so the contrast factor is written:

$$K_X = \frac{N_A e^4}{m_o^2 \mu^2 c^4} \left(Z_o - \frac{v_o}{v_m} Z_m \right)^2 = 7.9 \times 10^{-2} \frac{N_A}{m_o^2} \left(Z_o - \frac{v_o}{v_m} Z_m \right)^2 \quad (2.61)$$

where e and μ are the charge and the mass of one electron, c the speed of light, and Z_o and Z_m the number of electrons per basic unit of species o and m, respectively. N_A is again Avogadro's number and m_o the molecular weight of the basic constituent of the object o.

Here, the contrast depends upon the difference of the number of electrons between the two components. Unlike neutrons, the single-chain behaviour cannot be investigated unless labelling with heavy atoms is considered. This type of labelling is, however, liable to alter drastically the thermodynamics of the system, and eventually its structure.

Note that, thanks to the low, mostly negligible incoherent scattering, X-ray scattering may prove as advantageous as, or even better than, neutron scattering, especially when using synchrotron radiation. For instance, contrast of agarose in water is high enough to permit investigation of the structure of its gels. In addition, X-ray scattering offers two decisive advantages: (1) no correction of the incoherent background scattered by the hydrogenous agarose is necessary; and (2) deuterium exchange, which reduces considerably the contrast of agarose in heavy water (4 protons out of 18 are exchanged with the environing aqueous solvent), is inherently absent as light water is used instead.

In addition, the high photon flux provided by synchrotron sources allows time-resolved investigations to be carried out, a possibility still out of reach of neutron scattering.

Apart from the contrast, X-ray diffraction does not differ from neutron diffraction. Unlike with neutron, only one structure factor is observed unless heavy atoms labelling is employed.

Light scattering

As a rule, light-scattering studies are now carried out with laser devices that provide a polarized beam with a well-defined wavelength. In most cases two wavelengths are used: $\lambda = 430$ nm and $\lambda = 635$ nm. Note that the tranfer momentum in a medium characterized by a refractive index n_D is written:

$$q = \frac{4\pi n_D}{\lambda_o} \sin \theta/2 \quad (2.62)$$

where λ_o is the wavelength in a vacuum.

In a medium where $n_D = 1.5$, a typical accessible q-range is:

$$0.01 < q \text{ (nm}^{-1}) < 0.01$$

Light scattering is mostly used for characterizing aggregates (clusters), i.e. objects of a finite size as opposed to gels that are infinitely large networks. In this case two parameters can be obtained by the technique: (1) the radius of gyration of the clusters (the *z-average*, noted R_z for short); and (2) their molecular weight (the weight-average, noted M_w). They are determined through the Zimm relation:

$$I^{-1}(q) \propto \frac{1}{M_w} \times \left[1 + \frac{q^2 R_z^2}{3}\right] \tag{2.63}$$

Note that this relation also holds for neutron and X-ray provided that $qR_z < 1$ is always fulfilled. The light wavelength allows studies of large systems, typically from 25 to 200 nm, whereas neutron and X-ray are restricted to particles usually smaller than 25 nm.

Visible photons interact with the dipoles of molecules. The contrast is therefore related to the refractive index, and particularly to the difference in refractive indices between the different components. The contrast factor is expressed as the refractive index increment dn/dC, which is determined from refraction measurements between the pure solvent and the solution. In some cases this parameter is close to zero, so no scattering can be detected.

One point is worth mentioning in relation to the visual aspect of gels exposed to polychromatic daylight. Thermoreversible gels are usually transparent yet slightly turbid. In addition, through a careful observation of a gel one can notice their slightly bluish aspect. This effect arises from their fibrillar architecture wherein the fibre cross-section, unlike the mesh size, is much smaller than the light wavelength.

Consider the scattering by cylindrical objects under the condition $qr < 1$, where r is the fibre cross-section. The scattered intensity reads when introducing Rayleigh's effect, namely for a polychromatic wave $I \sim 1/\lambda^4$:

$$I(q) \propto \frac{1}{q} \times \frac{1}{\lambda^4} \propto \frac{1}{\lambda^3} \tag{2.64}$$

Relation (2.64) indicates that the intensity scattered by short wavelengths will predominate, hence the bluish aspect.

Conversely, suspensions of spherulites are usually whitish. These objects possess spherical symmetry and in most cases $qR > 1$, where R is the spherulite radius. Under these conditions, the intensity is written:

$$I(q) \propto \frac{1}{q^4} \frac{1}{\lambda^4} \propto \frac{1}{\lambda^o} \tag{2.65}$$

Relation (2.65) does not depend upon the wavelength of light, hence the whitish aspect of spherulitic suspensions.

Highly transparent gels are systems of special interest. Their visual aspect probably gave birth to the fringed-micelle model wherein the physical junctions are said to be so tiny as not to give off any forward scattering. While the fringed-micelle model may be encountered (e.g. see Fazel et al. 1994), it turns out that in many cases the match between the refractive index of the polymer and that of the solvent accounts quite naturally for transparency. The case of PVC/bromobenzene is a typical example (see Yang and Geil 1983).

CHAPTER 3

NMR Spectroscopy

NMR spectroscopy is a powerful experimental tool for characterizing both the structure and the dynamics of complex systems at the molecular level. Several monographs which describe in details this technique are available (e.g. see Atkins 1986). Here, it is only intended to present a few aspects that are particularly relevant to the study of polymer–solvent complexes. Modern NMR spectrometers are based on so-called Fourier transform NMR (FT NMR). The sample placed in a static magnetic field B_0 is exposed to a short pulse of the radiofrequency magnetic field B_1. It is shown that, in the so-called rotating frame due to radiofrequency pulse, the vector of nuclear magnetization M_0 (originally along the z-axis, i.e. in the direction of the static magnetic field B_0) rotates by an angle θ given by:

$$\theta = \gamma B_1 t_w \tag{3.1}$$

where γ is the gyromagnetic ratio of the atom experiencing the radiofrequency pulse of intensity B_1 and of length of time t_w. Most often $\pi/2$ pulses are used, with the result that:

$$\gamma B_1 t_w = \pi/2 \tag{3.2}$$

$\pi/2$ pulse rotates the nuclear magnetization **from the z-axis into the x,y-plane** where it is detected immediately by a coil as an oscillating signal which decays as a function of time (so-called FID, i.e. *Free Induction Decay signal*). The NMR spectrum is then obtained from the FID signal by Fourier transformation.

If $S(t)$ is the signal, the NMR spectra, namely $I(\nu)$ where ν is the frequency, is given through:

$$I(\nu) = 2 \operatorname{Re} \int_0^\infty S(t) e^{2\pi i \nu t} \, dt \tag{3.3}$$

Fourier transformation is performed on a computer that is normally an integral part of the NMR spectrometer.

The use of NMR in studies of molecular dynamics is based on relaxation processes. Owing to nuclear magnetic relaxation, the equilibrium population of nuclear states (which is perturbed by the radiofrequency pulse) is restored. The rate of this relaxation process, as described by relaxation times, reflects the internal dynamics of the sample under study. There are two basic relaxation times: *spin–lattice relaxation time* T_1 (also called longitudinal relaxation time because it is connected with the behaviour of the z-component of the nuclear magnetization) and *spin–spin relaxation time* T_2 (also called transverse relaxation time because it is connected with the behaviour of components of nuclear magnetization in the x, y-plane). Relaxation time T_2 is directly connected with the linewidth $\Delta \nu$ through:

$$T_2 = (\pi \Delta \nu)^{-1} \tag{3.4}$$

This relation shows the simplest way to determine the spin–spin relaxation time T_2. The better way is to use more sophisticated techniques based on the so-called *spin-echo* phenomenon (e.g. see Atkins 1986).

For measurements of the *spin-lattice relaxation time* T_1 the inversion-recovery pulse sequence is generally used. This is a two-pulse sequence, a 180° pulse followed by a 90° pulse separated by a variable time τ. The 180° pulse orientates the magnetization vector M_o along the z-axis, which gives $-M_o$, while the 90° pulse rotates the magnetization vector $M(\tau)$ into the *x,y-plane* where the detection coils are sensitive. When the relaxation behaviour of the magnetization follows a single exponential dependence (as is most often the case), the measured intensity I of a line at a given frequency obeys the equation (Figure 3.1):

$$I = I_o [1 - 2 \exp(-\tau/T_1)] \tag{3.5}$$

where I_0 is the equilibrium intensity. For $\tau = 0$, immediately after the second 90° pulse the relation $I = -I_0$ holds, i.e. the magnetization is inverted. As time interval τ increases, I values gradually increase, passing through zero for:

$$\tau = (\ln 2) T_1 \tag{3.6}$$

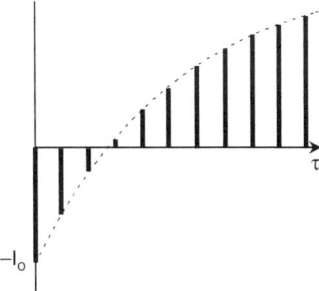

Fig. 3.1. Schematic representation of the increase of intensity as a function of the time τ between the 180° and the 90° pulses. The dotted line stands for equation (3.5).

Therefore, in T_1 measurements, a series of spectra always has to be measured as a function of the time interval τ.

Usually T_1 measurements are performed *nonselectively*; i.e. due to the first 180° pulse in the inversion-recovery sequence all lines in the spectrum are simultaneously inverted and then each of these lines tends to go to equilibrium with individual relaxation time T_1. However, it is possible by selective irradiation to invert only selected signals in the spectrum, while other signals remain unperturbed (Valensin et al. 1982).

An example is shown in Figure 3.2, where *selective* partially relaxed ^{1}H NMR spectra of bromobenzene in sPMMA gels are shown (Spěváček and Suchoparek 1997). Only signals of protons numbered 2 and 6 of bromobenzene are selectively irradiated, while signals of the remaining protons 3,4,5 are unperturbed. For selective irradiation in principle 'soft' pulses (i.e. pulses with low intensity B_1) might be used; generally, some special pulse sequences (e.g. DANTE sequence, see Morris and Freeman 1978) or *Gaussian-shaped* pulses are used for this purpose. Measurements of *selective* relaxation times T_1 represent an important tool for investigating polymer/solvent complexes, as recently shown for gels of sPMMA, iPMMA and PVC (Saiani et al. 1998; Spěváček et al. 1998; Spěváček and Brus 1999).

Assuming a dipolar relaxation mechanism, the *selective*, $R_1(SE)$, and the *nonselective*, $R_1(NS)$ relaxation rates for any proton i are given by:

$$R_1(SE) = (T_1^i)^{-1}(SE) = \sum \rho_{ij} \qquad (3.7a)$$

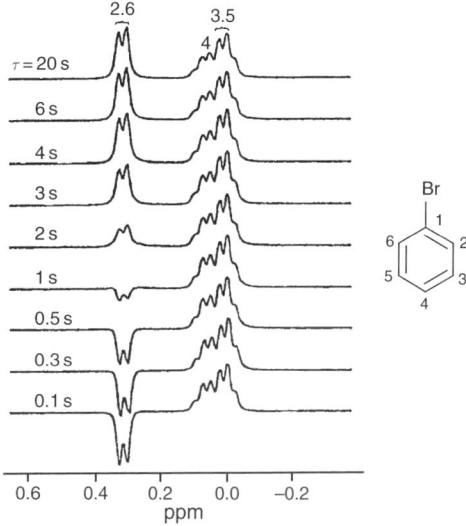

Fig. 3.2. Selective partially relaxed ^{1}H NMR spectra of bromobenzene in s-PMMA gel (Intensity vs chemical shift). **Right:** numbering of the hydrogens on the bromobenzene molecules. (From Spěváček and Suchoparek 1997. Reprinted with permission from ACS.)

and

$$R_1(NS) = (T_1^i)^{-1}(NS) = \sum \rho_{ij} + \sum \sigma_{ij} \qquad (3.7b)$$

In these equations, ρ_{ij} and σ_{ij} are the direct relaxation term and the cross-relaxation term, respectively, for a pair of protons i and j:

$$\rho_{ij} = \frac{h^2 \gamma_H^4}{10\pi^2 r_{ij}^6} x \left[\frac{3\tau_c}{1+(\omega_o \tau_c)^2} + \frac{6\tau_c}{1+4(\omega_o \tau_c)^2} + \tau_c \right] \qquad (3.8a)$$

and:

$$\sigma_{ij} = \frac{h^2 \gamma_H^4}{10\pi^2 r_{ij}^6} x \left[\frac{6\tau_c}{1+4(\omega_o \tau_c)^2} - \tau_c \right] \qquad (3.8b)$$

where r_{ij} is the interproton distance, ω_o the resonance frequency, τ_c the motional correlation time, and other constants have their usual meanings.

From equation (3.6) it follows that *nonselective* relaxation rate $R_1(NS)$ and *selective* relaxation rate $R_1(SE)$ exhibit different dependence as a function of the correlation time τ_c; a marked difference exits in the slow-motion region, i.e. for correlation times $\tau_c > \omega_0^{-1}$ (Figure 3.3). While the dependence of the *nonselective* relaxation rate $T_1^{-1}(NS)$ shows a maximum for $\omega_0 \tau_c \approx 1$, resulting in the fact that $T_1(NS)$ values for the case of the high mobility ($\omega_0 \tau_c \ll 1$) and low mobility ($\omega_0 \tau_c \gg 1$) need not be too different, selective relaxation rate $T_1(SE)^{-1}$ monotonically increases with increasing correlation time τ_c. This makes $T_1(SE)$ values more sensitive to slow motions when, for example, a part of the solvent molecule is bound to polymer chains forming a polymer/solvent complex.

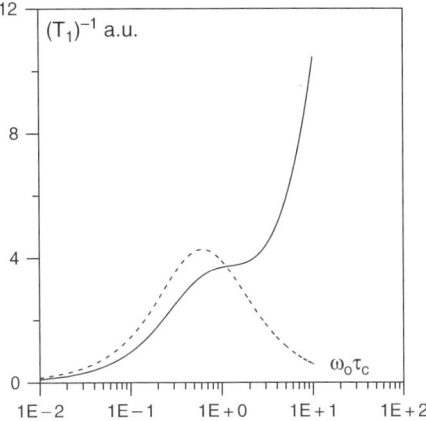

Fig. 3.3. Selective (full line) and non-selective (dotted line) variation of the relaxation rate (in arbitrary units) as a function of the correlation time τ_c rescaled by the resonance frequency ω_o. (From Valensin et al. 1982.)

For the determination of the correlation time τ_c it is convenient to use the ratio $R_1(NS)/R_1(SE)$, where the interproton distance r_{ij} is eliminated; the limiting values for this ratio are 1.5 for $\omega_o\tau_o \ll 1$ and 0 for $\omega_o\tau_o \gg 1$. If it is assumed that a fraction X of the solvent is bound while the remaining fraction $(1 - X)$ is free, and also that there is a rapid exchange between bound and free solvent molecules, the following relation holds.

$$(T_{1\,obs})^{-1} = (T_{1\,free})^{-1} + (T_{1\,bound})^{-1} \tag{3.9}$$

If bound molecules experience slow motion ($\omega_o\tau_c \gg 1$), one can expect to observe, as highlighted in Figure 3.3, a stronger effect on $T_1(SE)$ in comparison with $T_1(NS)$.

NMR also allows determination of the fraction p of polymer engaged in the framework of a gel. Those monomers belonging to this fraction display very low or even no mobility at all, so the corresponding NMR signals are so broad that they escape detection (e.g. see Spěváček and Schneider 1975, 1987). Under these conditions, the fraction of so-called associated polymer p is given through:

$$p = 1 - \frac{I}{I_o} \tag{3.10}$$

where I and I_o are the integrated intensities of the resonance peaks for associated and for non-associated states, respectively.

CHAPTER 4

Some theoretical approaches

Some recent theories, primarily developed for other types of systems, may be relevant to describing the behaviour of *fibrillar thermoreversible gels* after making the appropriate assumptions.

A fractal view of thermoreversible gels

Here, it should be clear that the fractal view developed below is totally at variance with that used as part of the percolation theory (de Gennes 1976; Stauffer 1976). In the percolation approach the fractal dimension of the aggregates formed below the critical degree of conversion is considered. As can be seen from Figure 4.1, the gel is an assembly of fibrils. As a result, it is convenient to define a geometrical parameter, the *longitudinal fractal dimension of the fibrils* connecting at the junctions, which will be of further use in what follows, particularly in accounting for some of the rheological properties of gels using appropriate theory (Jones and Marquès theory for instance).

This parameter can be easily defined by expressing the mean-square radius of gyration of the basic fibril portrayed in Figure 4.2, simply given through:

$$R_z^2 = R_L^2 + r_\sigma^2 \tag{4.1}$$

where R_L^2 is the squared radius of gyration of the long axis. R_L^2 can be expressed by using a relation derived by Loucheux et al. (1958) for polymer chains:

$$R_L^2 = \frac{L_F^{2/D_f}(2l_p)^{2-2/D_f}}{(1+2/D_f)(2+2/D_f)} \tag{4.2}$$

where l_p is the persistence length with respect to the long axis, L_F the contour length of the long axis. D_f is the longitudinal fractal dimension of the fibril, and is defined in the classic way as follows.

$$<S_F^2> \propto L_F^{2/D_f} \tag{4.3}$$

where S_F is the measure of the end-to-end distance of the long axis.

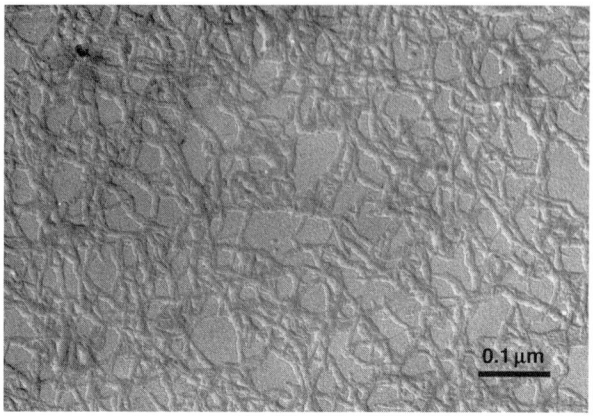

Fig. 4.1. Electron micrograph of an iPS xerogel prepared from 1-chlorododecane. (Guenet, unpublished.)

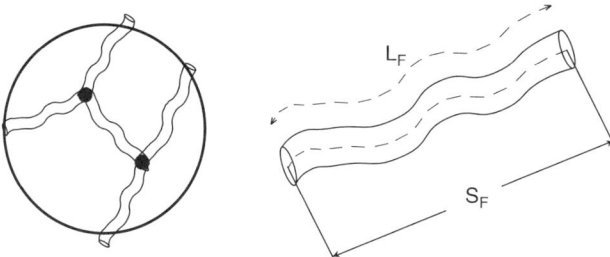

Fig. 4.2. Left: schematic blow-up of a gel of the type shown in Figure 4.1. **Right:** sketch of a basic fibril constituting a gel. S_F is the end-to-end distance of the long axis, while L_F is its contour length.

Finally, r_σ^2 is the cross-section radius of gyration which reads:

$$r_\sigma^2 = r^2/2 \qquad (4.4)$$

where r^2 is the quadratic radius of gyration of the cross-section.

Cross-terms are absent because they cancel out by virtue of the local circular symmetry (Reinecke et al. 2000).

As a rule, fibrils possess a length significantly longer than their cross-section, so the dominant term in equation (4.1) turns out to be R_L^2. The effect is considerably amplified for polydispersed systems. In the case of light scattering, the mean radius of gyration corresponds to the 3rd moment, namely the z-average R_z^2. If there is no correlation between length polydispersity and cross-sectional polydispersity, then R_z^2 simply reads:

$$R_z^2 = <R_L^2>_z + <r_\sigma^2>_z \approx <R_L^2>_z \qquad (4.5)$$

Fibril fractal dimension vs chain persistence length

Intuitively speaking, the longer the chain persistence length, the straighter the fibrils produced by a *simple bunching process of the chains*. For instance, chain persistence length of agarose chains is rather large and agarose gel fibrils are very straight with a longitutinal fractal dimension D_f close to unity. Conversely, iPS chain persistence length is much shorter, and, correspondingly, the fibrils are distorted with a longitudinal fractal dimension D_f larger than 1 (Figure 4.3).

The relation between chain persistence length and fibril longitudinal fractal dimension can be derived on a qualitative basis by using simple arguments. The approach developed here holds only for fibrils where the chain molecular order is nematic-like. As above, let L_F be the contour length of the longitudinal axis, and S_F its end-to-end distance. For the sake of simplicity, we shall assume that the chain end-to-end distance equals S_F; this means that all the chain ends are located within the same disc, which is the fibril cross-section. Chains are assumed to be Brownian and characterized by the same contour length, L and persistence length l_p.

The following relations hold true:

$$S_F^2 \propto N^{2/D_f} a_F^2 \qquad (4.6)$$

assuming that the contour length is constituted of N elements of length a_F, and:

$$s^2 \propto 2Ll_p \qquad (4.7)$$

Equating $s = S_F$ and keeping in mind that the ratio L/a_F is a constant, equation (4.6) eventually yields:

$$l_p/a_F \propto N^{2/D_f} \qquad (4.8)$$

which can be rewritten:

$$Log\left(l_p/a_F\right) \propto 2/D_f \qquad (4.9)$$

Relation (4.9) therefore suggests that the larger the l_p, the straighter the fibrils.

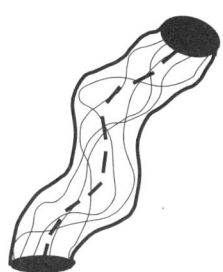

Fig. 4.3. Schematic representation of a fibril made up by bunching chains (thin threads) together without disturbing their worm-like conformation. The dashed line stands for the fibril long axis.

The critical gelation concentration C_{gel}

The temptation is great to relate the critical gelation concentration C_{gel} to the chain overlap concentration C^*. As has already been emphasized above, this would be meaningful if below C_{gel} individual chains would constitute the sol state, which is now known to be wrong (e.g. see Candau et al. 1987). Below C_{gel} aggregates are already formed, so C_{gel} simply corresponds to the concentration at which the average size of these aggregates diverges to infinity. Evidently, the way C_{gel} can be expressed mathematically depends upon the morphology of the aggregates. By taking a naïve approach to this problem by considering that the aggregates are assemblies of basic fibrils as described in Figure 4.2, C_{gel} can be written:

$$C_{gel} = \frac{6M}{\pi S_F^3} \qquad (4.10)$$

where M is the fibre molecular weight and S_F the end-to-end distance with respect to the long axis. Mathematically, this is the same expression as that used for deriving C^*.

The molecular weight is further expressed as:

$$M = \pi \rho r^2 L_F \qquad (4.11)$$

where ρ is the density. This eventually gives for C_{gel}:

$$C_{gel} = \frac{6\rho r^2}{L_F^{(3/D_f - 1)}} \qquad (4.12)$$

Equation (4.12) shows that, for r being a constant, the lower the fractal dimension, the lower the C_{gel}. This is effectively what happens if one compares agarose ($D_f \approx 1$) and PVC ($D_f \approx 1.5$). Also, if r is too large, gelation may not occur at all, and only a suspension of fibres can be produced, not to mention the probable branching.

The variation of C_{gel} with chain molecular weight is probably an irrelevant question. Indeed, fibres are certainly not built up by chains associating either side-to-side or end-to-end, or both. Most probably splicing processes come into play, giving rise to branching, and, as already underlined above, chains may display loops that render the comparison between chain length and fibre length meaningless.

In the case of a polydisperse system, assuming that polydispersity in lengths and in cross-sections are uncorrelated, $<C_{gel}>$ can be written:

$$<C_{gel}> = \frac{6 \sum_i N_i r_i^2 \sum_j N_j L_j}{\sum_i N_i \sum_j N_j L_j L_j^{(3/D_f - 1)}} = \frac{6\rho <r^2>_n}{<L_F^{(3/D_f - 1)}>_w} \qquad (4.13)$$

where $<>_n$ and $<>_w$ stand for the number average and the weight average, respectively. $<C_{gel}>$ is therefore more sensitive to the short cross-section fibres of significant length.

Thus far we have assumed that the fibre morphology and the fibre length remain unchanged up to C_{gel}. Yet, as observed with PVC aggregates, this is not true, so deriving a mathematical expression for this parameter becomes exceedingly difficult. While relations (4.12) and (4.13) are useful on a qualitative basis, they are no longer sufficient to derive a quantitative approach of the gelation threshold.

Another snag concerns the gel state, as its corresponding C_{gel} may be a function of the preparation concentration C_{prep}. For instance, fibre cross-sections have been observed to vary with C_{prep} (see PVC gels in Chapter 6). Clearly, values derived from the dilute side (determining the divergence of the aggregate size for instance) may not be useful for the concentrated side, namely the gel.

Finally, C_{gel} can be very dependent upon ageing processes such as those taking place in PVC or gelatin.

Fibril stability

In the case of a fibril consisting of crystallites alternating with disordered domains (Figure 4.4), cross-sectional growth is limited if chain folding is absent because of the appearance of instabilities at the domain interface (Guttmann et al. 1979). Fibrils of this type are encountered in PVC gels.

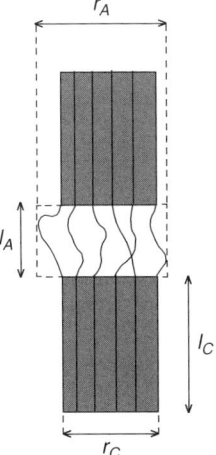

Fig. 4.4. Sketch of a fibril with alternating disordered and crystalline domains.

The conditions for stability can be derived from simple arguments. As a rule the density in the disordered domains is lower than that in the crystallites ($\rho_a < \rho_c$). ρ_a and ρ_c can be expressed as:

$$\rho_a = \frac{n_a \mu L_a}{\pi r_a^2 l_a} = \frac{n_a \mu}{\pi r_a^2} \left(\frac{L_a}{l_p}\right)^{1/2} \quad \text{as } l_a = L_a l_p$$

$$\rho_c = \frac{n_c \mu l_c}{\pi r_c^2 l_c}$$

(4.14)

in which μ is the mass per unit length of the chains, n with the appropriate subscript the number of chains in the different domains ($n_a = n_c$ in the absence of chain-folding), and r and l with the appropriate subscripts the cross-sectional radius and the length of the different domains (a = disordered domain, c = crystalline domain) respectively. L_a is the contour length of the chains in the disordered domain; l_p is their persistence length. It is assumed that μ is independent of the domain the chains cross, which usually holds true to a good approximation. Here, it is assumed that disordered chain portions that connect two adjacent crystalline domains display a Gaussian behaviour, hence $l_a = L_a l_p$.

This following inequality is finally obtained:

$$r_a \geq r_c \left(\frac{\rho_c}{\rho_a} \times \frac{n_a}{n_c}\right)^{1/2} \left(\frac{L_a}{l_p}\right)^{1/4}$$

(4.15)

This clearly shows that the cross-sectional radius of the disordered domain must exceed that of the crystalline domain. Yet, r_a cannot be physically larger than a certain value related to the chain contour length L, namely:

$$r_a \leq r_c + L_a/2$$

(4.16)

This eventually gives the following relation for the maximum value of the crystalline domain cross-section r_c^{max}:

$$r_c^{max} \propto \frac{L_a}{\left(\frac{\rho_c}{\rho_a} \times \frac{n_a}{n_c}\right)^{1/2} \left(\frac{L_a}{l_p}\right)^{1/4} - 1}$$

(4.17)

For $n_a = n_c$ and $L_a > l_p$ equation (4.17) suggests that r_c^{max} increases when increasing L at constant l_p. Also, r_c^{max} is supposed to become infinite for:

$$\frac{n_a}{n_c} = \frac{\rho_a}{\rho_c} \times \sqrt{\frac{l_p}{L_a}}$$

(4.18)

keeping in mind that $\rho_a \approx \rho_c$ in real cases, then:

$$\frac{n_a}{n_c} \approx \sqrt{\frac{l_p}{L_a}}$$

(4.19)

Some theoretical approaches

This shows that the more flexible a chain is, the higher the content of chain-folding that should be introduced to make the fibril stable, and vice versa. These outcomes bear out those conclusions drawn by Guttmann et al. (1979) derived from numerical simulation.

The arguments used to derive relation (4.17) are purely steric and do not take into account thermodynamic stability. As has been discussed in the section on Gibbs phase rules in Chapter 1, the minimum values for r_c^{min} and l_c^{min} are given through thermodynamic parameters:

$$\left(\frac{2}{r_c^{min}} + \frac{2}{l_c^{min}} \right) = \left[1 - \frac{T_m}{T_m^o} \right] \times \frac{\Delta h_f}{\sigma} \tag{4.20}$$

in which T_m^o is the melting temperature of an infinite crystal and T_m the temperature at which the actual crystal stands. Δh_f is the melting enthalpy of the crystal and σ its surface free energy. l_c^{min} is determined by the length of the crystalline sequence and as a result directly governs the value of r_c^{min}. Clearly, if r_c^{min} as calculated from equation (4.20) becomes larger than r_c^{max} then the fibril cannot exist.

Rheological behaviour: modulus vs concentration

No specific theory on the mechanical properties of thermoreversible gels has been so far developed. In recent writings (Guenet 1992; Guenet 2000) the author has argued that the Jones and Marquès theory on the elasticity of rigid network can be used to account for the behaviour of the elastic modulus as a function of the polymer volume fraction of the network (Jones and Marquès 1990). A brief account of this theory is given here.

Two different processes describe the elastic behaviour of polymer systems: one is of *entropic* origin and/or the other one is of *enthalpic* origin. **Entropic elasticity** occurs with flexible objects and involves a conformational change. The best known example is rubber elasticity where the variation of free energy under deformation arises essentially from a change of entropy of the chains. For instance their end-to-end distance is significantly altered.

Conversely, *enthalpic elasticity* occurs in rigid systems. It consists essentially in bending objects without conformational change. The dominant term in the free energy is therefore the enthalpic contribution.

A network is an array of objects connecting at junctions. On this basis, these two types of elasticity can be defined in a more general fashion. If we consider the mean square fluctuation $<\delta\psi^2>$ of the angle ψ between the vectors joining the ends of each of two objects connecting at the same junction, then entropic elasticity occurs when $<\delta\psi^2> \neq 0$ and enthalpic elasticity for $<\delta\psi^2> = 0$.

Note that entropic elasticity may occur for rigid yet freely hinged objects connecting at the same junction since the condition $<\delta\psi^2> \neq 0$ is fulfilled.

Jones and Marquès theory is intended for polymer networks where the objects are individual chains. Their theory can be applied to thermoreversible gels provided that the size of the junctions be negligible compared to S_F, and also that $S_F \gg r_\sigma$.

a) Enthalpic elasticity

Consider a network where junctions are separated by a distance R at which n prolate objects are connected (n is commonly designated as the *functionality*). If F is the force acting on each element the stress σ is written:

$$\sigma \approx \frac{nF}{R^2} \tag{4.21}$$

If X is the deflection of the element by the force F, the strain ε is X/R if affine deformation is assumed. Young's modulus is then simply written:

$$G_r = \frac{\sigma}{\varepsilon} = \frac{nF}{XR} \tag{4.22}$$

For rigid objects, the force is $F = kX$ where k is a spring constant, which is expressed as:

$$k \approx \frac{Ba}{NaR^2} \tag{4.23}$$

where a is the elementary length of the prolate object. Ba is its bend constant, which is written:

$$Ba = er_\sigma^4 \tag{4.24}$$

in which e is the intrinsic modulus of the object. Introducing the fractal dimension R is written:

$$R \propto N^{D_f} a \tag{4.25}$$

For an arbitrary fractal dimension, Young's modulus reads:

$$G_r \propto \frac{er_\sigma^4 n}{N^{(3+D_f)/D_f} a^4} \tag{4.26}$$

Introducing the network volume fraction φ_{net}:

$$\varphi_{net} = \frac{nNar_\sigma^2}{R^3} \tag{4.27}$$

G_r becomes:

$$G_r \propto \frac{er_\sigma^4 n}{a^4} \times \left(\frac{\varphi_{net} a^2}{nr_\sigma^2}\right)^{(3+D_f)/(3-D_f)} \tag{4.28}$$

Note that for $D_f = 1$ (cylindrical objects), G_r does not depend upon the cross-section whereas it does for higher values of the fractal dimension.

In the case of straight cylinders G_r varies as φ^2 for a constant n whereas at constant R but varying n leads to $G_r \sim \varphi$.

b) Entropic elasticity

Entropic elasticity varies as kT and assumes that Young's modulus is proportional to the number of junctions per unit volume, namely to the reciprocal of the cube of the mesh size l_n. Scaling arguments yield (Daoud et al. 1975):

$$l_n \propto \varphi_{net}^{1/(D_f-3)} \tag{4.29}$$

hence:

$$G_e \propto kT \times \varphi_{net}^{3/(3-D_f)} \tag{4.30}$$

Here, freely hinged cylinders would give $G_e \propto \varphi_{net}^{1.5}$, an exponent significantly differing from that occurring for enthalpic elasticity.

Whatever the type of elasticity, the network volume fraction is usually not equivalent to the polymer volume fraction φ_{pol}. This is especially so in the case of biphasic systems where the network volume fraction is only related to the polymer-rich phase. By applying the lever rule, φ_{net} is written:

$$\varphi_{net} = \frac{\varphi_{pol} - \varphi_{poor}}{\varphi_{rich} - \varphi_{poor}} \tag{4.31}$$

where φ_{poor} is the polymer-poor phase volume fraction and φ_{rich} the polymer-rich phase volume fraction (namely the stoichiometric concentration for compounds). These two parameters are constant at constant temperature and can be determined from the phase diagram.

If φ_{poor} is negligible, a case often encountered in thermoreversible gels, then:

$$\varphi_{net} = \frac{\varphi_{pol}}{\varphi_{rich}} \tag{4.32}$$

In the case of compounds, φ_{rich} can differ considerably from one solvent to another. Such a case occurs for isotactic polystyrene gels (see Chapter 15).

On the definition of thermoreversible gels

A system that possesses the mechanical properties of a solid is the simplest definition that can be put forward for ascertaining the gel status of a two-phase system where one component is a liquid. This implies that an elastic modulus at either zero frequency or infinite time can be measured. This definition is merely based on the mechanical aspect of the system. It usually applies to chemical gels, but fails in most cases when thermoreversible gels are dealt with. For instance, Guenet and McKenna (1986) have reported considerable relaxation in iPS/decalin gels (about 10% per decade) although the gel status of these systems appears offhand to be unquestionable. These systems can be handled in the same way as chemical gels, and are therefore not highly viscous solutions.

Other definitions consider that a gel is a system of infinite viscosity. Simple experiments, such as tube tilting, are then used. If the system does not flow, it is regarded as a gel

50   Polymer-Solvent Molecular Compounds

(e.g. see Prasad and Mandelkern 1990). Yet, many systems do not display flow behaviour while not being gels (e.g. assemblies of spherulites produced from crystallization in solutions, humid sand).

Unlike chemical gels, where *covalent bonds* form the cross-links between chains, *van der Waals interactions* or *hydrogen bonds* are involved in the network formation process of physical (thermoreversible) gels. These bonds are potentially labile under the action of a stress. Clearly, the total number of junctions will not vary but their positions with respect to one another will. Accordingly, rheological definitions based on enduring bonds are worthless.

Instead of considering a definition based only on rheological properties, Guenet (1992) has suggested using two main criteria:

1. A *topological criterion*: a gel is a *network*. According to the accepted definition, a network is *a large system of lines, tubes, wires, etc. that cross one another or are connected with one another* (Longman Dictionary of Contemporary English, 1987). Assemblies of spherulites do not fulfil this definition.
2. A *thermodynamic criterion*: gel formation and gel melting should proceed via a *first-order transition*.

Later, Daniel et al. (1994) devised a simple mechanical test (not a definition) to differentiate a gel as defined above from other systems that may be considered gels on the basis of the tube tilting method. These authors reason that, if the relaxation under a given compressive deformation arises only from the lability of the junctions, there should be some reversibility of the deformation process. The experimental test is described in Figure 4.5.

First a deformation λ_1 is applied to the sample, and relaxation is allowed to proceed for a given time. The deformation is suddenly altered to λ_2 with $\lambda_2 > \lambda_1$. For a gel, what occurs at the microscopic level can be viewed as follows: under the deformation λ_1

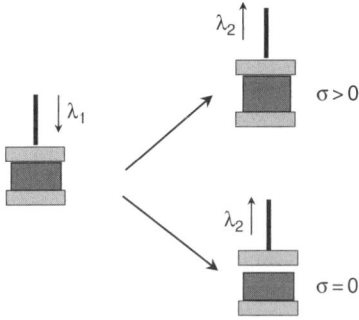

Fig. 4.5. Sketch of the test used by Daniel et al. (1994) for establishing the gel status of a system. Whether the system is irreversible or partly reversible the resulting stress σ is either zero or differs from zero. (See text for details.)

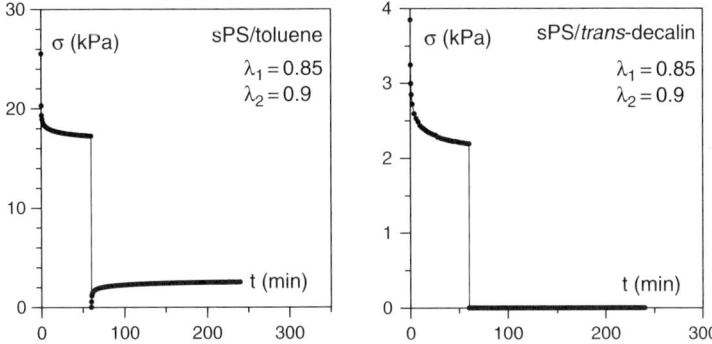

Fig. 4.6. Use of the test of Daniel et al. (1994) with sPS/toluene (fibrillar gel) and sPS/*trans*-decalin (assembly of spherulites). In both cases the polymer concentration is $C = 0.1 \text{ g/cm}^3$. (After Ramzi and Guenet, unpublished.)

the fibrils between junctions are bent but the bending gradually decreases owing to the rearrangement of the labile junctions, hence the occurrence of a relaxation. After altering the position of the piston to λ_2 the sample, and, correspondingly, the junctions, are no longer submitted to a deformation, so the fibrils can unbend reversibly. This entails partial recovery of the initial shape of the sample to such an extent as to produce the reappearance of a stress as soon as the sample pushes onto the piston (see Figure 4.5). Clearly, if there exists a deformation λ_2 for which such a behaviour can be observed, then there is some degree of reversibility of the process. Conversely, in the case of an assembly of spherulites produced from crystallization in solution, the cohesion between spherulites is usually very weak, so the deformation is virtually irreversible. As a result, no stress reappears after moving the piston of the compression device from λ_1 to λ_2 (Figure 4.6).

Chain conformation

In this section is given a succinct account of the theory developed by Frederickson (1993) for linear polymer chains complexed with oligomeric surfactants such as polyaniline. This theoretical approach is also useful for explaining qualitatively the behaviour of solvated polymer chains as occurs in gels from stereoregular polymers.

Consider a linear chain containing binding sites on to which surfactant molecules of an average size R_M can attach. Different regimes can be observed depending upon the degree of coverage σ ($\sigma \sim 1/R_M$). Frederickson has studied in detail the effects in a good solvent. If M is the molecular weight of the surfactant molecules it is assumed that $R_M \sim M^{3/5}$, although in the case of short surfactant molecules this is not strictly true. In what follows R_M will be used so that no assumption is made as to the surfactant conformation.

In the low-coverage limit, namely $\sigma \ll 1/R_M$, Flory's theory suffices to estimate the coil expansion arising from complexation. The free energy in a good solvent is simply written:

$$\frac{F}{kT} = \frac{R^2}{Na^2} + a^3 \frac{N^2}{R^3} + R_M^3 \frac{(\sigma N)^2}{R^3} \tag{4.33}$$

where R is the chain radius of gyration, and N the number of elements of statistical length a in the chain. The first two terms stand for the classic Flory relation (Flory 1953). Equation (4.33) requires minimization with respect to R in order to determine the equilibrium dimension of the complexed chain. In the very low coverage limit $\sigma \ll R_M^{-3/2}$, one retrieves the Flory radius $R_F \sim N^{3/5}$. The effect of complexation comes into play when the third term in equation (4.32) becomes larger than the second term, i.e. for $R_M^{-3/2} \ll \sigma \ll R_M^{-1}$. Under these conditions the radius of gyration becomes:

$$R_c \propto R_F \sigma^{2/5} R_M^{3/5} \tag{4.34}$$

In the high coverage limit, namely for $\sigma > 1/R_M$, a bottle-brush structure is formed, which considerably enhances the persistence length of the complexed polymer. Frederickson calculates the persistence length of the bottle-brush structure by considering a toroidal configuration where the curvature $C = 1/\rho$ greatly exceeds the cross-sectional radius R_M. The change in free energy from that of the undeformed state is determined by expanding to the second order (the first order being nought as the free energy does not depend upon the sign of the curvature). The persistence length is simply identified as the coefficient of this quadratic term in the free energy times ρ^2. For the free energy of the unperturbed state, the expression derived by Wang and Safran (1988) is used:

$$\frac{F}{kT} = a^{-1} \sigma^{13/8} R_M^{5/8} \tag{4.35}$$

which gives, for the case of the toroidal configuration:

$$\frac{F}{kT} = a^{-1} \sigma^{13/8} R_M^{5/8} \left[1 + \frac{113}{490\pi^2} c^2 + .. \right] \tag{4.36}$$

in which c is a reduced curvature, i.e. $c = R_{mo}/\rho$, and R_{mo} is the cross-sectional radius for the rod-like structure. This radius will be altered on bending, so:

$$R_{Mo} \propto a\sigma^{1/4} R_M^{5/4} \tag{4.37}$$

The persistence length l_p finally scales as:

$$l_p \propto a\sigma^{17/8} R_M^{25/8} \tag{4.38}$$

It is interesting to examine the results obtained for iPS/*cis*-decalin and iPS/*trans*-decalin systems in the SOL state in the light of this theory (see Chapter 15). By neutron scattering, it is found that the persistence length of the chain in the former system is about twice that found in the latter. Phase diagrams indicate that the stoichiometric composition is 1.75 *cis*-decalin/monomer against 1.15 *trans*-decalin/monomer. The stoichiometry is nothing but σ, and R_M can be regarded as the radius of the solvent molecules, which is virtually identical for *cis*-decalin and *trans*-decalin. Relation (4.38) indicates that the ratio of persistence lengths should be about 2.4, in good agreement with the experimental result.

Scale-free networks

As has been discussed in Chapter 2, Guenet (1994) observed that the scattering curve of fibrillar network in the condition where $ql_n > 1$, where l_n is the average mesh size, could be accounted for by considering fibril cross-sections characterized by a large polydispersity, where this polydispersity obeyed a power-law distribution. The size of the fibril cross-section is directly related to the number of chains participating in a given fibril, and the biggest cross-sections are the result of fibrils interacting side by side (Figure 4.7). Clearly, there are few fibrils of very large cross-section and many of small cross-section, hence the power-law distribution. This type of structure has been observed in the so-called scale-free networks; it characterizes such networks as airline networks where there exist a few large hubs and a wealth of small hubs (Albert et al. 1999; Baraba'si and Albert 1999). Schematically speaking, in thermoreversible, fibrillar gels the 'big hubs' are those domains where fibrils are interacting with one another and the 'small hubs' are the smallest fibrils.

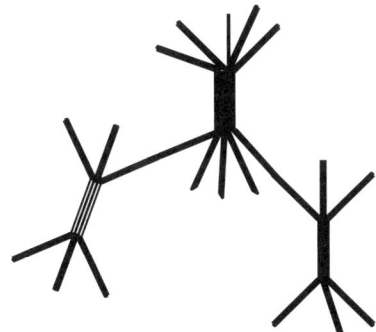

Fig. 4.7. Schematic representation of a fibrillar, thermoreversible network where interacting fibrils create 'big hubs'. As shown by scattering experiments, these networks are characterized by a distribution of cross-sections obeying a power-law. (Guenet 1994.)

Water taken in moderation cannot hurt anybody

Mark Twain (notebook, 1935)

PART II

Biopolymer complexes

As a rule, biopolymers can establish hydrogen bonds either between themselves or with polar molecules such as water. In many cases biopolymers such as polysaccharides are relatively rigid chains that possess the propensity to form regular helical structures. As a result, these helical chains tend to form more organized structures, and, if water is present, tend to generate molecular compounds with this special solvent through hydrogen-bonding interactions. This happens very often in living organisms, and is one way of storing water molecules. Other polar molecules, including amines, can also form molecular compounds with these biopolymers.

In view of the two types of molecular compounds discussed in the introduction, those formed with biopolymers are of the *enthalpic* type and should, rather, be designated as complexes, although some also have the characteristics of intercalates. This term will be, however, restricted to *entropic* molecular compounds as detailed in Part I.

In Part II most of the systems discussed are polysaccharides (cellulose, agarose, amylose, chitin), but some data are also presented for DNA.

CHAPTER 5

Cellulose

Cellulose is the most abundant naturally occurring polymer on earth. It is mainly found in all higher plants acting as a structural material in the cell wall. It is the major constituent of wood, cotton, etc. and, as such, has a wide range of industrial applications. Cellulose is a polysaccharide consisting of $1 \rightarrow 4$ linked β-D-glucose residues (Figure 5.1). The chemical repeat unit consists, however, of two residues, the spatial placement of the second one being obtained by a simple rotation. The morphology of cellulose is microfibrillar as revealed by electron microscopy and diffraction techniques (Manley 1964; Nieduszynski and Preston 1970; Chanzy 1975; Chanzy and Roche 1976).

As is the case for all polysaccharides, cellulose crystals cannot be melted prior to chemical degradation, and there is but a small number of solvents in which cellulose chains are readily dissolved. These solvents are usually highly polar, such as DMSO-paraformaldehyde, N-methylmorpholine N-oxide and hydrazine at high temperature (Johnson 1970; Litt and Kumar 1970; Johnson et al. 1976).

Cellulose I and II

Meyer and Misch (1937) first derived the crystalline structure of cellulose in its native state: it consists of a *centred monoclinic crystalline lattice* with parameters $a = 0.835$ nm, $b = 0.79$ nm, $c = 1.03$ nm (fibre axis) and $\gamma = 96°$ (so-called *cellulose I*; Figure 5.2). Later work on a highly crystalline cellulose obtained from the sea alga *Valonia ventricosa* revealed a number of weak reflections which could not be indexed by the Meyer and Misch lattice (Honjo and Watanabe 1958; Nieduszynski and Atkins 1970). In order to index these reflections the different groups of workers considered a unit cell four times the size of that proposed by Meyer and Misch: $a = 1.634$ nm, $b = 1.572$ nm, $c = 1.038$ nm (fibre axis). According to Gardner and Blackwell (1974) the larger 8-chain unit cell could be due to small shifts of H-bonded sheets past each other along the a axis or in the ac plane. Note that the chains take on an extended conformation, namely a 2_1 helix (as is apparent from Figure 5.2), and also that the crystal structure is stabilized by a network of hydrogen bonds within the 020 crystallographic plane only, as highlighted

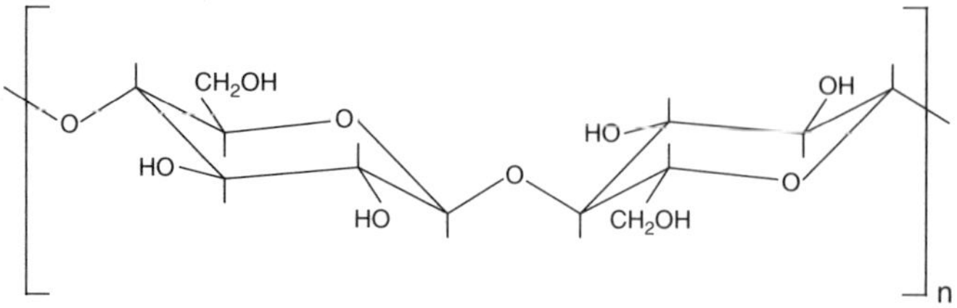

58   Polymer-Solvent Molecular Compounds

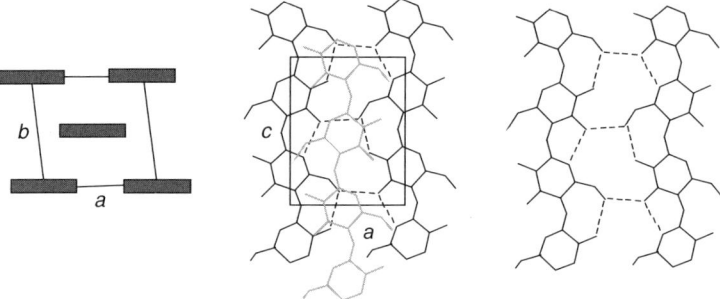

Fig. 5.1. Chemical repeat unit of cellulose.

Fig. 5.2. Left: projections of the unit cell of cellulose I as seen parallel to the chain axis (rectangles stand for the cross-section of the chains), and perpendicular to the chain axis (**middle**). Dashed lines indicate hydrogen bonds within the 020 plane (**right**). (After Gardner and Blackwell 1974.)

in Figure 5.2. No hydrogen bonding can be established along the unit cell diagonals or along the *b* axis (Gardner and Blackwell 1974).

Note 5.1   Crystallographic planes indexing

In some papers the *b* crystallographic axis is taken as the fibre axis while in others the *c* direction is. For the sake of consistency, we shall systematically consider, as is customary, the *c* direction as the fibre axis. As a result, *hkl* values mentioned in this chapter MAY differ from those used in the original publication. For instance, the 101 plane referred to in some publications corresponds here to the 110 plane.

It has been further shown by means of a refinement procedure on the same type of cellulose sample that the chains are definitely parallel in the *cellulose I* lattice, and that there is a so-called *quarter stagger* of the central chain with respect to the corner chains (Gardner and Blackwell 1974; Sarko and Muggli 1974). Whether the same holds true for the less crystalline native cellulose remains under discussion, but seems highly probable as its diffraction pattern is nearly identical to that of *Valonia* cellulose albeit being less well resolved. The parallel arrangement of the chains in the crystal unit cell implies the

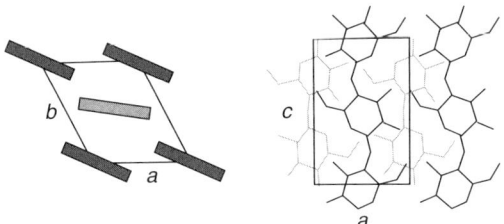

Fig. 5.3. Projections of the crystalline unit cell of cellulose II as seen **left**, parallel to the chain axis (rectangles stand for the cross-section of the chains, and the different shadowing schematize parallel and anti-parallel chains) and **right**, perpendicular to the chain axis. (After Kolpak and Blackwell 1975.) Stipanovic and Sarko have derived a model close to this one although differing in some details (Stipanovic and Sarko 1976).

absence of chain-folding, and therefore cellulose chains are completely extended within the microfibrils.

The 1-4 linkage between anhydroglucose residues in the cellulose chemical repeat unit makes it probable from the viewpoint of diffraction that a 2_1 helix is present. As a result the first meridional reflection should occur for $l = 2$, i.e. for the 002 plane (the space group is therefore $P2_1$). There are, however, some very weak 001 reflections that, according to some authors, probably arise from defects along the chain.

Cellulose I can be converted into *cellulose II* (see Figure 5.3) by allowing native cellulose to swell in a concentrated alkali solution followed by a thorough washing with water, or while crystallizing from solutions. Swelling and transformation in concentrated alkali is designated as *mercerization* after J. Mercer, who invented the process (Mercer 1850). There is so far no known procedure for reverting *cellulose II* back to *cellulose I*. *Cellulose II* seems to be entirely artificial although there is at least one report in the literature of its occurrence in one form of algae (Sisson 1941).

Similarly, *cellulose II* can be grown from dilute solutions, and in such a circumstance *cellulose II* is designated as *regenerated cellulose* (Maeda et al. 1970; Ramesh et al. 1973).

The crystalline unit cell of *cellulose II* was first determined by Andress (Andress 1929). It also consists of a *monoclinic crystalline lattice* of parameters $a = 0.814$ nm, $b = 0.974$ nm, $c = 1.03$ nm (fibre axis) and $\gamma = 118°$.

The major change with respect to *cellulose I* is the anti-parallel packing of chains (Stipanovic and Sarko 1976; Kolpak and Blackwell 1975). Anti-parallelism implies a change of chain direction: there is an up-chain and a down-chain that can be superimposed by a rotation of 180° (see Note 5.2). The appearance of an anti-parallel chain in the crystalline unit cell could have arisen from chain-folding, which provides the necessary rotational transformation. However, as will be shown below, transformation of *cellulose I* into *cellulose II* proceeds by another mechanism wherein the formation of compounds plays a major role.

> **Note 5.2 Parallel and anti-parallel chains of cellulose**
>
> As shown below (*left*) one can define a direction in cellulose chains (usually designated as *polarity*, a term somehow misleading for the layman as there is no electric charge involved).
>
> A mere 180° rotation allows transformation of up-chains into down-chains, and vice versa. Chain-folding, as occurs in lamellar crystals, can do the trick (*right*). Chain *polarity* utterly differs from chain chirality because chains of differing handedness are mirror-image, and cannot therefore be transformed into one another by a simple rotation. Note that chain polarity also exists with synthetic polymers such as stereoregular polystyrenes.

Although the dimensions of the crystalline unit cells have been known for years, determination of the fine details of the unit cell, namely the positioning of atoms and the chain direction, has only recently been made possible thanks to computer-assisted stereochemical structure refinement together with improvement of the diffraction data by using highly crystalline samples (Stipanovic and Sarko 1976); Kolpak and Blackwell 1975). Note that, unlike *cellulose I*, there is also a network of hydrogen bonds both in the 020 plane and in the 110 plane.

While *cellulose I* is the naturally occurring form of cellulose, *cellulose II* is energetically more stable. It is thought that *cellulose I* usually arises from the simultaneous action of biosynthesis and crystallization.

Complexes of cellulose

Cellulose has the propensity to form complexes with a large variety of polar molecules, among which alkalis and amines are particularly noteworthy for the existence of systematic and in-depth studies thereon. Other systems are known to produce complexes, in particular, inorganic acids such as nitric acid and perchloric acid.

Complexes with soda. Cellulose I–cellulose II transformation

As mentioned above, mercerization or regeneration of cellulose, namely the formation of *cellulose II*, is achieved through the action of concentrated *caustic soda* (sodium hydroxide) solutions (in most cases above 2N; e.g. see Chédin and Marsaudon 1955; Warwicker and Wright 1967; Rousselle et al. 1976; Rousselle and Nelson 1976).

Cellulose then forms well-defined compounds with sodium hydroxide whose stoichiometry is dependent upon the caustic soda concentration. It is suggested that clusters of water molecules surrounding the metal ion is the active agent responsible for the swelling of cellulose in aqueous alkali solutions (Dobbins 1970; Warwicker 1971). The following stoichiometries are accepted (see Warwicker 1971 for a review):

$$\begin{array}{ll} \text{Soda cellulose 1} & C_6H_{10}O_5/NaOH/3H_2O \\ \text{Soda cellulose 4} & C_6H_{10}O_5/NaOH/2H_2O \\ \text{Soda cellulose 2} & C_6H_{10}O_5/NaOH/1H_2O \\ \text{Soda cellulose 5} & C_6H_{10}O_5/NaOH/5H_2O \end{array}$$

These compounds are not affected by a change of temperature but are unstable when washing with water. Under these conditions a water-cellulose complex is formed, which contains only water molecules with a tentative stoichiometry of 1 water molecule per anhydroglucosic residue (Sakurada and Okamura 1937).

The various stoichiometries can be explained in terms of the solvated structure of the ions, namely the amount of water carried by the ions on adsorption within the cellulose crystalline lattice (Heuser and Bartunek 1925; Saito 1939). Chédin (1952, 1955) and Chédin and Marsaudon (1954, 1955, 1956) have developed the ideas of soda hydrates as active in the swelling process. A reaction is postulated between pairs of solvated ions and cellulose following the equation:

$$Na^+OH.nH_2O + R_{cell}(OH)_3 \rightarrow R_{cell}(OH)_3.NaOH.(n-3)H_2O + 3H_2O$$

In other words, the three hydroxyl groups per anhydroglucose residue replace three molecules of water from the pair of solvated ions, and one molecule of soda is fixed per anhydroglucose unit. The value of n decreases with increasing the sodium hydroxide fraction, so the stoichiometry of the cellulose/soda complex should contain fewer and fewer water molecules when exposed to soda solutions of higher and higher concentrations.

Compound formation occurs without going through a sol state, namely without complete dissolution of cellulose, but only through a swelling process. Warwicker and Wright (1967) have shown by X-ray diffraction studies on various cellulose samples that the distance associated with the 110 plane increases quite markedly, unlike the distance corresponding to the 1$\bar{1}$0. These authors therefore rightly conclude that complexation occurs essentially by insertion of soda molecules between the 110 plane of the *cellulose I* crystalline lattice (see Figure 5.4). Such a scheme implies the breaking up of the hydrogen bonds that have been shown to stabilize the *cellulose I* structure (Gardner and Blackwell 1974). Only van der Waals interactions held the chains together in the sheets formed by the 110 planes. The absence of hydrogen bonding gives the possibility of chain-folding with adjacent re-entry to take place within these sheets. After removal of the intercalated sodium hydroxide *cellulose II* can then be obtained.

As has been emphasized by Warwicker and Wright, considerable disturbance and disorientation of the cellulose crystalline structure results from the action of caustic soda. Rousselle and Nelson (1976) have reported a subsequent drop of the degree of crystallinity in cotton fibres from 75% down to 50%. Similarly, the morphology is strongly

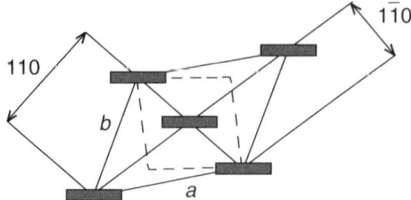

Fig. 5.4. Representation of the swelling of the initial *cellulose I* crystalline lattice (dashed line) after treatment in sodium hydroxide. The distance associated with the 110 plane is modified whereas that corresponding to the 1$\bar{1}$0 plane remains virtually unaltered. (After Warwicker and Wright 1967.)

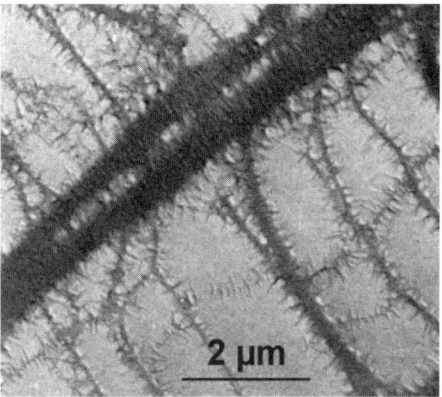

Fig. 5.5. Electron micrograph of cellulose of *Valonia macrophysa* after transformation from *cellulose I* into *cellulose II*. (After Chanzy and Roche 1976. Reprinted with permission from Wiley Interscience.)

affected, as shown by Chanzy and Roche (1976) on the cellulose material obtained from *Valonia macrophysa*. A shish-kebab structure grows at the expense of the original fibrillar structure (see Figure 5.5). This suggests the formation of chain-folded crystals as occurs for synthetic polymers such as polyethylene (see Pennings and Kiel 1965). Indeed, the conformational energy map established by Sarko shows that a tight fold is energetically allowed, which makes adjacent re-entry possible for cellulose chains (Sarko and Muggli 1974; Sarko 1976).

The mechanism for the transformation of *cellulose I* into *cellulose II* is, however, not thought to proceed in this way. A series of experiments by Sarko and co-workers (Okano and Sarko 1984, 1985; Nishimura and Sarko 1987) has revealed a mechanism where as many as five intermediate *Na-cellulose* forms are involved (*Na-cellulose* meaning complexes with soda). The interconversion scheme of these *Na-celluloses* is shown in Figure 5.6.

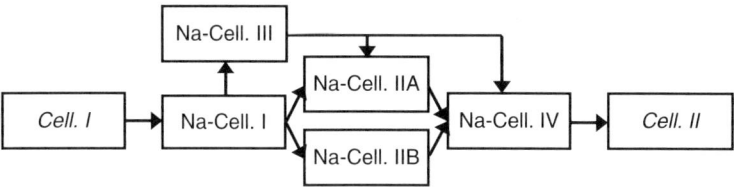

Fig. 5.6. Interconversion scheme of Na-celluloses during mercerization. (After Okano and Sarko 1984.)

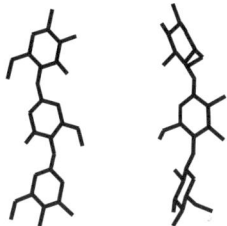

Fig. 5.7. Right: threefold helix for Na-cellulose IIB (left-handed). **Left:** the twofold helix for the sake of comparison. (After Okano and Sarko 1985.)

In *Na-cellulose I, III* and *IV* the chains still adopt a 2_1 helical conformation characterized by a 1.03 fibre repeat, whereas in *Na-cellulose IIA* and *IIB* Okano and Sarko have observed a fibre repeat unit of 1.5 nm. They have assigned this fibre repeat distance a threefold helix, as shown in Figure 5.7.

Otherwise the main differences lie in the size of the unit cell (Table 5.1). The high values for *a* and *b* simply reflect the swelling due to soda intercalation.

Okano and Sarko suggest that *Na-celluloses* fall into three distinct classes: (1) in the first class, *Na-celluloses I* and *III* that have similar unit cells, with chains adopting a twofold conformation, and the same stoichiometry; (2) in the second class, *Na-celluloses IIA* and *IIB* whose unit cell markedly differs from that of the previous *Na-celluloses*; (3) in the third class, *Na-cellulose IV* whose unit cell resembles that of *cellulose I* except that it is probably hydrated.

Table 5.1. Crystallographic characteristics of Na-celluloses. (After Okano and Sarko 1985.)

Na-cellulose	a (nm)	b (nm)	c (nm)	γ	Chains/unit cell	Density (g/cm^3)	% NaOH
I	1.17	2.57	1.012	94	6	1.8	33
IIA	1.29	1.05	1.548	112	2		
IIB	1.494	1.494	1.539	120	2	1.4	65
III	1.254	2.64	1.028	115	6	1.8	34
IV	0.988	0.965	1.028	126	2	1.5	0

A mechanism whereby *cellulose I* transforms irreversibly into *cellulose II* has been proposed by Okano and Sarko on the basis of these findings. This mechanism relies on the following observations:

- Cellulose fibres are usually composed of a large number of small crystallites (crystalline fibrils wherein chains are either all up or down) of less than 10 nm in diameter.

- Transformation occurs more easily in semi-crystalline native cellulose. Highly crystalline native cellulose is more difficult to transform.

- The 'amorphous' interface regions contain a mixture of up and down chains still under an extended conformation.

Okano and Sarko suggests that the amorphous domain swells first under the action of soda, thus giving birth to *Na-cellulose I*, a crystalline form which already contains up and down chains. Further absorption of soda in the *cellulose I* domains leads to the formation of *Na-cellulose IIA* and *IIB*, where, thanks to the threefold helical form, contact between adjacent cellulose chains is removed. This favours a random reshuffling of the stems, eventually producing the conditions for the formation of *cellulose II* after subsequent removal of soda.

Complexes with liquid ammonia

Several authors have observed that complexes can be formed at low temperature with *liquid ammonia* (NH_3) (Hess and Trogus 1935; Barry et al. 1936; Clark and Parker 1937). The use of liquid NH_3 in industrial applications is an alternative to soda processing (for a review see Herrick 1983).

As with soda solutions the main effect is the swelling between 110 planes while leaving the distance between cellulose stems within the 110 plane virtually unaffected. The resulting distance is strongly dependent upon the way the sample is prepared and studied, as highlighted by conflicting results (Barry et al. 1936; Clark and Parker 1937). The same problem is encountered as to the stoichiometry. It varies from 1/1 to 6/1 NH_3 *per anhydroglucose residue*. The former is obtained from weighing the sample after evaporation of the free NH_3 while the latter is calculated on the basis of the total amount of absorbed ammonia. Clearly, the former value is probably too low as the compound may not be totally stable in the open atmosphere, while the latter value may be too high because amorphous or disordered domains may absorb more than the crystalline phase.

As will be detailed in a forthcoming section, removal of ammonia does not give rise to *cellulose II*, unlike the case with soda, but to a new polymorph *cellulose III*.

Complexes with amines

The formation of complexes between *cellulose I* and amines was reported as early as 1931 by Trogus and Hess for diamines (Trogus and Hess 1931), and Davis et al. for

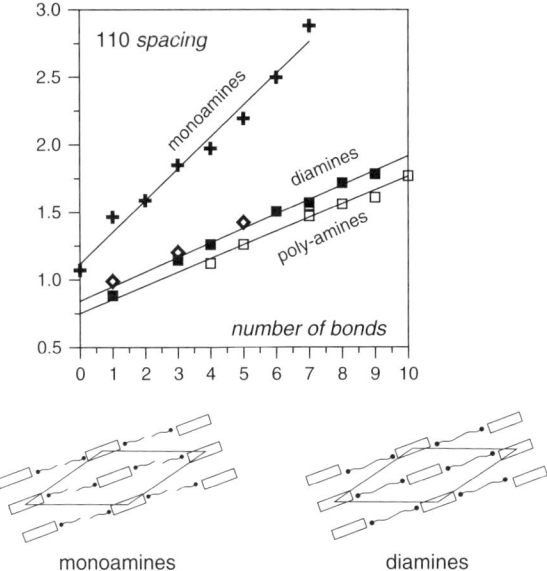

Fig. 5.8. Upper: Variation of the 110 spacing as a function of the number of bonds in the amine molecules (this is truly proportional to the length of the extended molecule). (□) After Creely et al. 1959; (+) after Davis et al. 1943; (◆) after Trogus and Hess 1931; (■) after Creely and Wade 1978b.) Note that the case where the number of carbons is 0 corresponds to hydrazine. **Lower:** probable placement of the monoamines (left) and the diamines (right) between the 110 planes.

monoamines (Davis et al. 1943). These authors observed that, after complex formation, the 110 reflexion was shifted towards smaller angles, which meant that the interplanar distance subsequently increased (Figure 5.8). Further confirmation of this effect was later given by Creely et al. on a series of diamines (Creely et al. 1959). Interestingly, the variation of the interplanar distance depends on whether monoamines or diamines are dealt with. As a rule, the distance depends upon the length of the amines but there is a ratio of about 2 between diamines and monoamines (see Figure 5.8). According to Howsmon and Sisson the bifunctional diamine molecules form hydrogen bond cross-links between two cellulose molecules (Howsmon and Sisson 1954). Monoamines also form hydrogen bonds but with only one cellulose molecule and are said to form a bilayer, as is illustrated in Figure 5.8. This is probably so because the aliphatic tail of the monoamines cannot mutually interdigitate due to steric hindrance. These models account nicely for the ratio 1:2 between monoamines and diamines. It is worth mentioning that this type of model has been also suggested for PVC–ester complexes (see The Sol State in Chapter 6).

Further work on amine type has been carried out by Creely and Wade, in particular with cyclic amines such as *p*-phenylene-diamine (Creely and Wade 1978a). They, however, dwell upon the fact that formation of a complex with cyclic amines requires pretreatment with ethylamine. According to these authors, ethylamine disrupts the hydrogen bonds between the 110 planes, and thus opens the way to cyclic amines, the compounds eventually being formed. Indeed, these compounds are unstable on exposure to air as decompostion takes place within 1–3 hours. In the case of p-phenylene-diamine/*cellulose I* compound, these authors find a stoichiometry of *two p-phenylene-diamines per cellulose* repeat unit (or one p-phenylene-diamine per anhydroglucose residue).

Creely and Wade have also shown that polyamines can form different complexes depending upon the way the differing amine functions interact with cellulose chains (Creely and Wade 1978b). For instance, in the case of diethylene triamine (DETA; $H_3N(CH_2)_2NH(CH_2)_2NH_3$), two complexes can be identified. One complex involves the terminal primary amine and the secondary amine function. Each amine function interacts with cellulose so as to bridge two chains located in two adjacent 110 planes (designated as the contracted form; Segal and Loeb 1960). The other complex occurs with both terminal primary amines (designated as the extended form complex). This effect is clearly evidenced in Figure 5.8 where the 110 interspace distance is plotted as a function of the length of the part of the polyamine molecule involved in the complex (i.e. the number of bonds in the polyamine molecule).

Blackwell and co-workers have made an extensive study of the molecular arrangement of the cellulose chains and the solvent within these complexes. The crystal lattices are dependent upon the type of cellulose as expected (*cellulose I* or *cellulose II*) but also upon the content of water in the hydrazine sample, as well as upon the morphology of the cellulose sample. The discovery of the occurrence of different complexes with the two different polymorphs of cellulose was first made by Trogus and Hess in the early thirties (Trogus and Hess 1931). For the *cellulose I*/hydrazine complex a monoclinic crystalline lattice was proposed with $a = 0.968$, $b = 0.996$ and $\gamma = 54.8°$ (Halle 1934; Creely et al. 1959). However, as was underlined by Lee and Blackwell, all these results were obtained with hydrazine containing 40% water, so a ternary complex may form cellulose/hydrazine/water (Lee and Blackwell 1981). In the case of hydrazine samples containing only 3% water, Lee and Blackwell have determined the crystalline unit cell of three compounds, the values of which are given in Table 5.2:

Table 5.2. The crystal unit cell parameters for the different complexes between cellulose and hydrazine. (After Lee and Blackwell 1981;*after Lee et al. 1983 once structure refinement has been used.)

sample	a (nm)	b (nm)	c (nm)	γ
cellulose I (ramie)	0.919	1.639	1.037	97.4
cellulose II (fortisan)*	0.937	1.988	1.039	120
cellulose II (merc. ramie)	0.948	1.649	1.037	96.4

Mercerized ramie is obtained by swelling ramie cellulose in caustic soda, while fortisan is produced from solutions (regenerated cellulose). The fortisan/hydrazine complex possesses the highest stoichiometry with regard to mercerized ramie/hydrazine (three hydrazine molecules per cellulose repeat unit for the former against 0.5 for the latter). Lee and Blackwell suggest that the different morphologies of the cellulose samples together with the different degrees of crystallinity may be responsible for this effect.

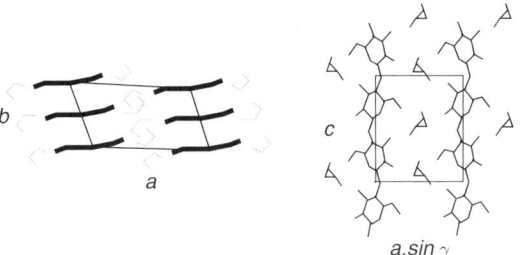

Fig. 5.9. Crystalline unit cell of *cellulose I*/diethylamine as seen parallel to the chain axis (**left**), and its projection down the b axis (**right**); there, the aspect of the solvent molecules arises from the superposition of two of them. $a = 1.287$ nm, $b = 0.952$ nm, $c = 1.035$ nm, $\gamma = 118.8°$. (After Lee et al. 1984.)

Cellulose I/diethyl diamine is one of the most crystalline compounds that are stable in the absence of water. As a result, Lee et al. have been able to carry out a more detailed analysis of the crystalline unit cell by means of the refinement procedure of the X-ray data together with stereochemical criteria (Lee et al. 1984). Owing to the observed meridional réflexion for $l = 2$, a 2_1 helical model of repeat unit 1.035 nm containing two glucose residues has been chosen. The crystalline unit cell is monoclinic with $a = 1.287$ nm, $b = 0.952$ nm, $c = 1.035$ nm and $\gamma = 118.8°$ (see Figure 5.9). The stoichiometry of the compound is 2:1 (two ethylene diamines per cellulose repeat unit, or one per 1 anhydroglucose residue). Note that the chains are parallel as in the parent *cellulose I*, but that the quarter stagger is lost. The occurrence of such a crystal structure implies the scission of the intermolecular hydrogen bonds together with the disruption of the stacks of the quarter staggered chains.

Henrissat et al. (1987) have given further support to this structure by means of solid state ^{13}C NMR, in particular as to the placement of the methyl hydroxyl group.

A detailed study of *cellulose II/hydrazine* compound (fortisan, i.e. regenerated cellulose) has been carried out by Lee et al. (1983). By the same approach as that used for *cellulose I*/diethyl amine compound they have arrived at the structure portrayed in Figure 5.10.

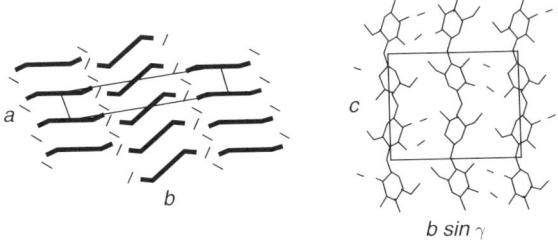

Fig. 5.10. Crystalline unit cell of *cellulose II*/hydrazine as seen parallel to the chain axis (**left**) and its projection down the b axis (**right**). $a = 0.937$ nm, $b = 1.988$ nm, $c = 1.039$ nm, $\gamma = 120°$. Chain polarity varies from one 020 plane to the next. (After Lee et al. 1983.)

Again, a 2_1 helix of fibre repeat 1.039 nm is considered. Lee and colleagues stress that uncertainty remains as to the exact hydrazine positions and the hydrogen bonding network. They also contemplate the possibility that some water molecules could be present in addition, or even replacing some of the hydrazines. Despite these minor points the compounds formed either in hydrazine or diethylamine highlight the ability of cellulose to form highly stable stacks of unstaggered chains, which may arise from the hydrophobic forces between surfaces of the glucose rings, as pointed out by Lee et al. Note that the distance between cellulose chains within the stacks is invariably the same independent of the complex considered (about 1 nm).

As will be discussed below, a new polymorph, namely *cellulose III*, is produced on removing amine molecules from the complex similar to that obtained with liquid ammonia.

Complexes with inorganic acids

The complex formation with nitric acid has been studied by several authors although chiefly limited to the crystalline unit cell and the stoichiometry (Katz and Hess 1927; Andress 1928; Chédin and Marsaudon 1954). Andress reports a crystalline unit cell with parameters $a = 1.22$ nm, $b = 0.973$ nm, $c = 1.028$ nm and $\gamma = 53.7°$. The complex involved is said to be $C_6H_{10}O_5/HNO_3$ (Katz and Hess 1927; Trogus 1934).

Andress and Rheinhardt (1930) found that a complex occurs between native cellulose and perchloric acid. From the X-ray diffraction pattern, they derived the following parameters for the crystalline unit cell: $a = 1.65$ nm, $b = 1.07$ nm, $c = 1.03$ nm and $\gamma = 93°$. The stoichiometry is reported to be $2C_6H_{10}O_5/HClO_4$, a value confirmed by Lieser and Fichtner (1941).

Cellulose III

As has been highlighted in the section devoted to cellulose complexes prepared from ammonias, a new polymorph can be obtained on removing the solvent. Instead of reverting to *cellulose I* a new structure is obtained which is designated as *cellulose III$_I$*. The index simply indicates that this polymorph is obtained from the complexed *cellulose I*. Indeed, there exists a *cellulose III$_{II}$* which is produced from *cellulose II* complexes. Both polymorphs can be prepared from cellulose/amine complexes. Another polymorph exists which is obtained by a special heat treatment (*cellulose IV*). The crystalline unit cells of *cellulose III$_I$* and *cellulose IV* are shown in Table 5.3 (Marchessault and Sarko 1967; Sarko et al. 1976).

The *cellulose III$_{II}$* unit cell is nearly identical to that of *cellulose III$_I$*. The only difference lies again in chain polarity: parallel packing in *cellulose III$_I$* and antiparallel packing in *cellulose III$_{II}$*.

Sarko et al. (1976) have resolved the structure of the crystalline unit cell of *cellulose III$_I$* (see Figure 5.11) by combined stereochemical structure refinement and X-ray diffraction

Table 5.3. Unit cell dimensions of cellulose polymorphs III_I and IV.

Polymorph	a (nm)	b (nm)	c (nm)	γ
Cellulose III_I	1.025	0.778	1.034	122.4°
Cellulose IV	0.79	0.811	1.03	90°

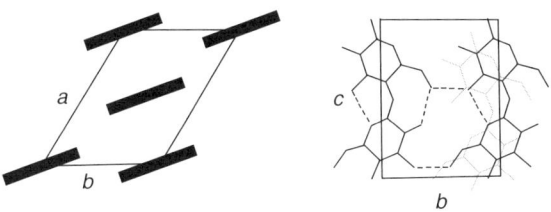

Fig. 5.11. The crystalline unit cell of *cellulose III_I* as determined by Sarko et al. (1976). $a = 1.025$ nm, $b = 0.778$ nm, $c = 1.034$ nm, $\gamma = 122.4°$. The dotted lines represent the hydrogen bonds.

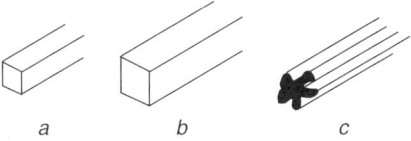

Fig. 5.12. Schematic drawing of a *Valonia* microfibril (a) undergoing complexation, and correspondingly swelling, with an amine (b) followed by removal of the amine for producing *cellulose III_I* and after reversion to *cellulose I* (c). As can be seen, despite the same crystalline structure the fibril in (c) has been considerably damaged. (After Chanzy et al. 1986.)

analysis. As can be seen, the network of hydrogen bonding is re-established in the 100 plane, as with *cellulose I*, yet a reminder of the complexed state is in evidence through the still-existing stacks along the 110 direction.

Immersing *cellulose III* in hot water allows reversion of *cellulose III_I* to *cellulose I*. Yet, as shown by Chanzy and co-workers, this reversibility is only at the level of the crystal unit. Fibrils that originally had a square cross-section are subsequently damaged after reversal to *cellulose I* (Figure 5.12) (Roche and Chanzy 1981; Chanzy et al. 1986).

CHAPTER 6

Agarose

Agarose is a polysaccharide occurring in agar, a constituent of cell walls in red seaweeds. It is an alternating copolymer consisting of 1,4-*linked* 3,6-*anhydro-α-L-galactose* and 1,3-*linked β-D-galactose* (Figure 6.1).

Agarose has the propensity to form thermoreversible gels in water and in binary aqueous solvents. The gels display a fibrillar morphology (see for instance Hickson and Polson 1968; Griess et al. 1993a, b; Sugiyama et al. 1994). Under no circumstances were spherulitic structures observed for this polymer. Thanks to their controllable porosity, agarose gels are used extensively in biology as a medium for electrophoresis and for DNA sequencing.

Early studies performed by means of polarimetry experiments led one to conclude that agarose gelation proceeded through a *coil–helix transition* (e.g. see Rees et al. 1969). Feke and Prins (1974) further suggested, on the basis of a time-resolved light-scattering study, that a *spinodal decomposition* was involved. As will be discovered throughout this book, spinodal decomposition may interfere with gelation but is rarely the driving mechanism. Also, recent neutron scattering investigations of the sol state by Guenet et al. (1993) and Rochas et al. (1994) have called into question the *coil–helix* transition and they rather favour a *loose helix–tight helix* transition, a notion put forward by Atkins (1986).

The helical structure involved in the gel structure is still a matter of controversy. Arnott et al. (1974b) have suggested considering the occurrence of a double helix made up by intertwining two threefold left-handed helices of pitch 1.9 nm (Figure 6.2). Historically speaking, the agarose double helix was contemplated at a time when double helical structures as those observed with carrageenans seemed to be a characteristic shared by polysaccharides. Admittedly, diffraction patterns obtained on gels produced from ι-carrageenans and/or κ-carrageenans contain a large number of reflections that are unambiguously accounted for by means of double-stranded helices (e.g. see Anderson et al. 1969; Arnott et al. 1974a; Millane et al. 1988). Conversely, the paucity of the diffraction patterns recorded on agarose gels opens possibilities (Arnott et al. 1974b). In fact three layer lines are usually detected, the first occurring at a spacing of 0.95 nm,

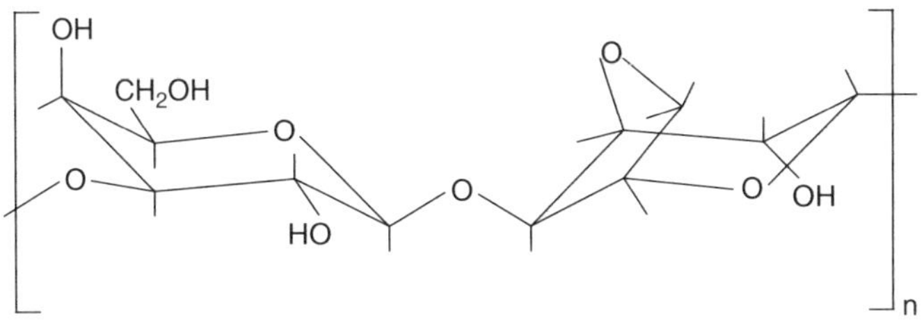

Agarose

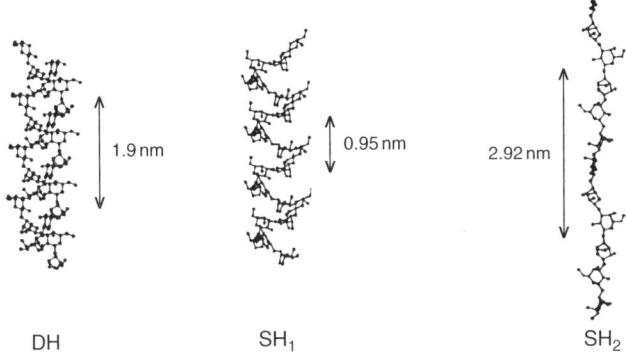

Fig. 6.1. Agarose repeat unit.

Fig. 6.2. The double helix (DH, **left**) as proposed by Arnott et al. (1974b) and two typical single helices: 3_1 form of pitch 0.95 nm and axial rise 0.317 nm, SH_1 (**centre**); and pitch 2.92 nm and axial rise 0.973 nm, SH_2 (**right**) proposed by Foord and Atkins (1989).

the first meridional reflection corresponding to a spacing of 0.317 nm. These layer lines have been interpreted by Arnott and colleagues as even orders of a fundamental threefold helix of pitch $P = 1.9$ nm and axial rise of 0.634 nm respectively. The odd orders are expected to be cancelled due to the intertwining of these helices with a precise axial stagger of $P/2$ (the first meridional therefore corresponds to $l = 6$). Note that the diffuse diffraction pattern indicating limited order is oddly at variance with the supposedly precise stagger P/2 strictly required for cancelling the odd layer lines. Recently, Foord and Atkins re-examined the helical structure in the light of experiments performed on stretched and dried gels. The preparation procedure tremendously improves the diffraction patterns that are then only consistent with single helices. Unless disruption of the double helices occurs while stretching through a mechanism yet to be invented, these experiments certainly cast doubt on the very existence of double-stranded helices in agarose gels. As a result, Foord and Atkins suggest considering single helices instead. Two typical examples, which are consistent with the diffraction pattern, are given in Figure 6.2 (Foord and Atkins 1989). In particular, single helix SH_1 is consistent with the diffraction pattern observed on gels stretched at room temperature. This time the meridional reflection corresponds to the third layer line. Ab-initio calculations further show that these helices are energetically feasible (Foord and Atkins 1989; Jimenez-Barbero et al. 1989; Kouwijzer and Pérez 1998) unlike what was stated by Arnott et al. (1974b).

Although it is suspected that water is probably occluded in the gel fibres, no mention of complex formation has been explicitly made until recently. Neutron scattering experiments carried out on samples in the sol state turned out to be instructive on that point (Guenet et al. 1993; Rochas et al. 1994). A thermodynamic study of gelation in aqueous binary solvent also provided information in connection with complex formation (Ramzi et al. 1996).

The sol state

As mentioned above, agarose chains were thought to take on a flexible, Gaussian coil conformation in the sol state. This statement relied upon polarimetry experiments that showed the loss of the helical structure present in the gel state. Guenet (1993) and Rochas (1994) have investigated the short-range chain conformation by small-angle neutron scattering in a series of water/DMSO mixtures. These authors have observed that agarose chains are in fact very rigid and can be described by means of worm-like statistics. The chains are therefore globally Brownian, but locally rigid, i.e. rod-like. The distance l_p below which agarose chains can be regarded as rod-like (the persistence length, see Part I) is estimated to be larger than $l_p \approx 9$ nm (Figure 6.3). This conformation is found independent of the binary solvent: from pure water to pure DMSO, and also in water/NaSCN medium. Clearly, this conformation is remote from the so far supposed flexible coil.

Of incidental note, the fact that agarose chains are quite rigid in the sol state accounts for why no chain-folded crystals can be produced from this biopolymer. The re-entry

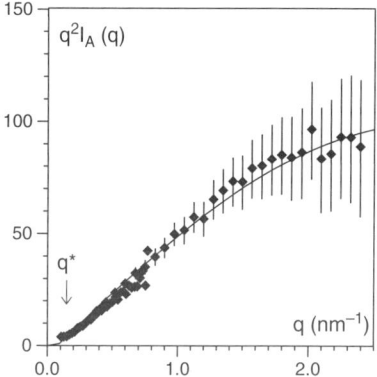

Fig. 6.3. Scattering curve for agarose in DMSO, $C_{agarose} = 30$ g/l. q* highlights a possible departure from the 1/q behaviour at low q. This may stand for the cross-over between the 1/q and $1/q^2$ regimes expected for worm-like chains in which case $l_p \approx 9$ nm. Alternatively, if the solution is semi-dilute, then this cross-over does not correspond to the persistence length but to the screening length. In this case l_p may be larger than 9 nm.

length would be too large, making the crystals highly unstable. Conversely, a fibrillar structure is far more stable.

At large q the departure from the 1/q behaviour arises from a cross-section effect (see Part I for details). The cross-section of the chains can be evaluated through the following relation:

$$I(q) = KC_{agarose} \frac{\pi \mu_L}{q} \times \frac{4 J_1^2(qr_H)}{q^2 r_H^2} \tag{6.1}$$

in which r_H is the cross-section radius and μ_L the mass per unit length (see Part I for details). Values experimentally determined in water are: $r_H = 0.35 \pm 0.15$ nm and $\mu_L = 315 \pm 30$ g/nm/mol.

These results taken together suggest that the chains take on a *loose helix* conformation in the sol state quite close to the single helix of type B as described by Foord and Atkins. This result led Guenet and Rochas to wonder how worm-like chains with such persistence length could intertwine to give rise to the double helix put forward by Arnott et al. Clearly, these results provide further support to the views put forward and defended by Foord and Atkins.

Guenet and Rochas have also noticed that the mass per unit length varies with the DMSO content in water/DMSO binary solvents (Rochas et al. 1994). As can be seen in Table 6.1, this parameter can even be lower than that the agarose chains would have should they take an all-extended conformation ($\mu_L = 300$ g/mol/nm). As this situation is clearly impossible, the mass per unit length as determined by neutron scattering must be considered apparent. According to Rochas and Guenet this effect can be accounted for by contemplating the existence of an agarose/solvent complex. This means that solvent molecules are tightly bound to the agarose chains, so a new entity, the complex, should be considered for calculating the contrast factor K of equation (6.2). For a complex containing n solvent molecules, K is then proportional to:

$$K \propto \left[a_{agarose} + na_s - \frac{\lambda(v_{agarose} + nv_s)}{v_s} a_s \right]^2 \tag{6.2}$$

in which a stands for the scattering amplitude, v for molar volume and m for molar weight, with subscript s corresponding to the solvent (here only one type of solvent is considered for the sake of simplicity). λ is an adjustable parameter which takes into account the fact that the molar volume of a complex is seldom the sum of the molar volumes of each species involved in the complex. As can be seen, if $\lambda \neq 1$, the value

Table 6.1. The agarose mass per unit length as a function of the solvent composition in the sol state. (From Rochas et al. 1994.)

water/DMSO (V/V)	100/0	70/30	50/50	30/70	0/100
μ_L (g/mol/nm)	360 ± 36	341 ± 36	293 ± 36	256 ± 36	230 ± 36

of K can be lower than it would be in the absence of a complex thus yielding lower scattered intensities, and ultimately lower values for the mass per unit length. Ramzi has determined by the variation contrast method the fraction of hydrogenous DMSO needed to match the coherent scattering of the agarose/DMSO complex (Ramzi 1996). This method consists in determining the fraction of hydrogenous DMSO required for having $K = 0$ (see Part I for details). In the event of no complex forming this volume fraction should be $X_{DMSOH} = 0.41$. Experimentally, Ramzi has obtained $X_{DMSOH} = 0.5$, which is a clear indication of complex formation. In the present case it is difficult to determine λ precisely without knowing n, yet it is likely that $\lambda < 1$ as $a_s > a_{agose}$. Experiments have been carried out in this connection by Rochas et al. for water/DMSO binary mixtures, these researchers coming up tentatively with $n = 2$ and $\lambda = 0.975$ as probable, sensible values (Rochas et al. 1994).

Admittedly, this effect can originate in preferential adsorption of one solvent on to the agarose chains in the case of a binary solvent. However, this mechanism cannot be invoked in the case of pure DMSO. Gamini et al. (1997) have came up with the same conclusions, namely the occurrence of a complex, from NMR investigation. Also, Ramzi et al. have used the contrast variation method for the gel state in binary solvents and found a discrepancy between theoretical and experimental values, thus giving support to the complex effect (Ramzi [thesis] 1996; Ramzi et al. 2000a). That a neutral polysaccharide, as is agarose, can form a complex with polar molecules is not surprising in itself. As will be discussed in Part II, amylose is known to co-crystallize with a large number of solvents.

Ramzi et al. have obtained additional circumstantial evidence on the occurrence of complexes in the ternary system agarose/water/DMSO by studying the dynamic behaviour of aggregates by means of electric birefringence (Ramzi et al. 2000b). This technique allows one to access the rotational diffusion coefficient D_{rot} by application of a pulsed electric field.

They have observed that the rotational coefficient is strongly dependent on the applied electric field (Figure 6.4). This behaviour is most unusual, and therefore suggests that the size of the aggregates decreases with increasing the intensity of the electric field. Clearly, application of an electric field breaks up agarose aggregates. In the range of electric field values studied the variation of D_{rot} is linear, so Ramzi et al. have analysed their results by considering the derivative dD_{rot}/dE. This parameter together with D_{rot} decreases with increasing agarose concentration, which therefore suggests that aggregates are larger and are less affected by the electric field. It turns out that, at constant agarose concentration, dD_{rot}/dE goes through a maximum at $f_{DMSO} \approx 0.2$. This led Ramzi et al. to suggest that ternary agarose complexes are formed, and that these complexes can be destroyed through the application of an electric field. Possibly, the electric field affects the orientation of the solvent molecules within the complex, and thus destabilizes it, hence the break-up of the aggregates. As will be shown below, the DMSO fraction at which the effect is maximum ($f_{DMSO} \approx 0.2$) corresponds to the stoichiometric composition of the ternary complex. It is worth mentioning that similar effects have been observed recently for syndiotactic polystyrene compounds irradiated with light (Itagaki et al. 2005). Probably, the disorientation of the solvent by some external field entails the destabilization of the complex, and correspondingly its 'melting'.

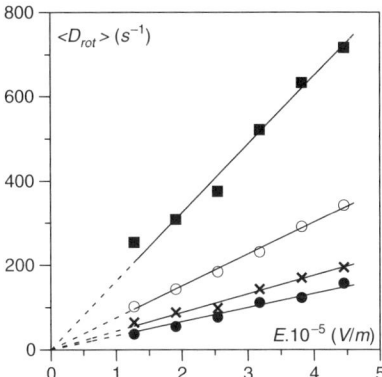

Fig. 6.4. Variation of the rotational diffusion coefficient of agarose aggregates with the intensity of the electric field. (■) $C_{agarose} = 0.03$ g/l; (○) $C_{agarose} = 0.05$ g/l; (×) $C_{agarose} = 0.08$ g/l; (●) $C_{agarose} = 0.1$ g/l; $f_{DMSO} = 0.2$. (Data from Ramzi et al. 2000b.)

The gel state

Thermodynamics

The temperature–composition phase diagram for the ternary system agarose/water/DMSO was first established by Watase and Nishinari (1988) and later confirmed by Ramzi et al. (1996).

As can be seen, both the gelation and the melting temperatures display a maximum for a molar ratio of about 5H$_2$O/1DMSO (Figure 6.5). Watase and Nishinari have interpreted this effect by considering the strong interaction that is known to exist between water and DMSO (Cowie and Toporowski 1961; Rasmussen and MacKenzie 1968). This effect can be expressed through Flory's relation for ternary systems (see Part I for details):

$$\frac{1}{T_m^{1,2}} - \frac{1}{T_m^1} = -\frac{R V_p}{\Delta H_p V_1} \times G_{\varphi_p}(\varphi_i, \chi_{ij}) \qquad (6.3)$$

in which $T_m^{1,2}$ is the polymer melting temperature in a binary solvent, T_m^1 the polymer melting temperature in the reference solvent (water for instance) and $G_{\varphi_p}(\varphi_i, \chi_{ij})$ a function of the different interaction parameters involved. This function can exhibit a maximum when there exists a strong interaction between the solvents. The binary solvent may then behave as a poorer solvent than the poorest of the two solvents. As a result, a maximum in the melting temperature may occur for a given composition of the binary solvent.

To account for the simultaneous increase of the associated enthalpies, Watase and Nishinari consider that the degree of crystallinity of the gel has increased, which means that more agarose is participating in the network or there are fewer and fewer amorphous

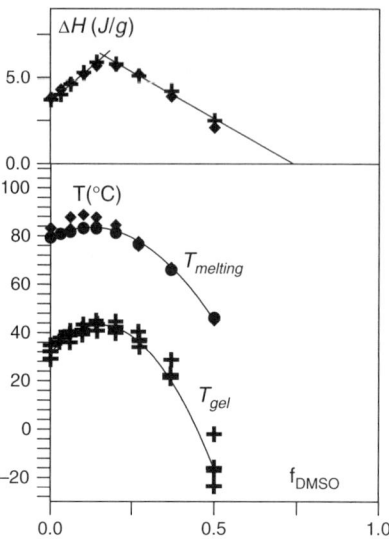

Fig. 6.5. Temperature–composition phase diagram for agarose/water/dimethyl sulphoxide system. Gel formation (T_{gel}, +), gel melting ($T_{melting}$, ◆ = $C_{agarose}$ = 5 g/l; • = $C_{agarose}$ = 10 g/l) and the associated enthalpies display a maximum at $f_{DMSO} \approx 0.17$. The gel melting temperature depends little upon agarose concentration. (Data from Ramzi et al. 1996.)

domains in the fibrils. In both cases, however, one should observe a corresponding increase in the gel elastic modulus. As shown by Ramzi et al. such is not the case: the elastic modulus remains virtually unchanged; meanwhile the associated enthalpies increase by about 60%. Evidently, this interpretation is not self-consistent.

Ramzi et al. have suggested that the phase diagram may be interpreted in terms of agarose/solvent complexes instead. This means that within the gel fibres agarose and solvent(s) form crystallo-solvates. To fulfil the rules for phase diagrams they have to consider the coexistence of a *binary complex agarose/water* and a *ternary complex agarose/water/DMSO* in the composition domain where the melting temperature is seen to increase. That agarose forms a binary complex with water does not come as a surprise since it is known that agarose powder always contains a non-negligible fraction of water. Conversely, a ternary complex is a new notion for this polysaccharide. Incidentally, the probable occurrence of such a complex in the gel state is rather consistent with the conclusions drawn from neutron scattering experiments and birefringence studies in the sol state that have been discussed above. The ternary complex in the sol state is simply kept in the gel state. Note that the increase of associated enthalpies is, this time, consistent with and even required for the occurrence of a complex. The total stoichiometry of the ternary complex cannot be determined due to the limited range of agarose concentrations explored. Only the composition of the binary solvent within the complex can be derived from the maximum, which yields a ratio of about 5 *molecules of water for* 1 *molecule of DMSO*. Now, if one considers a stoichiometry of two solvent molecules by agarose residue, as derived by Rochas et al., this would give a possible stoichiometry of 1 DMSO/5 H$_2$O/3 *agarose residues* (i.e. per turn of helix if one considers a 3$_1$ form of type *B* and *C* as shown in Figure 6.2).

Ramzi et al. have further established phase diagrams in other ternary systems: *agarose/water/dimethyl formamide*, *agarose/water/methyl formamide* and *agarose/water/formamide*.

The phase diagram in *agarose/water/dimethyl formamide* systems is quite similar to that in DMSO. Here again it can be interpreted the same way. A more interesting case, by far, is for the system *agarose/water/methyl formamide*. While the gel melting temperature constantly decreases with increasing methyl formamide content, the associated enthalpies display a maximum (Figure 6.6). Here, the explanation by invoking only the strong interaction between both solvents is totally inconsistent with the simultaneous increase of the enthalpies and decrease of the melting temperature. In Watase and Nishinari's view, improving solvent quality certainly could not lead to an increase of the degree of organization. Alternatively, the notion of a complex is still consistent with this behaviour. ΔH is proportional to the amount of ternary complex formed, hence its increase up to the binary solvent stoichiometry, while the melting temperature is monitored by the interaction between the complexes and the environing solvent.

Finally, there is a conspicuous absence of ternary complex in the case of *agarose/water/formamide* gels (see Figure 6.6). Here, the associated enthalpies are virtually constant in the whole range of investigated compositions. This invariance implies that only one type of entity is formed (the binary complex *agarose/water*).

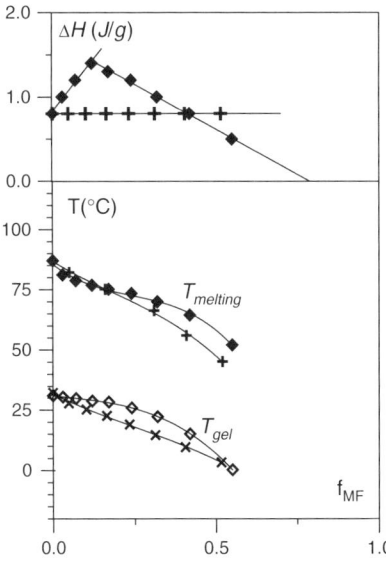

Fig. 6.6. Temperature–composition phase diagram for agarose in water/methyl formamide (♦) and in water/formamide (+) with the associated enthalpies. (Data from Ramzi et al. 1996.)

These binary solvents considered on their own in the solid state show interesting features. *DMSO/water* and *DMF/water* produce two congruently melting compounds, *MF/water* one singular-point melting compound and *formamide/water* a eutectic system only (Kessler et al. 1981, 1982; Ramzi et al. 1996). As emphasized by Ramzi et al., the propensity of DMSO, DMF and MF to form compounds with water, i.e. to establish privileged interaction with OH groups, gives further support to the conclusions drawn from the phase diagrams. Indeed agarose, which contains four OH groups per residue, is potentially liable to produce complexes with DMSO, DMF and MF. Conversely, the only occurrence of eutectic melting in the solid state between *formamide* and *water* is to be compared with the absence of ternary complex, as deduced from the above phase diagrams.

Surprisingly, complex formation does not seem to be correlated with hydrogen bonding. In fact *formamide*, the solvent with the highest capacity to establish such bonds, turns out to be unable to form compounds. It seems that electrostatic interactions may play an important role (CH_3 groups repel electrons, thus bearing a positive fractional charge while a negative fractional charge of the same magnitude appears on another site of the molecule, the oxygen atom in DMSO for instance).

As was discussed above, more evidence of compound formation was obtained by the contrast variation method in neutron scattering (see part I). Ramzi et al. (2000a) have observed conspicuous discrepancies between the calculated proportions of deuterated and hydrogenous solvent needed for matching the coherent scattering of agarose in the gel state. A typical graph of such a study is shown in Figure 6.7 while the results are summarized in Table 6.2.

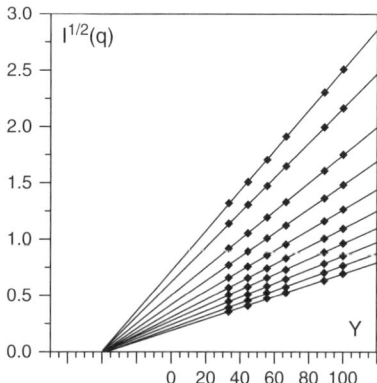

Fig. 6.7. Variation of the square root of the scattered intensity as a function of the fraction Y (%) of deuterated DMSO ($DMSO_D$) in a 9/1 mixture of water/DMSO (70/30 in w/w). The data points at constant Y stand for the different values of q. The extrapolation to $I^{1/2}(q) = 0$ gives the value that would be required for matching the coherent scattering of agarose in gels, namely $Y_o = -0.36 \pm 0.04$ while the calculated value is $Y_{theo} = 0$. (Data from Ramzi et al. 2000a.)

Agarose

Table 6.2. Values of the calculated and the theoretical fractions of deuterated vs hydrogenous solvent needed for matching the coherent scattering of agarose in gels. Y_o and Y_{theo} stand for the experimental and the theoretical values of the fraction of $DMSO_D$ in the mixture; X_o and X_{theo} stand for the experimental and the theoretical values of the fraction of D_2O in the mixture. (After Ramzi et al. 2000a.)

water/DMSO fraction (mol/mol)	X_o	X_{theo}	Y_o	Y_{theo}
9/1 D_2O/DMSO ($DMSOD_6$/$DMSOH_6$)			-0.36 ± 0.04	0
9/1 H_2O (H_2O/D_2O)/$DMSOD_6$	0.16 ± 0.03	-1.00		

Gel nanostructure

As shown by several investigations by electron microscopy on dried gels, the mesoscopic structure is fibre-like (Griess et al. 1993a, b; Sugiyama et al. 1994). Fibrils seen by this technique are about 7–10 nm thick. The mesh size l_n is of the order of several nanometers and decreases with increasing agarose concentration. A quantitative analysis of this parameter as a function of agarose concentration has been achieved by Righetti et al. (1981) by determination of the mobility of calibrated latex spheres:

$$l_n = 705 \times C_{agarose}^{-0.7} \qquad (6.4)$$

Typical values range from 229 nm for $C_{agarose} = 5\,g/l$ to 36 nm for $C_{agarose} = 70\,g/l$.

Recently, Fernandez and Guenet have obtained AFM pictures of 1% agarose gels in water (Fernandez and Guenet, unpublished). They find a similar fibrillar structure with fibrils cross-section of about 10–15 nm diameter (see Figure 6.8).

Ramzi et al. (1998) have studied gel nanostructure as a function of agarose concentration (from 5 g/l to 70 g/l), and the nature of the binary solvent and its composition. Two typical

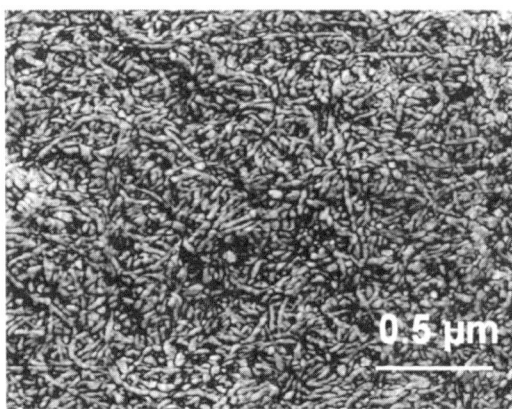

Fig. 6.8. AFM picture of a 1% w/w agarose gel in water. (Data from Fernandez and Guenet, unpublished.)

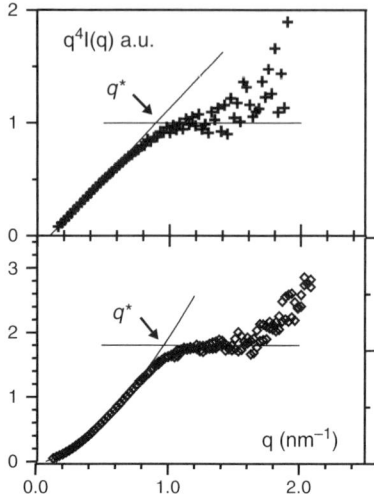

Fig. 6.9. $q^4 I(q)$ vs q representation for gels prepared in water/DMSO ($f_{DMSO} = 0.1$), $C_{agarose} = 5$ g/l (**upper diagram**), for $q < q^*\lambda = 1$; water/dimethyl formamide ($f_{DMF} = 0.35$), $C_{agarose} = 20$ g/l (**lower diagram**), for $q < q^*\lambda = 1.6$. (Data from Ramzi et al. 1998.)

scattering curves are shown in Figure 6.9 by means of a $q^4 I(q)$ vs q representation. Note that, according to equation (6.4) the condition $q l_n > 1$ is always fulfilled. The shape of the scattering curves is very reminiscent of what has been observed with PVC gels, and can be analysed through Guenet's approach (Guenet 1994). In particular a *transitional domain* can be identified (see Part I for details). In this domain the fibril cross-section distribution of the type $r^{-\lambda}$ can be characterized. It is observed that the characteristic exponent λ is dependent upon the agarose concentration and the binary solvent composition. Typically, for water-rich compositions and low agarose concentration $\lambda \approx 1$, while increasing the concentration and the fraction of organic solvent entails an increase of this exponent up to $\lambda \approx 1.6$. The increase of λ implies that the fraction of fibrils of low cross-sectional radius increases. The *Porod domain* always occurs at about the same scattering vector, i.e. $q \approx 0.9$–1.0 nm^{-1}. Analysis of the scattering curves by means of equations detailed in Part I gives the following values for the cut-off radii.

$$r_{min} = 0.8\text{--}0.9\,\text{nm} \quad \text{and} \quad r_{max} = 6\text{--}9\,\text{nm}$$

The values of r_{max} correspond to those observed by electron microscopy. Direct imaging by EM misses, however, the fibrils of lowest cross-sectional radius that are in the majority in view of the distribution considered for the analysis.

As with PVC gels, departure from the Porod regime is observed. Unlike PVC, however, this behaviour vanishes beyond $C_{agarose} = 50$ g/l, and the *Porod-regime* is seen all along for q > q*. Clearly, this behaviour is not related to the short-range molecular structure. Ramzi et al. have suggested that it may be due to the presence of free and/or dangling chains (referred to as *loose chains* by Ramzi and colleagues) possessing the same conformation as in the sol state. The resulting intensity can be approximated in the q-range of interest to:

$$\frac{q^4 I(q)}{C} \approx X \frac{4\pi\rho}{r_n} + (1-X) 4\pi q \mu_L \frac{J_1^2(qr_H)}{r_H^2} \tag{6.5}$$

where X is the volume fraction of the network and $1-X$ the volume fraction of *loose chains*, and r_H the chain cross-section. These chains scatter approximately as $1/q^2$ in this q-range (Figure 6.9) and therefore their intensity obliterates that of the network. The fraction of these chains is said to decrease as the agarose concentration increases in order to account for the gradual vanishing of this behaviour. As will be seen in the section devoted to gel rheology, this interpretation is consistent with the variation of the elastic modulus as a function of agarose concentration.

About the sol–gel transition

The so-far-admitted picture of the events taking place at the sol–gel transition is the formation of double helices out of a solution containing flexible coils. As convincingly highlighted by the experiments of Foord and Atkins (1989), the occurrence of double helices relies on a shaking basis, and the formation of single helices seems to be a more promising scenario. Further, the notion of flexible coil is certainly not consistent with neutron scattering data reported by Guenet et al. (1993) and Rochas et al. (1994). These authors suggest that the stiffness of the agarose chains, with a conformation close to a single helix of type SH_2, will tend to lead to a parallel association of chains under this conformation (the *loose helix–tight helix transition*). However, the diffraction data obtained from gels are difficult to reconcile with this helix. In particular, the first meridional reflection is not consistent with the required 0.973 nm axial rise. The characteristic diffraction pattern for SH_2 appears only when the gel is stretched to a sufficient extent. Two additional arguments against the simple alignment of SH_2 helices can be put forward: (1) why would there be a hysteresis between gel formation and gel melting? And (2) why and how would such helices accommodate solvent molecules to form binary or ternary complexes?

Another sol–gel mechanism is tentatively put forward (see Figure 6.10) which relies on the rationale that no double helix conformation occurs. The sol state is seen as a solution of worm-like agarose chains that locally take on a loose SH_2 helix conformation. On cooling, gelation is preceded by an SH_2–SH_1 transition such that the gel is made up of single SH_1 helices. This mechanism and its consequences appear to be consistent with:

- both neutron findings and Foord and Atkins helical structures;
- the existence of a hysteresis: SH_2 helices may not be able to associate, unlike the case with SH_1 helices. As a result, gelation can only take place as soon as the SH_2–SH_1 transition has occurred. Conversely, melting of the SH_1 gel can occur well above this transition;
- the fact that SH_1 helices may be more appropriate structures for accommodating solvent molecules, and for producing polymer–solvent complexes.

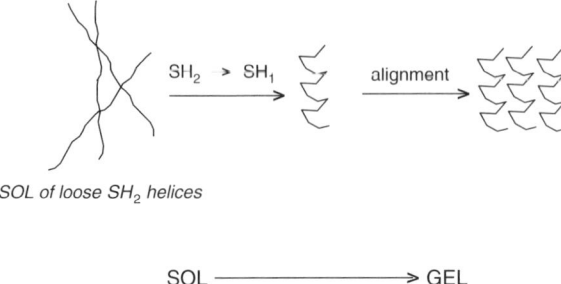

Fig. 6.10. Schematic representation of a possible scenario for the events at the sol–gel transition: transformation from an SH$_2$ loose helix into a tight SH$_1$ helix followed by immediate aggregation.

Gel rheology

Unlike synthetic polymers such as PVC, agarose gels are produced for concentrations as low as $C_{agarose} = 1$ g/l (against 22 g/l for PVC). Most of the studies aimed at establishing the modulus–concentration relation were restricted to concentrations below 20 g/l that showed a power law variation with an exponent of about $\nu = 2.2 \pm 0.1$ (Rochas et al. 1994). Yet, departure from this behaviour occurs at about $C_{agarose} = 20$ g/l (Figure 6.11). A cross-over is then observed and another power law with, this time, $\nu = 1.5 \pm 0.1$ is seen instead (Ramzi et al. 1998). This is a general behaviour observed independent of the solvent (water or binary aqueous solvents). These results ought to be examined in the light of the conclusions drawn from the small-angle scattering study, namely the presence of loose chains for low agarose concentrations. As a result, the network fraction, φ_N, which is the only part contributing to the gel elastic behaviour, differs from the agarose concentration. Conversely, at higher agarose concentration, no loose chains

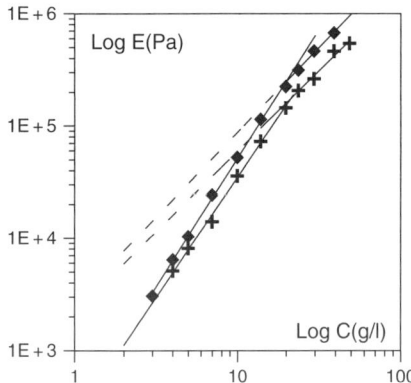

Fig. 6.11. Relation modulus–concentration for: (♦) agarose/water/DMF ($f_{DMF} = 0.19$) and (+) agarose/water/formamide ($f_{FOR} = 0.31$). The cross-over occurs for $C_{agarose} \approx 20$ g/l. (After Ramzi et al. 1998).

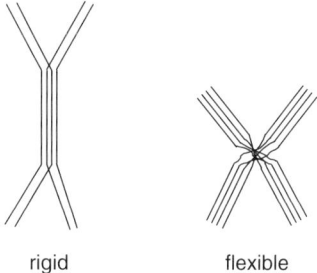

rigid   flexible

Fig. 6.12. Depiction of the two possible types of junction in agarose gels. The rigid junction is organized, unlike the flexible junction. Parallel association of chains occurs in both cases, a process which is directly responsible for the existence of the rigid junction while indirectly producing the flexible junction (note that the latter model had already been suggested by Arnott et al. 1974b).

are detected, which implies that $\varphi_N \approx C_{agarose}$. This implies that the exponent derived for $C_{agarose} < 20\,g/l$ is only apparent while that determined for $C_{agarose} > 20\,g/l$ better describes the network structure.

Agarose gels are random dispersions of fibrils, so Jones and Marqués theory (Jones and Marqués 1990) can be used to derive the exponent in terms of the longitudinal fractal dimension of the fibrils D_f (see equations (4.25) and (4.27)). These fibrils are close to cylinders which entails $D_f \approx 1$. The only way to account for $\nu = 1.5$ is to consider the occurrence of *entropic elasticity*. As fibrils are but rigid objects, entropic elasticity can only occur if the gel junctions are flexible (Figure 6.12).

Note that structural studies such as X-ray diffraction or NMR spectroscopy cannot distinguish between either type of junction. Rheology experiments therefore provide circumstantial evidence with regard to this issue.

It is worth stressing that the elastic modulii for systems displaying *entropic elasticity* should vary as kT (see Chapter 4, equation (4.28)). Results by Watase et al. (1989) and Zhang and Rochas (1990) do show in a given temperature range (typically from 15°C to 35°C) an increase of the modulus with increasing temperature of about the expected magnitude. Modulus decay is only seen with the onset of gel melting. Interestingly, this behaviour is not seen at all for carrageenan gels, for which only enthalpic entropy is liable to occur.

Gels from modified agarose

The chemical structure of agarose shown in Figure 6.12 is but theoretical. In practice, modification of the OH group can be carried out; OCH_3 is possible or even further modifications. Nature is capable of modifying agarose on specific sites, while chemists are only able to alter randomly both moieties of the agarose residue. This gives the structure shown in Figure 6.13, where those OH groups on carbons 2 and 4 of the *galactose* moiety and the OH groups on carbon 2 of the moiety *anhydro-galactose* are modified.

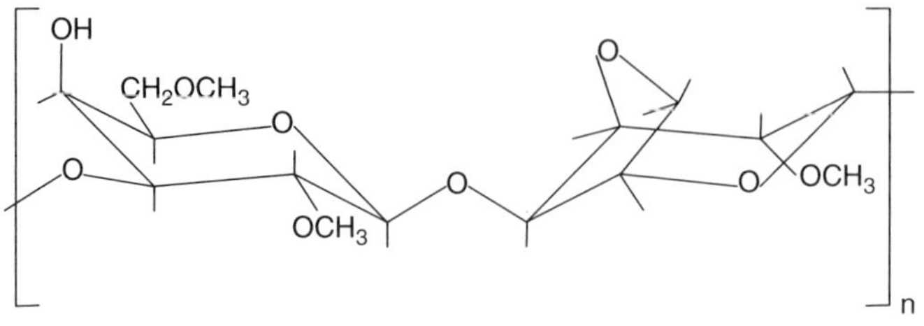

Fig. 6.13. Agarose residue where some OH groups have been replaced by OCH_3.

Dahmani et al. (2003) have carried out a systematic study of the thermodynamic properties and structure of gels prepared from agarose with increasing degree of modification. They have considered three samples, whose characteristics are as follows.

M_1 $C_{12}H_{17.46}(CH_3)_{0.54}O_9$
(0.24 CH_3 from the *anhydro-galactose* and 0.3 CH_3 from the *galactose*)

M_2 $C_{12}H_{17.32}(CH_3)_{0.68}O_9$
(0.4 CH_3 from the *anhydro-galactose* and 0.28 CH_3 from the *galactose*)

M_3 $C_{12}H_{18.07}(CH_3)_{0.93}O_9$
(0.5 CH_3 from the *anhydro-galactose* and 0.43 CH_3 from the *galactose*)

As has already been reported by several authors, the gel melting temperature drops drastically with the introduction of chemical defects (Guiseley 1970; Lahaye and Rochas 1989, 1991; Stevenson and Furneaux 1991; Miller et al. 1994; Takano et al. 1995; Falshaw et al. 1998). The temperature–composition phase diagram mapped out by Dahmani et al. for agarose/water/DMSO highlights this effect (Figure 6.14). Typically, the melting and formation temperatures are shifted downwards by approximately 25°C. This shift does not increase greatly with the degree of modification of the three samples considered above. As with 'pure' agarose, a maximum occurs for a DMSO molar fraction of about $f = 0.2$. Conversely, the value of the gel melting enthalpy depends significantly on the degree of modification, and drops to almost zero in pure water for the most modified sample. Replacing H atoms with CH_3 groups entails a higher solubility of the modified agarose in water. This is so because hydrogen bonds are therefore disrupted, which weakens interactions between agarose chains in the gel.

Modified agarose is therefore more soluble in water. By increasing the fraction of DMSO the gel melting enthalpies display a maximum for a DMSO mole fraction of about $f = 0.2$. Interestingly, the value of the melting enthalpies becomes virtually independent of the degree of modification. This clearly suggests that DMSO promotes gelation in spite of its expected behaviour as a 'good solvent' towards agarose. Here, the fact that water behaves as a better solvent with modified agarose than it does with agarose definitely rules out Watase and Nishinari's assumption as to the occurrence of a *good solvent/bad solvent* effect for the behaviour of the melting temperature in the water/DMSO phase diagram (Watase and Nishinari 1988). Conversely, these results give further support to the ternary complex model developed by Ramzi et al. (1996).

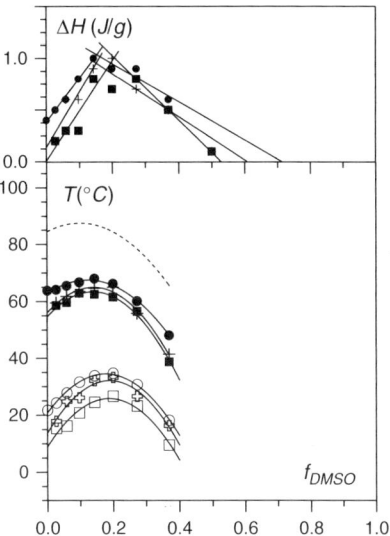

Fig. 6.14. Temperature–composition phase diagram for the ternary system agarose/water/DMSO. The composition is given in mol fraction. (●) = M1; (✦) = M2; (■) = M3. Filled symbols represent the gel melting temperature, while open symbols stand for the gel formation temperature. Polymer concentration $C_{pol} = 15$ g/l. (Data from Dahmani et al. 2003.)

Additional evidence for the occurrence of a complex is found from the study of the rheological behaviour, thus ascerting the role played by DMSO. Dahmani et al. have shown that the elastic modulus of the gels increases quite dramatically with increasing DMSO content (Figure 6.15). In the case of gels prepared from sample M3, the increase is about fivefold. The values go through a maximum at a DMSO fraction of about

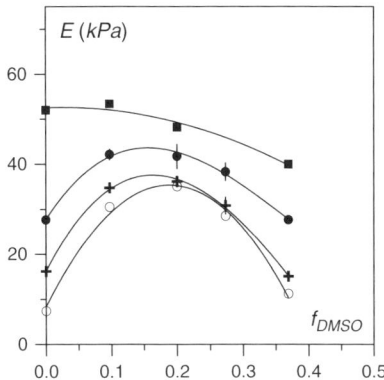

Fig. 6.15. Variation of the elastic modulus as a function of the DMSO mole fraction in ternary systems agarose/water/DMSO. (■) agarose; (●) modified agarose, M1; (✦) modified agarose, M2; (○) modified agarose, M3. Solid lines are a guide for the eye. (Data from Dahmani et al. 2003.)

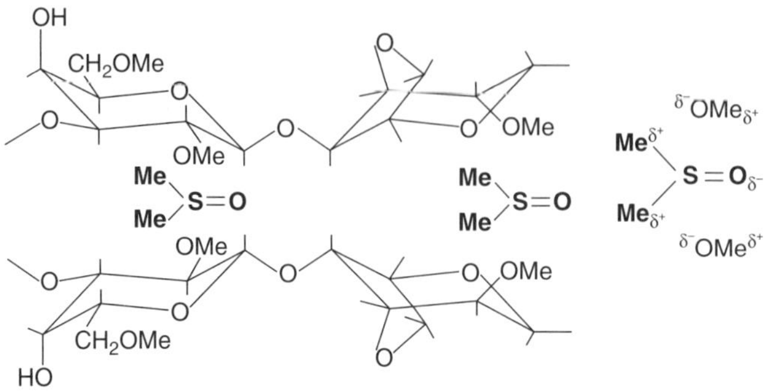

Fig. 6.16. Depiction of the possible way DMSO interacts simultaneously with two agarose chains through electrostatic interaction. (From Dahmani et al. 2003. Reprinted with permission from Elsevier.)

$f_{DMSO} = 0.2$, namely the stoichiometric composition of the ternary complex. Also, at this DMSO composition, the values of all the samples are rather close to that of 'pure' agarose, while they differ significantly in water.

As emphasized by Dahmani et al., everything happens as if DMSO were 'repairing' the missing hydrogen bonds originally present in the 'pure' agarose. This can be accounted for by considering the fractional charges that are created through bond polarization effects. In the case of DMSO, positive fractional charges are present on the methyl groups while negative charges appear on the oxygen atom. In the case of agarose, similar effects occur on the OCH_3 group. Dahmani et al. therefore reason that electrostatic interactions are established between opposite charges of modified agarose and DMSO (Figure 6.16). As DMSO can interact with two chains simultaneously, the absence of hydrogen bonds due to the replacement of H by CH_3 is therefore counterbalanced by an electrostatic interaction mediated by DMSO. In the case of 'pure' agarose, no effect is seen on the rheological behaviour simply because the hydrogen bond is this time replaced by the agarose/solvent electrostatic interactions.

CHAPTER 7

Amylose

Starch is a major component of green plants, and particularly of seeds, where its function consists in storing excess glucose. After extraction from plants and seeds, starch is employed industrially in food processing, and still finds use in laundry to stiffen shirt collars and ruffs of fine linen. Bread contains a large amount of starch, as it is prepared from grain. Clearly, starch is essential both for living systems and for human beings.

Starch is constituted of two main components: *amylose* and *amylopectin* in the ratio 2/8 to 3/7 depending upon the origin. Both components have the same chemical formula but differ in their structure: *amylose* is a linear biopolymer while *amylopectin* is a highly branched, tree-like biopolymer. Amylose and amylopectin are made up of $\alpha(1 \to 4)$-linked D-glucopyranosyl units. The chemical structure is shown in Figure 7.1.

Amylopectin molecules form the framework of the starch granules from which amylose can be extracted through the so-called gelatinization process. This process consists in swelling irreversibly in boiling water starch granules, from which amylose leaches out. This gives a kind of gel which is metastable and undergoes structural transformations, such as aggregation and recrystallization of amylose molecules designated as *retrogradation*.

Depending upon the origin of starch, amylose can contain from 50 to up to 500 $\alpha(1 \to 4)$-linked D-glucopyranosyl units (Hizukuri et al. 1989; Morrison and Karkalas 1990). For further reading on the composition, molecular structure and physicochemical properties of starch from various origins, the comparatively recent review by Hoover is quite informative (Hoover 2001).

Amylose and amylopectin are both crystalline biopolymers with four known crystalline forms, designated as A, B, C and V, all being molecular compounds whose occurrence depends upon the origin of the starch (Gallant et al. 1982; Dreher and Berry 1983; Soni and Agarwal 1983; Takeda et al. 1983; Yu et al. 1999). Only the first three forms are found as native starch from plants and/or seeds. Most of the studies have been carried out on amylose simply because this biopolymer is easier to re-crystallize on account

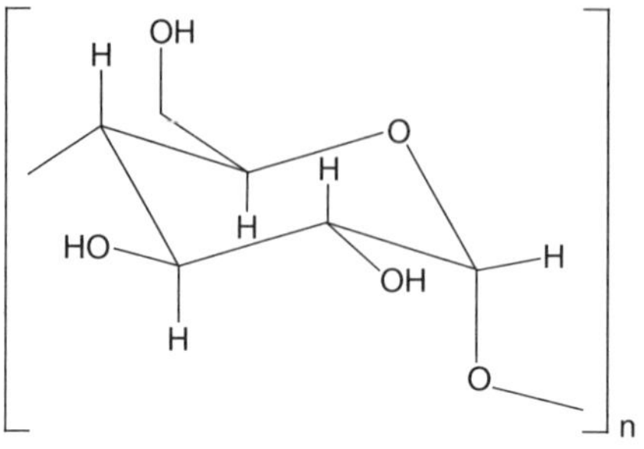

Fig. 7.1. Amylose repeat unit.

of its smaller molecular weight, and, also, it can be straightforwardly enzymatically synthesized. What follows is therefore a description of the different forms of amylose.

A-amylose tends to be found in cereals while B-amylose is encountered in tubers. Form C is thought to be a superposition of form A and form B as both allomorphs occur simultaneously in some starch granules (Sarko and Wu 1978; Buléon et al. 1998). Form V is produced only after some processing and extraction from the starch granules, as will be detailed below.

In solution the formation of A-amylose versus B-amylose depends upon crystallization temperature, concentration and amylose molecular weight (Hizukuri 1961; Buléon et al. 1984; Pfannemüller 1987; Gidley and Bulpin 1989). Long chains and low crystallization temperature favour B-amylose while high concentration and high crystallization temperature promotes the growth of A-amylose.

A-amylose

The very first report on the study of the crystalline structure of starch dates back from the early 1930s by Katz and van Itallie (Katz and van Itallie 1930). Their studies were carried out on non-oriented samples, with the result that the crystalline lattice derived then was later questioned, and in fact totally reconsidered.

Studies on oriented fibres by X-ray diffraction coupled with molecular modelling (Wu and Sarko 1978a, b), and then by means of electron diffraction investigations on in-vitro-grown single crystals (Buléon et al. 1984; Tran and Buléon 1987; Imberty et al. 1987, 1988), and finally by X-ray micro-diffraction with synchrotron radiation (Popov et al. 2006), have allowed the precise determination of the crystalline structure of A-amylose (Figure 7.2). A-amylose forms a compound with water with a monoclinic unit cell of parameters $a = 2.0874$ nm, $b = 1.146$ nm, $c = 1.055$ nm and $\gamma = 121°94$, and with a space group $B2$.

A double-stranded helix is formed, resulting from the association of 6-fold left-handed single helices of repeating distance $2c = 2.11$ nm. On the basis of the crystal density

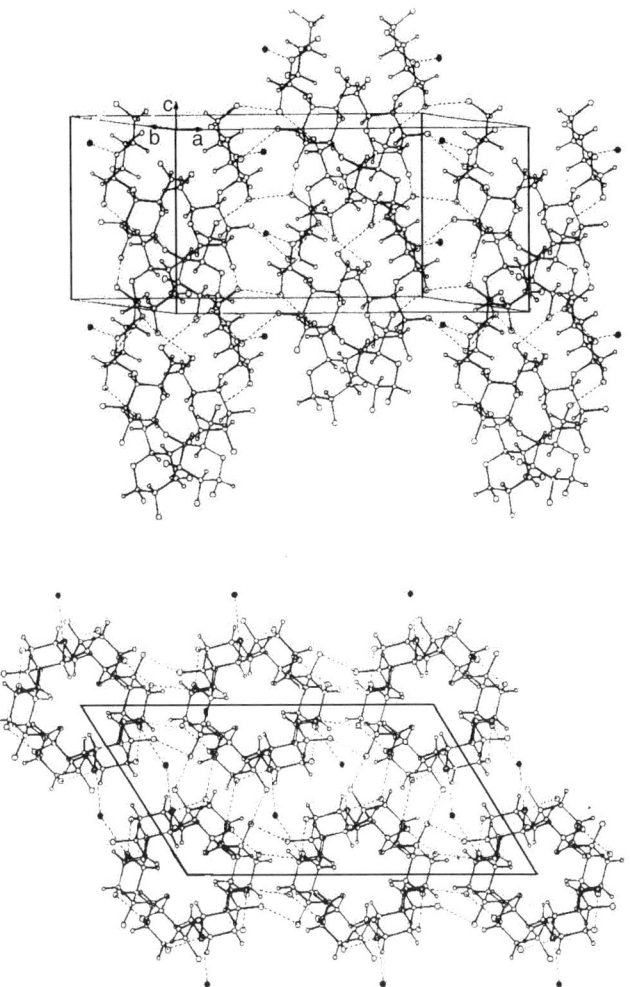

Fig. 7.2. Crystal structure of A-amylose. **Top:** as seen perpendicular to the helical axis; **bottom:** as seen parallel to the helical axis. The black dots represent water molecules, and dotted lines hydrogen bonds. Monoclinic unit cell $a = 2.0874$ nm, $b = 1.146$ nm, $c = 1.055$ nm and $\gamma = 121°94$. (From Imberty et al. 1988. Reprinted with permission from Elsevier.)

(1.48 g/cm^3), Imberty et al. suggest that the unit cell can accommodate 4 water molecules, which yields a stoichiometry of 1 H$_2$O/4 D-glucopyranosyl units.

It is worth stressing that these investigations, particularly those on single crystals, were carried out on amylose samples of low degree of polymerization ($13 < \mathrm{DP} < 27$) obtained either by mild hydrolysis of potato starch (Imberty et al. 1988; Robin et al. 1974) or by enzymatic synthesis from sucrose using amylosucrase (Potocki-Veronese et al. 2005; Popov et al. 2006). Also, preparation of A-amylose from amylose chains extracted from starch is a complex procedure (e.g. see Popov et al. 2006).

Fig. 7.3. SEM image of synthetic A-amylose crystals grown from recrystallization of synthetic amylose. (From Popov et al. 2006. Reprinted with permission from ACS.)

The morphology of the single crystals obtained from synthetic amylose is needle-like, as shown in Figure 7.3 (Popov et al. 2006).

B-amylose

As aforementioned, B-amylose is found in starches from tubers, such as potatoes. B-amylose was first described in a series of papers from data gathered by X-rays on starch granules by Katz and co-workers (Katz and Rientma 1930; Katz and van Itallie 1930; Katz and Derksen, 1930, 1933). As with A-amylose, its preparation from amylose chains extracted from starch is, however, a complex procedure (e.g. see Takahashi et al. 2004).

The crystal structure has been investigated by several authors (Rundle et al. 1944; Kreger 1951; Blackwell et al. 1969; Kainuma and French 1972; Cleven et al. 1978; Wu and Sarko 1978a, b; Imberty and Pérez 1988; Takahashi et al. 2004), yet it took more than 40 years to obtain the correct crystal unit cell. Incidentally, this highlights the difficulties in determining the crystal unit cell together with the group of symmetry with semi-crystalline polymers, and the more so when molecular compounds are dealt with.

The latest unit cell has been proposed by Imberty and Pérez, and later confirmed by Takahashi et al. Chains are arranged according to the hexagonal space group $P6_1$ with $a = b = 1.85$ nm; $c = 1.04$ nm. As with A-amylose the chains possess a 6_1 left-handed helical structure with repeat distance $2c = 2.08$ nm and associate to form double-stranded helices that are packed in a parallel fashion in the crystal unit cell (Figure 7.4). According to these authors the unit cell also contains 12 molecules of water (36 molecules per channel), which means that the stoichiometry of B-amylose is 1 H_2O/1 D-glucopyranosyl unit. Water molecules are essentially housed in channels created by six double-stranded helices (see Figure 7.4). Half of the water molecules are strongly linked though hydrogen bonds to the amylose chains, while the other half are only loosely bound.

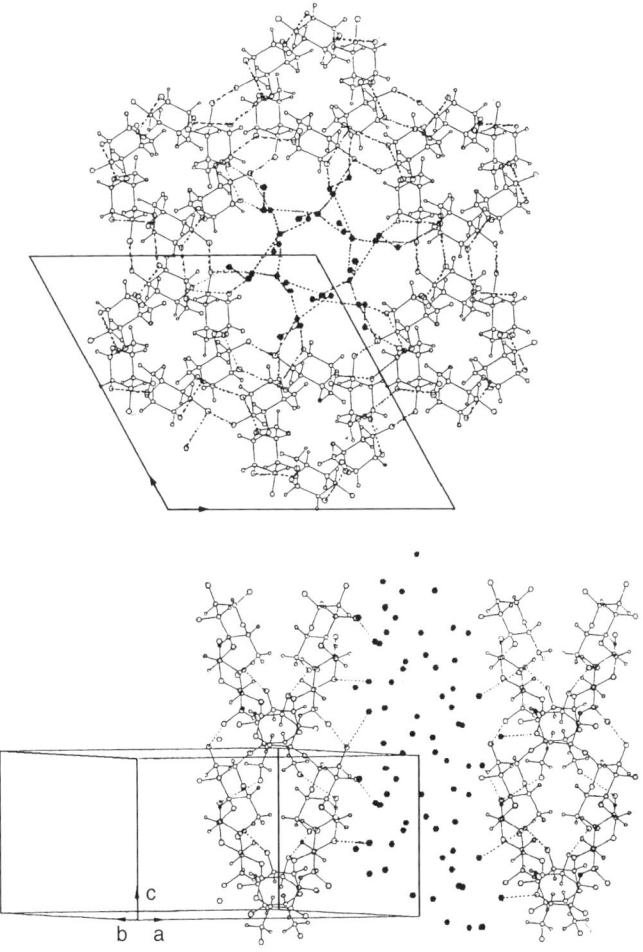

Fig. 7.4. Crystal structure of B-amylose. **Top:** as seen perpendicular to the helical axis; **bottom:** as seen parallel to the helical axis. Black dots represent water molecules, and dotted lines hydrogen bonds. Hexagonal unit cell $a = b = 1.85$ nm; $c = 1.04$ nm. (From Imberty and Pérez 1988. Reprinted with permission from Wiley Interscience.)

V-amylose

As has been emphasized above, V-amylose does not occur naturally in starch but is produced by crystallization of amylose from solutions. The name V-amylose can be somewhat misleading as it does not mean the *fifth* crystalline form of amylose. In fact, V-amylose was first isolated by Katz, who subsequently coined the name after the German word *Verkleisterung*, which means 'gelatinization'. Indeed, this form is produced only after extracting amylose from the starch granules through this process (Katz 1930; Katz and Rientsma 1930).

The formation of V-amylose results from complexation with small ligands such as alcohol and fatty acids. Amylopectin does not form such complexes, so formation of V-amylose also allows one to fractionate starch into its two components, and thus to recover amylose easily (e.g. see Bourne et al. 1948).

V-amylose is made up of single left-handed helices, unlike A-amylose and B-amylose. Thus far two helical structures have been observed, i.e. the 6_1 helix and the 8_1 helix. The 6_1 helix possesses an inner cavity which can house solvent molecules. The crystal unit cell depends upon the host molecule, yet four modifications have been identified and described to date. A series of identifying names was proposed by Helbert (1994): V6-I (also named V_h, the subscript h standing for 'hydrated') (Rappenecker and Zugenmaier 1981), V6-II (Helbert and Chanzy 1994), V6-III (Buléon et al. 1990) and V8 (Le Bail et al. 2005). The Arabic digits 6 and 8 stand for the number of D-glucopyranosyl residues per turn in the helical structure, while the Roman digits are related to the space available for the solvent molecules (III indicating the largest available volume).

Early studies by Sarko and Zugenmaier have shown that the complex formed with water possesses an orthorhombic unit cell, which is found to be with the parameters $a = 1.37$ nm, $b = 2.37$ nm and $c = 0.805$ (Sarko and Zugenmaier 1980; Rappenecker and Zugenmaier 1981).

The amylose/dimethyl sulphoxide complex has been investigated by means of X-ray diffraction on oriented fibres. The amylose occurs in a six-residue helix with alternate 'up' and 'down' chains packed in a pseudo-tetragonal unit cell (French and Zobel 1967; Winter and Sarko 1974). The parameters are $a = b = 1.917$ nm and c (fibre axis) $= 2.439$ nm with space group $P2_12_12_1$. This indicates that three turns of helix are necessary for the crystallographic repeat. Dimethyl sulphoxide is located inside the helix with one DMSO molecule for every three glucose residues. Four DMSO molecules together with eight water molecules are located between chains. According to Winter and Sarko, it is the interaction of these molecules with the helix that results in the pseudotetragonal chain packing. Also, this is probably the origin of additional layer lines in the diffraction pattern that are not consistent with the 0.813 nm helix repeat unit (Winter and Sarko 1972).

In-depth studies of the crystal structure of V-amylose have been achieved by means of electron diffraction on single crystals. Buléon et al. (1990) have determined the crystalline unit cell of V-amylose formed with isopropanol or acetone by this method (designated as V6-III). The unit cell is orthorhombic with parameters $a = 2.826$ nm, $b = 2.93$ nm, and $c = 0.801$ nm of space group $P2_12_12_1$ or $P2_12_12$. The helical structure is a left-handed 6_1 helix. With 1-butanol (also designated as V6-II), Helbert and Chanzy have also observed an orthorhombic unit cell with parameters $a = 2.74$ nm, $b = 2.65$ nm, $c = 0.80$ nm, again of space group $P2_12_12_1$ or $P2_12_12$ where chains take on a left-handed 6_1 helical conformation (Helbert and Chanzy 1994). With glycerol Hulleman et al. (1996) observed again an orthorhombic unit cell with parameters $a = 1.93 \pm 0.01$ nm, $b = 1.86 \pm 0.01$ nm, $c = 0.83 \pm 0.03$ nm, again of space group $P2_12_12_1$ and left-handed 6_1 helical conformation (see Figure 7.5).

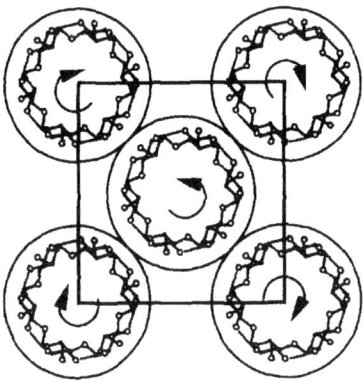

Fig. 7.5. Schematic representation of the structure of the amylose chains in the complex amylose/glycerol. As is illustrated by the arrows, there are up and down chains. (From Hulleman et al. 1996. Reprinted with permission from Elsevier.)

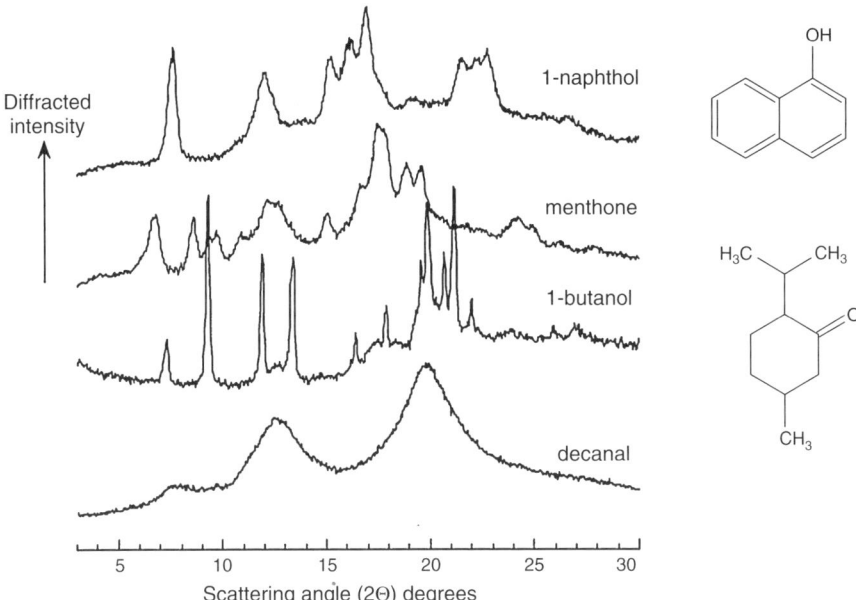

Fig. 7.6. Left: X-ray diffractograms for different amylose complexes. (From Le Bail et al. 2005. Reprinted with permission from Elsevier.) **Right:** chemical structure of 1-naphthol (**top**) and menthone (**bottom**).

A comparison of the X-ray diffraction patterns is shown in Figure 7.6 for different amylose/solvent complexes (Le Bail et al. 2005). These patterns highlight the different types of V-amylose complexes as defined by Helbert (1994).

Reparet et al. (2006) have studied the amylose/octanal system. Octanal is a solvent molecule which possesses an aliphatic tail and a polar head. The diffraction pattern can be indexed with an orthorhombic unit cell of the type V6-II with $a = 2.74$ nm, $b = 2.65$ nm, $c = 0.80$ nm and space group $P2_12_12_1$ or $P2_12_12$. Reparet et al. have, however, observed that the amylose/octanal complex also yields V6-I-type crystals. Clearly, this solvent triggers the growth of two crystalline types: this suggests that endogenous lipids complex with some part of the amylose without hindering the interactions with exogenous ligands. The authors could not definitely determine the location of the octanal molecules: between the helices or inside and outside the helices.

Many other systems have been studied with basically the same helical structure: amylose/palmitic acid and amylose/lauric acid (Le Bail et al. 2000), amylose/linalool (Rondeau-Mouro et al. 2004), amylose/1-decanol (Kowblansky 1985), amylose/stearic acid (Karkalas et al. 1995), n-pentanol and thymol (Kawada and Marchessault 2004).

It is worth mentioning that recent studies by means of solid state NMR have led the authors to suggest a model wherein the hydrophobic inner cavity of the 6_1 amylose helix houses the aliphatic part of fatty acid while the polar head is excluded (Le Bail et al. 2000). This entails the existence of pseudo-crosslinks between chains that, according to the authors, provides anti-staling properties (Figure 7.7).

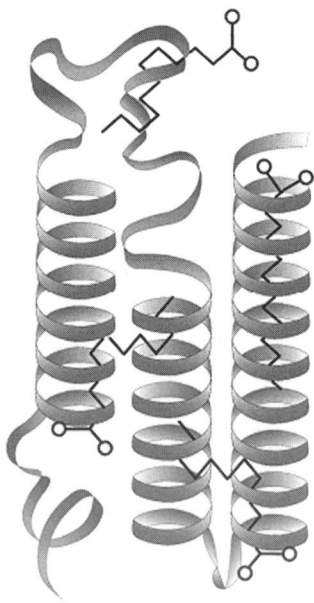

Fig. 7.7. Schematic drawing of molecules pervading several amylose 6_1 helices. These pseudo-crosslinks provide the anti-staling properties. (From Kawada and Marchessault 2004. Reprinted with permission from Wiley Interscience.)

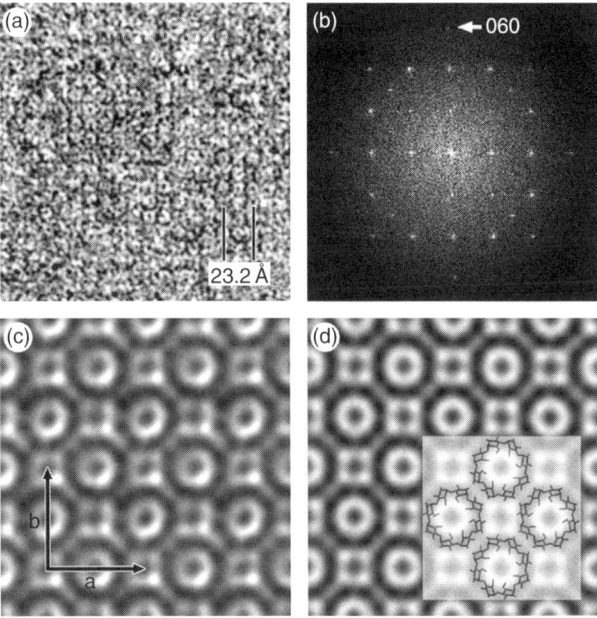

Fig. 7.8. (a) HREM image of a VR-naphthol crystal viewed along the chain axis c; (b) power spectrum of the lattice image; (c) real-space translational average; (d) 4-fold rotation average of the image in C. **Inset:** a molecular model corresponding to the projection of 8-fold amylose helices viewed end-on is superimposed on the average image. (From Cardoso et al. 2007. Reprinted with permission from ACS.)

The V-amylose complexes that have been described so far consist of 6_1 helices. Another type of helix has been observed for amylose when complexed with 1-naphthol, namely an 8_1 helix (Helbert 1994; Winter et al. 1998). The X-ray data can be indexed with a tetragonal unit cell of parameters $a = b = 2.33$ nm, $c = 0.79$ nm (Le Bail et al. 2005) and of space group $P4_32_12$ or $P4_12_12$. More recently Cardoso et al. have studied single crystals by electron diffraction and high resolution imaging (Figure 7.8) and came to the slightly different parameters $a = b = 2.2844$ nm, $c = 0.7806$ (Cardoso et al. 2007).

CHAPTER 8

Chitin

Chitin is, with cellulose, the most abundant natural polymer on earth. It is found in the exoskeleton of arthropods (crabs, shrimps, insects, etc.) but also in the cell walls of most mushrooms, in some bacteria and in some algae (Roberts 1992). From a structural point of view, chitin performs the same function in invertebrates as cellulose does in plants. The repeat unit is constituted of N-acetyl-D-glucosamine and β1,4-linked D-glucosamine (Figure 8.1).

Because of the existence of two biosynthesis mechanisms, chitin naturally occurs in two crystalline forms: α-chitin, which is by far the most abundant and β-chitin. The crystalline unit cell of α-chitin consists of two chains arranged in an *antiparallel* fashion for which the space group is $P2_12_12_1$ (Lotmar and Picken 1950; Minke and Blackwell 1978). β-chitin forms a monoclinic unit cell of space group $P2_1$, in which all the chains adopt a *parallel* arrangement (Dweltz 1961; Blackwell 1969; Gardner and Blackwell 1975).

β-chitin can transform into α-chitin after intracrystalline swelling in HCl (Lotmar and Picken 1950; Rudall 1963; Rudall and Kenchington 1973). As has been studied in more

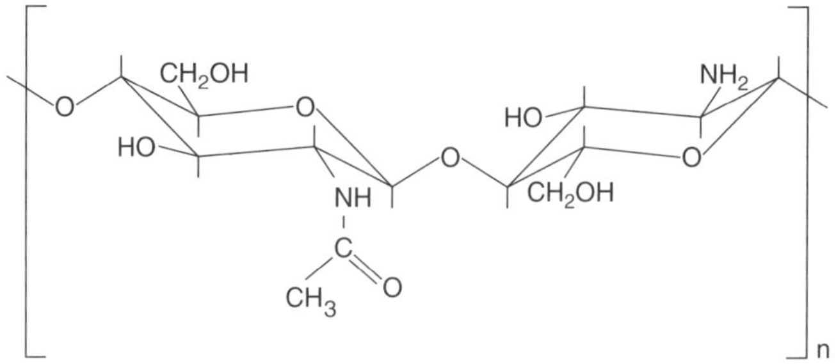

Fig. 8.1. Repeat unit of chitin.

detail by Saito et al. on chitin extracted from vestimentiferan tubes of *Tevnia jerichonana*, this intracrystalline swelling produces a solvated mesophase by a 'decrystallization' process (Saito et al. 1997a, b). In the range of HCl aqueous concentrations from 6N to 7N the chitin sample gives a mesophase with a nematic-like ordering. The diffraction pattern displays only three broad rings centred at 0.9 nm, 0.45 nm and 0.34 nm. According to Saito et al. the last two maxima correspond roughly to the areas of maximum diffraction for *β-chitin* while the maximum at 0.9 nm can be assigned either to chitin or to HCl. Note that in this range of HCl concentrations the mesophase can transform reversibly into *β-chitin*. The reversibility only occurs at the molecular level. At the mesoscopic level, striations can be observed from electron micrographs, suggesting the formation of longitudinal cracks leading to a series of smaller parallel subfibrils. For concentrations between 7N and 8N, decrystallization occurs together with chain scission and dissolution. The *β-chitin* gives birth to *α-chitin* by an epitaxial growth process. The *β-chitin* to *α-chitin* transformation is irreversible.

Saito et al. have further shown that, after swelling in HCl, 7N crystallo-solvates can be formed by immersion in linear alcohols (Saito et al. 1997a, b).

The unit cell is monoclinic with a $P2_1$ symmetry. The *a* and *c parameters* are those of the *β-chitin* unit cell while the *b parameter* increases with an increase in the size of the alcohol (Figure 8.2). This suggests that alcohol molecules are intercalated in

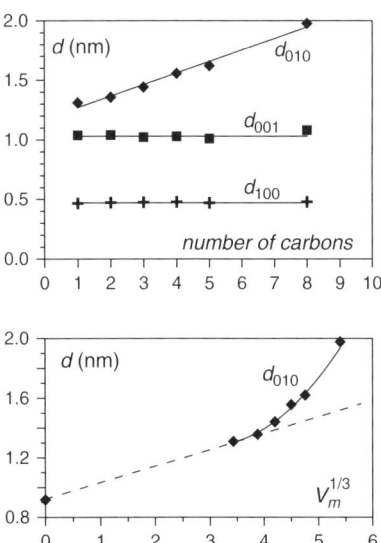

Fig. 8.2. Spacings d_{100} (+), d_{010} (♦), and d_{001} (■) of β-chitin/alcohol crystallosolvates in a series of linear alcohols. **Lower figure:** vs the cubic root of the alcohol molar volume; the dotted line stands for linear variation while the full line shows departure from linearity. **Upper figure:** vs the alcohol length (number of carbons). Monoclinic unit cell with $\gamma = 93° \pm 3$. (Plot after data from Saito et al. 1997a, b.)

the β-chitin unit cell and that chains form a sheet-like arrangement. Figure 8.2 also highlights the fact that the *b parameter* varies linearly as a function of the length of the alcohol molecules (number of carbons) but not as the cubic root of their molar volume. This implies that the conformation of the alcohols is most certainly extended. Note that after alcohol evaporation the unit cell reverts to the usual β-chitin unit cell.

CHAPTER 9

Deoxyribonucleic acid (DNA)

DNA (deoxyribonucleic acid) is a macromolecule, displaying a polyelectrolyte behaviour, that contains all the genetic information necessary for the growth of living cells. DNA occurs in the nucleus of cells or in viruses as a double-stranded helix, the structure of DNA being discovered some 50 years ago by Crick and Watson on the basis of the diffraction patterns collected by Franklin on DNA fibres (Watson and Crick 1953; Franklin and Gosling 1953a, b). Franklin identified two forms of DNA, which she called A-DNA and B-DNA. The B-DNA form was resolved by Watson and Crick. The double helix occurs because of the special interactions, namely hydrogen bondings, that are established between complementary bases, as shown in Figure 9.1.

Another form was discovered later, named Z-DNA, and its structure was solved rather recently (Wang et al. 1979). There exist two additional forms, C and D. In all cases double-stranded helices, where the phosphate groups are pointing outwards, are dealt with (Figure 9.2). Investigations by diffraction techniques on DNA fibres and single crystals grown from oligonucleotides, as well as by computer simulations, show that water molecules are located at well-defined positions and play a crucial role in the stabilization of the various conformations. Clearly, the formation of DNA/water molecular compounds is an important phenomenon for living systems. Several humidity-driven conformational changes have been reported (Franklin and Gosling 1953a; see also Mahendrasingam et al. 1986). Partial dehydration entails a change from B-DNA to A-DNA. Franklin soon realized the problem of water content and succeeded in obtaining the first diffraction patterns of DNA by mastering this parameter.

B-DNA is the commonest form occurring in the cell nucleus. It is made up of two right-handed, anti-parallel helices with a pitch of 3.32 nm and with an average number of 10 bases/turn. The diameter of the double helix is 2.0 nm. The double helical structure is not symmetric in that one helix is not half-staggered with respect to the other (this means $\varphi \approx 90°$ considering the helical section model drawn in Figure 2.2 in Part I). This

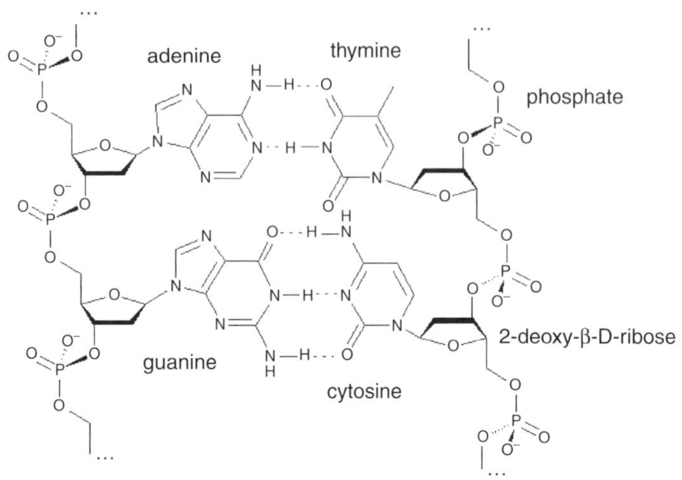

Fig. 9.1. Chemical structure of DNA showing the way adenine and thymine, on the one hand, and guanine and cytosine, on the other hand, interact.

creates structural features designated as minor and major grooves. The B-conformation is partly crystallizable within an orthorhombic unit cell of space group $P2_12_12_1$.

B-DNA is the helical form that contains the highest fraction of water molecules. It has been shown on oligomers that a 'spine' of hydration occurs in narrow regions of the minor groove (Drew and Dickerson 1981). The water content is larger, as water molecules interact more strongly with the phosphate groups that run along the inner edges of the major groove. The bases are involved in hydrogen-bonded pairing but are also capable of additional hydrogen-bonding links to water within the major or minor groove (Figure 9.3).

Water motion within the grooves is slowed down compared with the surrounding water, with the greatest reduction being found within the minor groove (Bonvin et al. 1998; Pal et al. 2005). However, there is a rapid exchange with bulk water, as residence times are smaller than 1 ns.

In order to obtain data on the placement of the water molecules, neutron diffraction was used by two independent groups (Fuller et al. 2004; Arai et al. 2005). This technique is best suited to placing molecules made up with hydrogen atoms thanks to the high values of the scattering cross-section of hydrogen and deuterium with respect to those of the other heavier atoms (see Chapter 2). Replacing light water with heavy water (D_2O) enhances the contrast of water molecules and allows minimization of the incoherent background without changing many of the DNA–water interactions. These studies were carried out both on DNA fibres under controlled moisture conditions (Fuller et al. 1989;

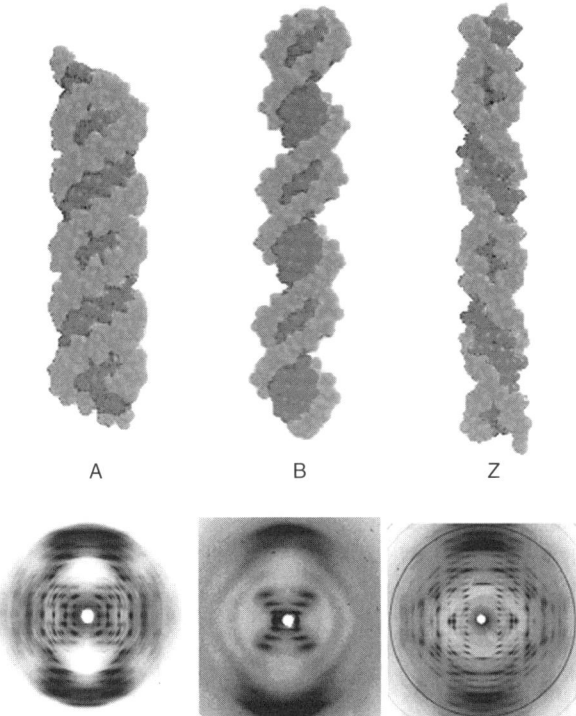

Fig. 9.2. Molecular representations of the helical forms of A-DNA, B-DNA and Z-DNA together with the corresponding fibre diffraction patterns. All forms are double-stranded helices. (Diffraction patterns from Fuller, W., Forsyth, T., Mahendrasingam, A. *Phil. Trans. R. Soc. B* **2004** *359* 1237. Reprinted with permission from The Royal Society.) (See Plate 1)

Shotton et al. 1998; Fuller et al. 2004), and on single crystals of oligomers (Arai et al. 2002, 2005).

Shotton et al. (1998) confirmed the previous X-ray findings and further concluded that two chains of ordered water can be identified in the minor groove of B-DNA with positions that suggest water bridging of sugar O4 and purine N3 or pyrimidine O2 atoms. Shotton et al. have then been able to identify four water sites on and within the A-DNA conformation (Figures 9.4 and 9.5).

The conformation of A-DNA is a 11_1 double right-handed helix with a pitch of 2.815 nm and a diameter of 2.6 nm (see Figure 9.2). Unlike B-DNA, the base pairs are tilted by about 20° with respect to the helix axis, and also shifted away from this axis, which creates a hollow core of about 0.6 nm in diameter. As with B-DNA, a deep major groove and a shallow minor groove occur. A-DNA crystallizes in the monoclinic space group $C2$ with parameters $a = 2.224$ nm, $b = 4.062$ nm, $c = 2.815$ nm and $\beta = 97°$. Here too, neutron diffraction studies have provided additional information on how the water molecules are arranged around and within the A-DNA double helix (Shotton et al. 1997).

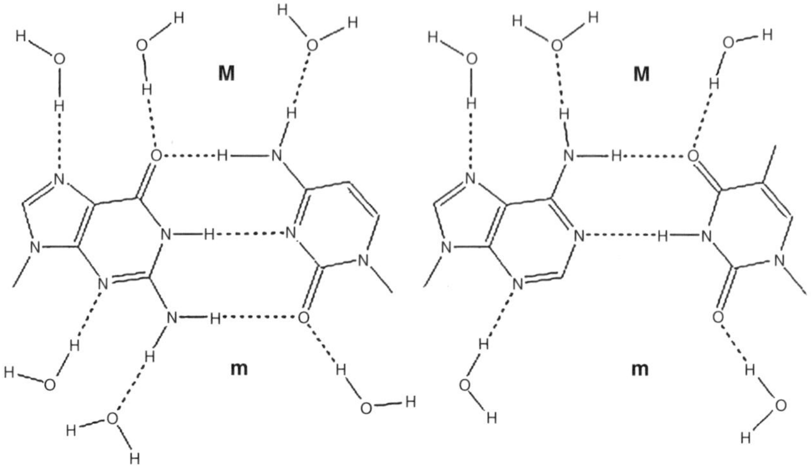

Fig. 9.3. The possible hydrogen bonding of water with the bases of DNA. M and m stand for major and minor groove, respectively.

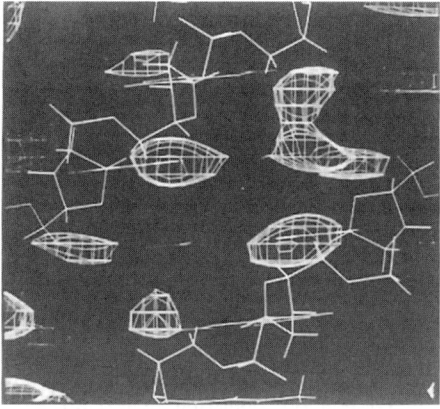

Fig. 9.4. Fourier difference synthesis map highlighting the location of two chains of ordered water in the minor groove of the B-DNA double helix. (From Shotton et al. 1998. Reprinted with permission from Elsevier.)

Shotton et al. (1997) have been able to identify four water sites on and within the A-DNA conformation. Site 1 is located in the major groove and contains the intra-strand bridging water molecules. Site 2 is located at the centre of the opening of the major groove at equal distances from the phosphates on either strand. Site 3 is found in the centre of the major groove at a smaller radius than site 2. Site 4 appears as a column of water molecules in the hollow central core.

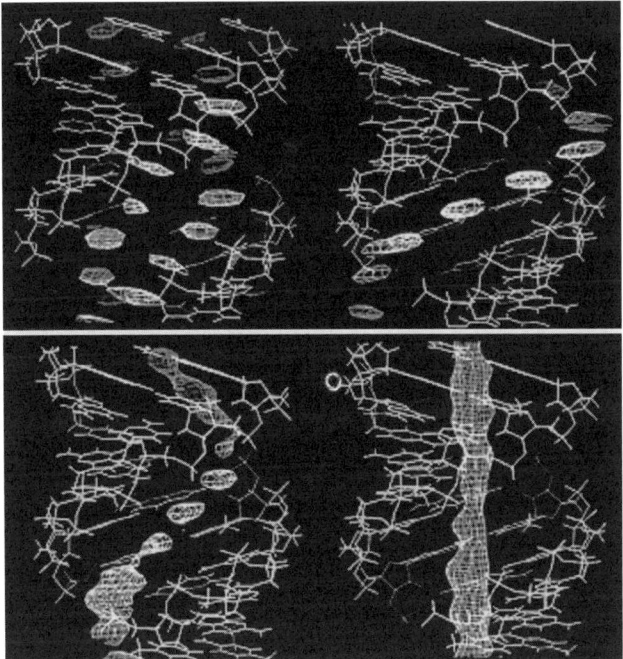

Fig. 9.5. Fourier difference synthesis map showing the four ordered water sites in the A-DNA double helix (locations of water molecules are in blue, while the DNA is in yellow). **Top left**, site 1; **top right**, site 2; **bottom left**, site 3; **bottom right**, site 4. (From Shotton et al. 1997. Reprinted with permission from Elsevier.) (See Plate 2)

The Z-DNA, where Z stands for *zig-zag*, the conformation taken on by the strands, has been discovered recently (Wang et al. 1979; Arnott et al. 1980). Its biological function, if any, is still under debate. Unlike the other DNA forms, Z-DNA helices take on a left-handed conformation. The pitch of the double helix is 4.3 nm with 12 base pairs per helix turn, while the diameter is 1.8 nm (see Figure 9.2). It has been shown on synthetic polynucleotides that the transition from B-DNA to Z-DNA, i.e. involving a change of handedness, can be achieved in a high-salt environment and as a function of the relative humidity (Mahendrasingam et al. 1990). This transition has been seen to be reversible.

Paréceme, Sancho, que no hay refrán que no sea verdadero, porque todos son sentencias sacadas de la misma experiencia, madre de las ciencias todas.

Miguel de Cervantes Saavedra (El ingenioso hidalgo Don Quijote de la Mancha)

PART III
Synthetic polymer complexes

Some synthetic polymers have the propensity to form hydrogen bonds, for example poly[vinyl oxide], or pseudo hydrogen bonds arising from electrostatic charges created through the polarization of chemical bonds, for example poly[vinyl chloride]. As has been discovered recently, some of these polymers form compounds with polar molecules. As with biopolymers, these compounds are therefore of the *enthalpic* type and are best described as polymer–solvent complexes. In Chapters 10–13, complexes involving polyaniline, poly[vinyl chloride] and poly[vinylidene fluoride], and some liquid-crystalline polymers, are detailed.

CHAPTER 10

Atactic poly[vinyl chloride]

The common version of commercial atactic poly[vinyl chloride] (PVC) is usually produced from radical polymerization at 50°C. Under these conditions, the triad composition as revealed by NMR is:

$$hetero = 0.49; \quad iso = 0.18; \quad syndio = 0.33$$

which makes it a chiefly atactic material.

Despite its atactic character, poly[vinyl chloride] samples can exhibit a low degree of crystallinity (about 5 to 10%; e.g. see Rehage and Hallboth 1968). This crystallinity has been shown to arise from the organization of syndiotactic material (e.g. see te Nijenhuis and Dijkstra 1975) as ascertained by comparison with the reflections observed by Krimm and co-workers on highly syndiotactic samples (Wilkes et al. 1973; see Figure 10.1 for the crystalline lattice). NMR studies yield a probability for a syndiotactic placement of about 0.55 to 0.57 by taking Bernouillian statistics. These figures entail the amount of long syndiotactic sequences being very low, and therefore in conflict with the degree of crystallinity. This paradox can be illustrated by means of Mutin's derivation of the analytical expression for the number and fraction of *syndiotactic sequences of a given length* as well as for the number and fraction of *syndiotactic sequences larger than a given length* (Mutin 1986). If α and β are the probability of isotactic and syndiotactic placements respectively ($\alpha = 1 - \beta$), then the probability of finding a syndiotactic sequence of N_s consecutive monomers is written:

$$P(N_s) = \alpha^2 \beta^{N_s - 1} \tag{10.1}$$

The average number of syndiotactic sequences containing N_s monomers for a chain made up with N monomers is written:

$$T_s(N_s) = (N - N_s + 1)\alpha^2 \beta^{N_s - 1} \tag{10.2}$$

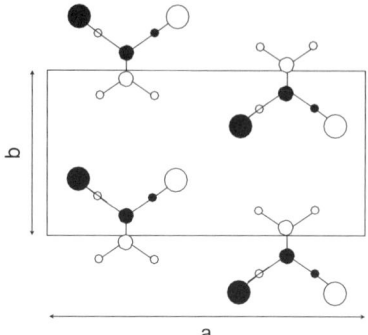

Fig. 10.1. Crystalline lattice of highly syndiotactic PVC as seen parallel to the chain axis (the chain takes on a planar zig-zag conformation). The full symbols stand above the open symbols. Largest symbols = chlorine atoms; medium-sized symbols = carbon atoms; and small symbols = hydrogen atoms. Crystal parameters: $a = 1.024$ nm, $b = 0.524$ nm, $c = 0.508$ nm. (After Wilkes et al. 1973.)

and the corresponding weight fraction:

$$W_s(N_s) = \frac{N_s}{N}(N - N_s)\alpha^2 \beta^{N_s - 1} \tag{10.3}$$

For large values of N, the weight fraction $W_s(N_s)$ is independent of molecular weight, in agreement with the experimental fact that tacticity is molecular weight-independent when polymerization is carried out under the same conditions. Conversely, $T_s(N_s)$ does depend upon molecular weight: the higher the molecular weight, the larger the number of syndiotactic sequences of N_s monomer units. The *total number of sequences of length larger than a given length* is then written:

$$T_s(Z_s > N_s) = \sum_{i=N_s}^{N} (N - i + 1)\alpha^2 \beta^{i-1} \approx \alpha N \beta^{N_s - 1} \tag{10.4}$$

and the corresponding weight fraction for large N:

$$W_s(Z_s > N_s) = \sum_{i=N_s}^{N} \frac{i}{N}(N - i)\alpha^2 \beta^{i-1} \approx \beta^{N_s - 1}[N_s \alpha + \beta] \tag{10.5}$$

Again the latter quantity is molecular weight-independent unlike $T_s(Z_s > N_s)$, which varies approximately as N. The degree of crystallinity should increase with molecular weight as the number of potentially crystallizable sequences also increases. This is what has been observed by Dorrestijn and te Nijenhuis (1990), although these authors have assumed the effect to arise from different syndiotactic contents.

As a rule, polymer crystals are of a thickness greater than 4 nm (e.g. see Wunderlich 1973) which, considering the crystal structure of syndiotactic material (see Figure 10.1), implies syndiotactic sequences larger than 16 monomer units. For a typical molecular weight of 1.2×10^5, equations (10.4) and (10.5) imply that the number of such sequences is below 1 *per chain*, and that the corresponding weight fraction amounts to 0.001.

In an attempt to resolve this crystallinity paradox, Lemstra et al. (1978) have suggested that some syndiotactic sequences are not recognized by NMR, which implies a non-Bernouillian distribution. Experimental results are, however, in very good agreement with this type of distribution. Juijn et al. (1973) have proposed a model wherein defects, such as short isotactic sequences, are incorporated into the syndiotactic crystalline lattice. Thanks to the small van der Waals radius of chlorine, this is sterically quite feasible as isotactic sequences can take on a planar zig-zag conformation, which is a prerequisite for being incorporated into the syndiotactic crystalline lattice without creating too much distortion. Standt (1983) has also considered that electrostatic interactions arising from the strong polarization of the H−C−Cl bond should favour this process.

As expected, the crystalline behaviour of bulky PVC has a direct bearing upon its propensity to form thermoreversible gels in a large variety of solvents (e.g. see Aiken et al. 1947; Stein and Tobolsky 1948; Alfrey et al. 1949; Walters 1954). These gels possess unexpectedly high elastic moduli, a fact that has remained a challenge to scientists for several years. In particular, an additional paradox has been pointed out by He et al., who studied a model multiblock copolymer containing 10% of crystallizable sequences alternating regularly with non-crystallizable sequences. Thermoreversible gels of such a copolymer possess a modulus a hundred-fold lower than that of PVC gels at the same concentration (He et al. 1988). Clearly, even if one succeeds in explaining the 10% crystallinity, comparison with model systems raises new questions.

On the basis of rheological experiments, Mutin and Guenet have suggested to account for this paradox by contemplating the existence of a *polymer–solvent complex*. These experiments have consisted in studying the gel modulus as a function of the degree of swelling in different solvents and by using two samples of PVC of differing tacticities (Mutin and Guenet 1989). Indeed, PVC samples containing a higher proportion of syndiotactic sequences can be obtained by performing the synthesis at lower temperature (in the present case −40°C). The fraction of syndiotactic triads increases up to 0.39 while the fraction of hetero triads remains virtually constant. For the same solvent, for example bromonaphthalene, gels from this highly syndiotactic PVC (*designated as LTPVC*) possess moduli of about 5 times larger than those from atactic PVC (*designated as HTPVC*), which is consistent with the fact that the degree of crystallinity is correspondingly larger. Interestingly enough, Mutin and Guenet have observed that changing the solvent type can produce the same effect as increasing the syndiotacticity. This is illustrated in Figure 10.2 where the evolution of the modulus for gels immersed in an excess of preparation solvent, and hence allowed to swell (or de-swell), is plotted as a function of concentration. All the gels are prepared at the same starting concentration ($C_o = 0.175 \, \text{g/cm}^3$) but display different types of behaviour. The HTPVC gel in bromonaphthalene swells markedly without significant alteration of the original modulus. Conversely, the LTPVC gel prepared from the same solvent de-swells considerably, while its modulus nearly doubles in size, which indicates an increase of crystallinity. Similar behaviour is seen with HTPVC in diethyl malonate (DEM), a fact that led Mutin

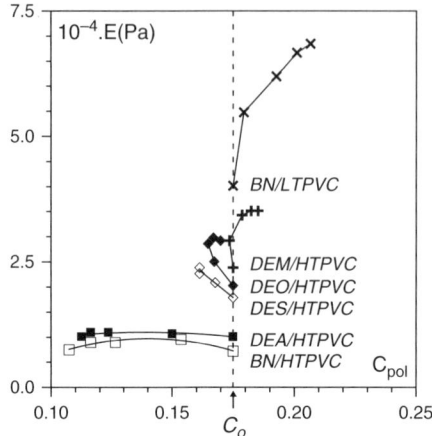

Fig. 10.2. Evolution of the modulus when the gel is placed in an excess of preparation solvent. All the gels have been prepared at a starting concentration $C_o = 0.175\,\mathrm{g/cm^3}$. C_{pol} is the concentration reached after some ageing time (maximum 20 days). BN = bromonaphthalene; DE, diethyl with M = malonate, O = oxalate, S = succinate and A = adipate. (After Mutin and Guenet 1989.)

and Guenet to state that some solvents can enhance the degree of organization, as does higher syndiotacticity. This effect can be designated as the *equivalence solvent-tacticity*.

The gel state

Formation and melting behaviour

As has been mentioned above, PVC forms thermoreversible gels in a large variety of solvents. As we shall discover in more detail in the section devoted to molecular structure, this is so because PVC cannot give birth to lamellae, and correspondingly to spherulites, but is restricted to the development of small crystallites that are stabler when building up fibrils (see Chapter 4). The occurrence of PVC–solvent complex does not necessarily pertain to all solvents. As will be apparent throughout this section, thermoreversible gelation of PVC originates above all in the crystallization of syndiotactic sequences (e.g. see Lemstra et al. 1978; Dorrestijn et al. 1981) while gel ageing arises essentially from the formation of PVC–solvent compounds.

Unlike many systems, gel formation does not produce any detectable exotherm in differential scanning calorimetry experiments. This may seem rather surprising as a freshly prepared piece of gel gives an unexpectedly huge endotherm on the first heating run. The magnitude of this endotherm has in actuality been shown to be meaningless by Mutin and Guenet (1989). Indeed, after melting and subsequent cooling, reheating the sample does not again produce any endotherm. While the position of the endotherm observed on the first heating cycle does correspond to the macroscopic melting of the gel, the associated enthalpy is therefore grossly overestimated. Mutin and Guenet have demonstrated it to arise from a mechanical effect triggered when a piece of gel of undefined shape collapses while melting. The sudden dropping of the liquid down to the bottom of the pan generates a parasitic effect foreign to the thermodynamics of the gel. When the liquid is cooled down again, a gel re-forms taking on the shape of the DSC pan. On reheating, *gel melting* occurs in a 'quieter' way so that the mechanical parasitic effect is considerably reduced, and correspondingly no endotherm is observed. The abnormally large endotherm can be made to reappear, admittedly to a lesser extent, by cutting the same gel sample into pieces. This is so because the sample collapse on gel melting can take place again. Incidentally, this indicates that the first heating cycle is liable to be smeared with all sorts of non-pertinent effects, especially when dealing with thermal events of low enthalpy. The best method to assess the validity of the DSC findings is to perform a study as a function of the heating rate. The enthalpy should not vary significantly with this parameter, or at least extrapolation to zero-heating rate should give a finite value (a case encountered with gels of higher melting enthalpy, i.e. iPS gels, but still subject to the artefact described above; see Chapter 14). In the case of PVC it has been shown that the value of the melting enthalpy does extrapolate to a value close to nought by carrying out this procedure (Mutin and Guenet 1989).

The virtual absence of an endotherm at the gel macroscopic melting probably originates in the low degree of crystallinity together with the broad distribution of fibril cross-section which spreads the melting process over a broad temperature range. This point will be further discussed in the section dealing with molecular structure.

When a gel, properly reformed within the sample pan, is allowed to age, an endotherm appears at $T \approx 50 \pm 10\,°C$ (designated as low-melting endotherm in what follows), whose associated enthalpy is rather low (0.4 to 1.0 J/g). This endotherm is clearly related to the ageing phenomenon which is known to take place in these gels (Walters 1954; te Nijenhuis and Dijkstra 1975), and which corresponds to the appearance of new physical links (Mutin and Guenet 1989; Garcia et al. 1990; see the section on mechanical properties).

A temperature–concentration phase diagram can be drawn (Figure 10.3) which highlights that the thermal event associated with the low-melting endotherm is a non-variant transformation. In view of the possible phase diagrams detailed in Part I, this type of invariant transition is compatible with an **incongruent melting**, which suggests the existence of a polymer–solvent complex. A similar diagram is obtained for *syndiotactic polystyrene/benzene gels* for which the occurrence of a polymer–solvent compound is unambiguously demonstrated (see Chapter 14). The *GEL II phase* consists of a mixture of junctions made up with syndiotactic microcrystals and with the organized form of the polymer–solvent complex (using the terminology of Mutin and Guenet, *the strong links* for the former and the *weak links* for the latter). At the low-melting endotherm, the polymer–solvent complex melts, so the *GEL I phase* contains only syndiotactic microcrystals, as will become explicit in the section on the molecular structure.

As has been pointed out by Yang and Geil (1983), the gel terminal melting depends markedly upon the molecular weight of the PVC sample. This, as has been discussed above, is related to the existence of longer syndiotactic sequences when increasing PVC molecular weight, so more stable syndiotactic crystals can be formed. This again hints

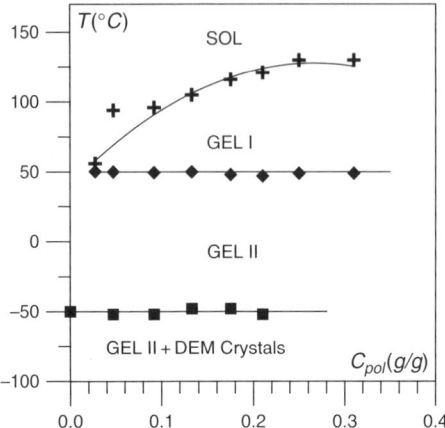

Fig. 10.3. Temperature–concentration phase diagram for HTPVC/diethyl malonate. + = macroscopic gel melting temperature as determined by DSC and the ball-drop method. The other thermal events were detected by DSC. (After Mutin and Guenet 1989.)

at the relation between the terminal gel melting and the melting of those syndiotactic crystals.

> **Note 10.1**
>
> As is shown in PART I, the melting temperature of a crystal is size-dependent. If the syndiotactic crystals of the PVC fibrils constituting the gels are regarded as cylinders of radius r and length l_c, then the actual melting point with respect to that of an infinite crystal is:
>
> $$T_m = T_m^o \left[1 - \frac{\sigma}{\Delta h_f} \times \left(\frac{2}{r} + \frac{2}{l_c} \right) \right]$$
>
> Thus, for r being constant, the longer the l_c, the higher the T_m.

The gel is a *two-phase system*: a polymer-rich phase, the network, embedded in a dilute phase, the latter being virtually pure DEM. The biphasic character is evidenced by the formation of DEM crystals within *GEL II* when the system is cooled to below the solvent melting point. The concentration at which DEM melting enthalpy becomes zero corresponds to 0.6 solvent molecules per monomer. Two populations of solvent molecules are therefore present in the gel: *free solvent* in the dilute phase and *bound solvent* in the polymer-rich phase (i.e. the gel framework). Similar solvent/monomer ratios have been measured in bromobenzene, bromonaphthalene and diethyl oxalate (Mutin 1986). This ratio may correspond to the stoichiometry of the complex although the actual value depends upon the 'crystallinity' and the solvent molecules trapped in the disordered domains. A value of about *one solvent molecule per two monomers* is probably a good estimate.

Although diethyl malonate is certainly not a good solvent to PVC, the biphasic character of the gel is not consequent to a *liquid–liquid phase separation*. The features of the phase diagram do not depend upon the cooling rate. Further, the same thermal event at 50°C is observed for PVC gels prepared from good solvents such as dibutyl oxalate. Kawanishi et al. (1986, 1987) have also noticed that liquid–liquid phase separation in PVC/butyrolactone is not a prerequisite for gelation to occur. In the phase diagram established by turbidimetry these authors recognize two 'gelling' regions: one involving a liquid–liquid phase separation, possibly by spinodal decomposition prior to crystallization, and another involving only crystallization.

Gel morphology

PVC gels possess a mesoscopic structure because they form via a phase separation process unfolding through two possible routes: *liquid–solid phase separation* only or *liquid–liquid phase separation prior to liquid–solid phase separation*. To date only arrays made up with fibrillar structures have been observed (Figure 10.4), whose average mesh size at similar polymer concentration seems to be dependent upon the route taken by the phase separation process: ranging from 0.1 to 1 μm when ***liquid–solid phase separation*** is involved (class I gels: see Yang and Geil 1983; Cho and Park 2001) ranging from 1 to 10 μm when ***liquid–liquid phase separation*** occurs at the early stage of gelation (class II gels: see Mutin and Guenet 1989).

Gels of class I are characterized by unexpectedly outstanding elastic properties whereas gels of class II are irreversibly deformed when subjected to a stress. This is so because the solvent is easily expelled under deformation. Also, gels of class II display strong heterogeneities in network density. The fibrillar structure of gels belonging to class I is very reminiscent of what is seen for isotactic polystyrene gels, i.e. no straight fibrils but "awry" ones instead (see Part IV). Also, fibril cross-section polydispersity is observed.

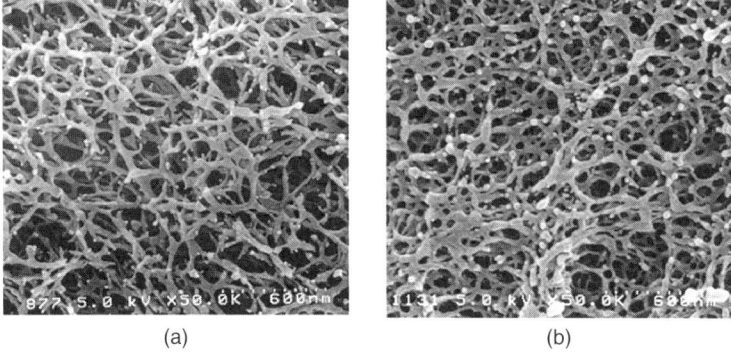

(a) (b)

Fig. 10.4. (a) Electron micrograph of PVC/dibutyl phthalate gel obtained by the critical point drying technique. ($C_{pol} = 0.2 \text{ g/cm}^3$). (From Cho and Park 2001. With permission from Wiley Interscience.) (b) Electron micrograph of PVC/butyl benzoate gel obtained by the critical point drying technique. ($C_{pol} = 0.2 \text{ g/cm}^3$). (From Cho and Park 2001. With permission from Wiley Interscience.)

In earlier papers Keller and co-workers observed a reflection at 0.52 nm on stretched gels that they assigned to the presence of platelet crystals essentially disconnected from the network (Lemstra et al. 1978; Guerrero et al. 1980; Guerrero and Keller 1981). Yet, even high-quality electron micrographs such as those obtained by Cho and Park through the critical drying procedure do not reveal such platelet crystals. As has already been the case with isotactic polystyrene gels (see Part IV), the interpretation of this unexpected reflection may not require postulating the existence of a morphology never observed to date.

Gel nanostructure

PVC/diethyl oxalate gels

The gel nanostructure of PVC/DEO gels has been extensively studied by Abied et al. (Abied et al. 1990) and, later, by Lòpez et al. (Lòpez et al. 1994) by means of small-angle neutron scattering in various solvents. A typical scattering curve is presented in Figure 10.5 for the PVC/DEO system.

As is apparent from this figure, three scattering regimes can be identified. At low q a transitional regime occurs (for further details see Part I, Chapter 2) where the intensity varies as:

$$q^4 I_A(q) \propto \pi q - \frac{2}{r_{max}} \tag{10.6}$$

where r_{max} is the largest cross-section radius that the PVC fibrils can reach. Such a behaviour therefore suggests that the fibrils are characterized by a high cross-section radius polydispersity consistent with a distribution of the type $w(r) \sim r^{-1}$ (see Chapter 2).

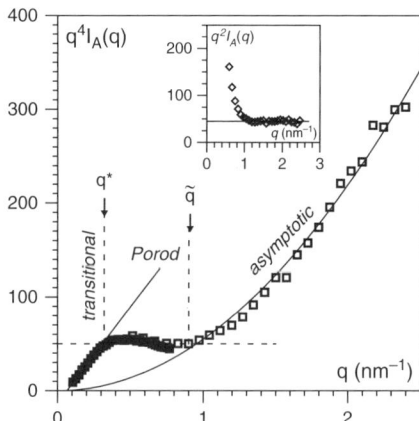

Fig. 10.5. Intensity scattered by a PVC_H/DEO_D gel ($C_{pol} = 0.05\,g/cm^3$). Three scattering regimes can be identified as indicated: in the transitional regime the variation is linear, the Porod regime appears as a plateau, and the asymptotic regime behaves as q^{-2} as shown by the best fit (solid line) and in the inset where $q^2 I_A(q)$ is plotted vs q. (After Lòpez et al. 1994.)

Note that this conclusion is in line with the conclusions from the DSC outcomes, which also hint at the existence of a large crystal size distribution.

Equation (10.6) allows calculation of r_{max} by determining the value $q = q_o$ where $q^4 I_A(q) = 0$ (see Part I):

$$r_{max} = \frac{2}{\pi q_o} \tag{10.7}$$

Lòpez and colleagues have derived $r_{max} = 11 \pm 1\,nm$, which, incidentally, is a value in close agreement with electron microscopy findings (Yang and Geil 1983; Cho and Park 2001). As is detailed in Part I, the value of r_{min} can be obtained from the cross-over q^* between the transitional regime and the Porod regime through:

$$\frac{r_{max}}{r_{min}} = \frac{q^*}{q_o} \tag{10.8}$$

Lòpez et al. have also derived $r_{min} = 1.8 \pm 0.2$ nm. Taking the chain density in syndiotactic crystals, a rough estimate can be made about the number of chains per fibril: about 50 chains in the thinnest fibrils to 750 in the thickest.

The type of cross-sectional distribution considered here implies that the large majority of fibrils in the gel have small cross-sections.

It is worth emphasizing that the $1/q^4$ behaviour is reached from above owing to the presence of $1/q^6$ terms in the Porod regime (see Chapter 2). The existence of these terms has been shown by Lòpez et al. to be directly related to the number of physical junctions between fibrils per unit volume.

The terminal asymptotic behaviour observed beyond $q = \tilde{q}$ varies as $1/q^2$. This hints at two possible interpretations: either there exists a *significant fraction of chains not connected to or partially disconnected from the network* (free or dangling chains as discussed for agarose gels, see Chapter 6), or this behaviour is related to the *short range structure of the fibrils*. The first situation may occur because low molecular weight chains contain only short syndiotactic sequences and are thus less liable to be incorporated into any organized structure as discussed above. The $1/q^2$ behaviour implies that the *free/dangling chains* obey Gaussian statistics, in which case the intensity scattered by the gel in the *Porod range* should be:

$$q^4 I_A(q) \propto a q^2 + b \tag{10.9}$$

or:

$$q^2 I_A(q) \propto a + \frac{b}{q^2} \tag{10.10}$$

in which a and b are constants.

Three arguments allow one to disregard this interpretation. The first argument derives directly from the scattering curve displayed in Figure 10.5 which indicates that the fit in the $1/q^2$ domains yields $b \approx 0$. This would imply the network fraction to be close to nought, which is clearly unrealistic. Second, the scattering curve obtained with the same PVC sample after molecular weight fractionation, whereby those low molecular weight chains have been subsequently extracted, does not display any noticeable modification in the asymptotic range (Reinecke et al. 1997a). Finally, the scattering curve remains unchanged when varying the PVC concentration unlike what is seen with agarose gels (Abied et al. 1990; Reinecke et al. 1997a).

Consequently, the second interpretation, i.e. a scattering behaviour related to the short-range structure of the fibrils, appears to be plausible enough. To account for the $1/q^2$ behaviour Lòpez et al. have come up with a model where *sheets of extended PVC chains* alternate with *layers of solvent molecules* as shown in Figure 10.6. This type of structure is very reminiscent of that put forward for cellulose/amines complexes, and has also been considered for the *p-phase* of the isotactic polystyrene/*cis*-decalin system (see Parts II and IV for details). This structure ought to possess a significant degree of organization, and Lòpez et al. consider it as nothing but the PVC/solvent complex evidenced by DSC experiments. This model, first considered by Abied et al. for PVC (Abied et al. 1990), is based on the propensity of a diester to interact with two different chains through electrostatic interactions (Tabb and Konig 1975; Monteiro and Mano 1984). Both the diester C=O bond and the PVC H−C−Cl bonds can be polarized so as to create a positive and a negative fractional charge, respectively (see Figure 10.6).

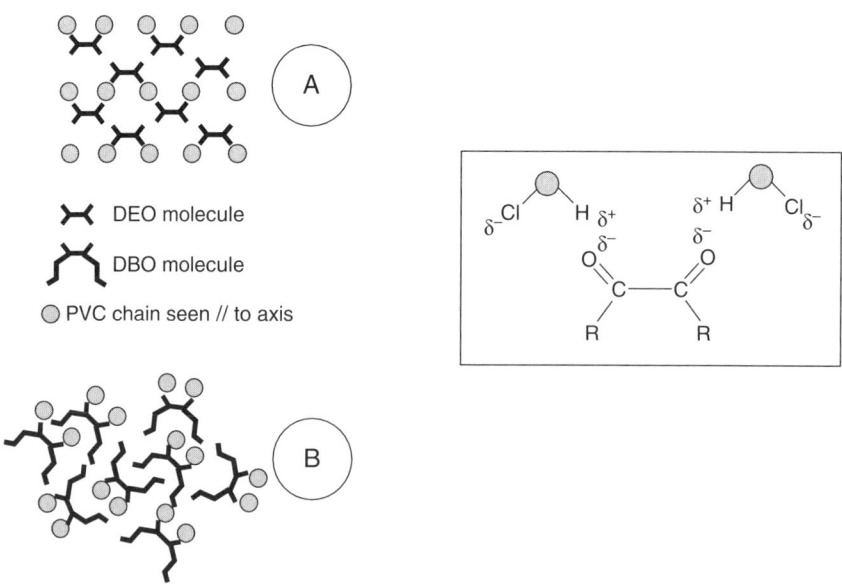

Fig. 10.6. Molecular models for the short-range structure of the PVC/solvent complex in two different cases: A = DEO, B = DBO. (After Lòpez et al. 1994 (A) and Reinecke et al. 1997a (B).) **Inset:** detail of how a diester molecule can interact electrostatically with two PVC chains. In all cases chains are seen parallel to their axis.

The fraction X of complex can be estimated from this asymptotic regime, keeping in mind that the syndiotactic crystals and the disordered domains still scatter as $1/q^4$ because their densities differ from the dilute phase. The total scattered intensity in the asymptotic range can be approximated, by neglecting interdomain intermolecular scattering, to:

$$I(q) \propto X^2 \times \frac{2\pi\mu_S}{q^2} + (1-X)^2 \times \frac{4\pi}{q^4 r_n} \ldots \quad (10.11)$$

where μ_S is the mass per unit area in the PVC sheets and r_n is the number-averaged cross-section radius. For large q only the first term is significant, which gives the sheet-like behaviour scaled by the complex fraction:

$$I(q) \propto X^2 \times \frac{2\pi\mu_S}{q^2} \quad (10.12)$$

Lòpez et al. have further suggested that μ_S should be about 450 g/nm per mol, which yields $X \approx 0.57$. This figure implies a high fraction of PVC/solvent complex, which makes sense in view of the high elastic modulus of these gels.

Lòpez et al. have also studied the gel structure at 70°C (i.e. *GEL I* with regard to the notation used in the phase diagram in Figure 10.3). They have found that at 70°C there is no significant change in relation to r_{max} and r_{min}. Conversely, in the asymptotic range, the scattered intensity appears to be of the form:

$$I(q) \propto \frac{A}{q} + \frac{B}{q^2} \quad (10.13)$$

a behaviour typical of a rod-like conformation (Luzzati and Benoit 1961).

Clearly, once the complex is molten, chains are sufficiently spaced within the fibril to enable them to be seen individually for $q > \tilde{q}$. Their rod-like scattering response entails that chains are locally extended (Figure 10.7).

The question of how fibrils interact has been examined by Lòpez et al. (1994). These authors suspect that an interdigitation process occurs between the PVC/solvent complex sheets of two different fibrils. There thus exists a type of junction where rows of PVC chains belonging to two different fibrils alternate. These junctions are most probably the *'weak'* links described by Mutin and Guenet (1989). The effect of complex formation would therefore be twofold: *to stiffen the fibrils* and *to create additional links between fibrils*. This statement will be further elaborated in the sections devoted to modified PVCs and to rheological properties.

The existence of rod-like sequences along the chain has been given further support through a small-angle neutron scattering study of gel samples wherein a few chains were

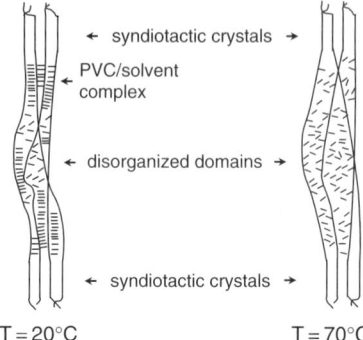

Fig. 10.7. Schematic representation of a PVC fibril according to Abied et al. 1990 and Lòpez et al. 1994. Abied et al. suggest that three types of domain are present: syndiotactic microcrystallites, PVC/solvent complex domains and disordered domains. After melting above 50°C the PVC/complex domains vanish, and the chains are locally extended (parallel dashes mimic the solvent in the organized domains while randomly oriented dashes simulate the solvent in the disorganized domains).

labelled (here the matrix is deuterated PVC and the labelled species are hydrogenous PVC; the reason for this is detailed in Part I).

The scattering curve recorded for two fractions of hydrogenous PVC at constant total polymer concentration is shown in Figure 10.8. This curve is, however, only meaningful at large q because of the upturn observed at a small angle, which depends upon the

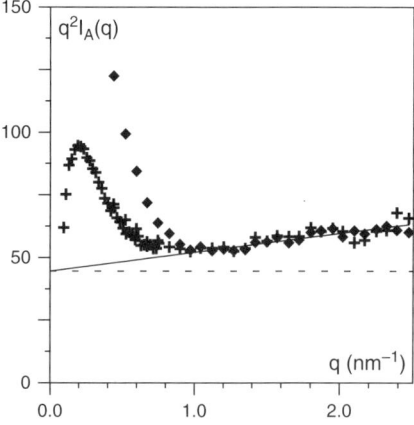

Fig. 10.8. Scattering curve plotted by means of a Kratky representation for PVC_H chains in a gel matrix composed of PVC_D + solvent. Total concentration $C_{pol} = 0.15\,\text{g/cm}^3$, (♦)$C_H = 0.06\,\text{g/cm}^3$, (✦)$C_H = 0.03\,\text{g/cm}^3$ (for the latter the intensity has been multiplied by 2 in order to rescale the concentration). (After Reinecke et al. 1997a.)

fraction of PVC_H chains. The latter effect may equally arise from isotopic segregation between PVC_H and PVC_D or from strong intermolecular scattering. Indeed, gel formation confines labelled chains to a smaller volume (the fibrils), so the interchain distances are considerably reduced and fall within the explored q-range. Similar effects had been observed for low-concentrated isotactic polystyrene gels (Guenet 1987). At large q both PVC_H fractions give the same scattering curve, which varies linearly with q in a Kratky-representation ($q^2I(q)$ vs q). Such behaviour may be accounted for by considering two types of sequences along the chain: *rod-like sequences* pertaining to the PVC/DEO complex domains as well as to the syndiotactic microcrystallites, and *random sequences* belonging to the disordered domains. Neglecting the intermolecular cross-terms between the different domains, the scattering curve can be approximated to:

$$q^2I(q) = C_H \mu_L \left[X^2 \pi q + (1-X)^2 \frac{12}{b} + .. \right] \qquad (10.14)$$

where X is the fraction of those rod-like sequences, μ_L the mass per unit length of the labelled chains, C_H their concentration and b the length of the statistical segment.

For PVC under an extended-chain conformation, i.e. planar zig-zag, $\mu_L = 250$ g/nm per mol, and $b \approx 1.5$ nm for a random coil. From these parameters X can be derived either from the slope or from the intercept with the $q^2I(q)$ *axis*. Reinecke et al. obtained $X = 0.41 \pm 0.03$, a value lower than that derived from the study of the gel structure ($X = 0.57 - 0.64$). Yet this value still conveys the idea that a significant part of the chain is engaged in the formation of the PVC/DEO complex.

PVC gels in other solvents

The gel nanostructure turns out to be solvent-dependent. Two systems studied by Reinecke et al. (1997a) are discussed in what follows: gels prepared from another diester, namely diester *dibutyl oxalate* (DBO), and gels produced from *bromobenzene*, both with HTPVC and LTPVC.

The scattering curve obtained for PVC/DBO gels is shown in Figure 10.9. Again, three regimes can be evidenced: the *transitional regime*, the *Porod regime* and the *asymptotic regime*.

In this solvent, the intensity does not show a linear dependence in the *transitional regime*. A distribution function of the type $w(r) - r^{-\lambda}$ with $\lambda = 0.33$ has to be considered instead (for further details see Chapter 2). This gives $r_{max} = 7.5 \pm 0.5$ nm, a figure close to that obtained in DEO at the same polymer concentration. The value of r_{min} as derived from q^* is lower than that obtained in DEO (1.2 ± 0.2 nm in DBO against 1.8 ± 0.2 nm in DEO).

Interestingly, the *asymptotic regime* markedly differs from that in DEO. It can be fitted here by means of equation (10.13), which indicates a rod-like behaviour. According to Reinecke et al. this behaviour can be accounted for by contemplating the molecular structure portrayed in Figure 10.6(B) where DBO molecules promote pairwise association of PVC chains through electrostatic interaction. This model is reminiscent of a

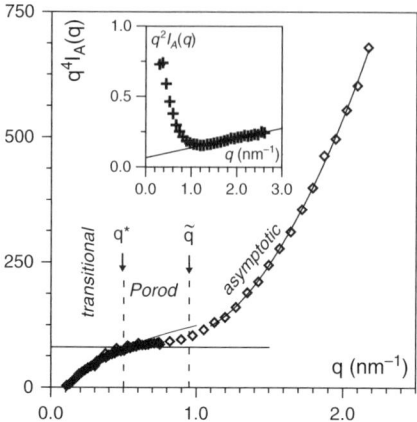

Fig. 10.9. Scattering curve ($q^4 I(q)$ vs q) obtained on PVC$_H$/DBO$_D$ gels. **Inset**: intensities from the asymptotic range plotted by means of a Kratky-plot representation ($q^2 I(q)$ vs q). (After Reinecke et al. 1997a.)

collection of zips when the fibril is viewed perpendicularly. Sheet formation, as with DEO, is not possible owing to the large molecular volume of DBO molecules. The cross-section of these zips is, however, sufficiently small that the departure from the $1/q$ behaviour is not seen in the explored q-range (see Chapter 2).

Surprisingly enough, the value of λ differs for a 2% system (i.e. only finite aggregates are formed). Reinecke et al. found $\lambda \approx 1$, which yields $r_{max} \approx 7.5$ nm, a value lower than that observed in DEO at the same concentration, but the same as in DBO at higher concentration.

The effect of PVC type has been examined in detail for bromobenzene for a sample synthesized at 50°C (HTPVC) and a sample obtained at −40°C (LTPVC). The scattering curves displayed in Figure 10.10 differ notably from one PVC type to the other. In the *transitional range* the values of λ differ considerably: $\lambda = 1$ for LTPVC gels against $\lambda = 0.34$ for HTPVC gels. The resulting values for r_{max} are $r_{max} = 6.7 \pm 0.5$ nm for HTPVC/bromobenzene and $r_{max} = 8.8 \pm 0.8$ nm for LTPVC/bromobenzene. These results are in line with those obtained from elastic modulus measurements, the analogy between tacticity and solvent type having already been highlighted (see Figure 10.2). LTPVC/bromobenzene structure closely resembles that of HTPVC/DEO gels in relation to mesoscopic structure. The same values for λ and for r_{max} are obtained.

Yet, this origin of this structural similitude arises from another type of short-range molecular organization. In fact, unlike *HTPVC/DEO*, *LTPVC/bromobenzene* gels still display a $1/q^4$ behaviour in the asymptotic regime. This implies that LTPVC fibrils are compact objects down to very short distances. This is so because the fibrils are composed here only of syndiotactic crystals and of disordered domains (the absence of any special organization in the latter makes them scatter as a compact medium with respect to the surrounding dilute phase). From the plateau regime Reinecke et al. have derived an

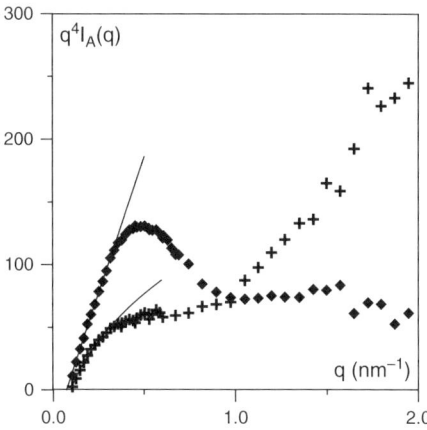

Fig. 10.10. Scattering curves ($q^4 I_A(q)$ vs q); (♦) = LTPVC/bromobenzene gel; (+) = HTPVC/bromobenzene gel. In both cases $C_{pvc} = 0.093\,\text{g/cm}^3$. Solid lines represent the best fit in the transitional range by using equation (2.36) (see Chapter 2). (After Reinecke et al. 1997a.)

estimate of the density within these disordered domains and found it to be about three time larger than the total polymer concentration ($C_{fibril} \approx 0.27\,\text{g/cm}^3$).

The analogy between solvent type and tacticity has also been reported for LTPVC/nitrobenzene gels, the scattering curve of which shows a close resemblance to that of HTPVC/DEO gels (Abied et al. 1990).

Nanostructure of gels from chemically modified PVCs

Hitherto the structure of the PVC/complex formed in diesters has relied upon the assumption that electrostatic interactions occur between the positively charged hydrogens located on the H−C−Cl bond and the negatively charged oxygens of the solvent molecules. Although it is known that such interactions exist (Tabb and Konig 1975; Monteiro and Mano 1984), no conclusive experiments have revealed their role in complex formation.

The studies carried out by Mijangos and co-workers on the chemical substitutions of PVC have proven to be of great help for testing this model (Mijangos et al. 1988, 1989, 1992). Methods devised by Pourahmady et al. (1992) allow the chemical modification of PVC chains by replacing chlorine atoms either with hydrogen or a bulkier group such as thiophenate. The modification can be carried out to the extent required, in particular to a very low degree, such that the resulting polymer can still be considered as a PVC containing some chemical defects. Another interesting outcome deals with the stereoselectivity of the chemical reactions. Lòpez et al. (1996, 1997) have shown by NMR investigations on the vinyl triads that only the heterotactic and the isotactic triads are altered while the syndiotactic triads remain virtually unchanged (Figure 10.11). This point is of great importance for testing the complex, which is said to occur only in less syndiotactic domains. Consequently, this type of chemical modification should

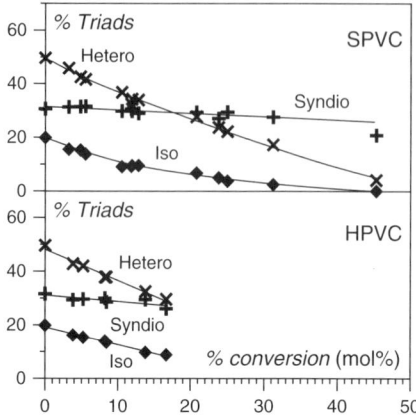

Fig. 10.11. Variation of the vinyl triads (VVV) as a function of the degree of modification. **Upper figure:** corresponds to substitution by benzene thiophenate (sample designated as SPVC); **lower figure:** corresponds to reduction by hydrogen (samples designated as EPVC). (After Lòpez et al. 1996.)

also alter the PVC propensity to form complexes with diesters. Sure enough, as will be discussed in the section devoted to rheology (see Rheological properties below), the elastic modulus drops markedly for a very low degree of modification.

Lòpez et al. (1994) have performed small-angle neutron scattering on these modified PVCs, in particular on those in which the chlorine atom is replaced by hydrogen (*EPVC samples*) and those in which it is replaced by a thiophenate group (*SPVC samples*). The rationale for selecting these types of modification relies upon two considerations: (1) replacing chlorine by hydrogen renders the opposite hydrogen atom no longer positively charged, so complex formation through electrostatic interaction is definitely hampered; (2) replacing chlorine atoms with benzenethiophenate is expected to produce the reverse situation, as DEO shows strong affinity for this group.

EPVC samples

Figure 10.12 shows a comparison of the intensity scattered by the unmodified PVC with scattering by *EPVC samples* (EPVC2 = 5.2% modified and EPVC3 − 11.1% modified). The main effect occurs in the *Porod domain* where the maximum arising from the $1/q^6$ terms gradually vanish with increasing the degree of chemical modification. As has been discussed in Chapter 2, the magnitude of these terms depends upon the density of contacts between fibrils. This means that the number of physical junctions diminishes, which is consistent with the decrease of elastic modulus. Hampering the complex formation therefore results in decreasing the interactions between gel fibrils. Lòpez et al. speculated on this point by suggesting that the *interdigitation process* between fibrils was strongly impeded by the formation of hypothetical bottlenecks owing to the crystallization of polyethylene-like sequences (those sequences originating from the replacement of chlorine by hydrogen). As we shall discover in the section devoted

Atactic poly[vinyl chloride]

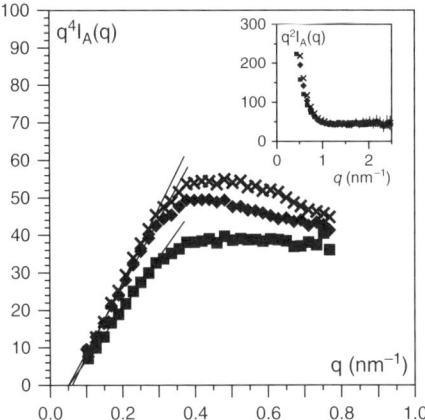

Fig. 10.12. Scattering curves in the transitional and the Porod ranges plotted as $q^4 I_A(q)$ vs q for PVC (×), EPVC2 (■) and EPVC3 (◆) (see Table 10.1 for details). Solid line stand for the best fit with $\lambda = 1$. Inset: intensities in the asymptotic range plotted by means of a Kratky representation. (After Lòpez et al. 1994.)

Table 10.1. Values of the different parameters obtained from the scattering curves ($C_{PVC} = 0.05$ g/cm^3). The cross-sectional radii are in nanometres. The bracketed values correspond to those data obtained at $T = 70°C$. (After Lòpez et al. 1994.)

sample	% modif	r_{max}	r_{min}	r_n	ρ g/mol/nm^3
PVC	0	10.6 [11.3]	2.2 [1.8]	4.4 [3.9]	318 [220]
EPVC2	5.2%	12 [13.0]	2.1 [2.0]	4.5 [4.4]	297 [180]
EPVC3	11.1%	11.3 [11.3]	1.95 [1.7]	4.1 [3.8]	260 [110]
SPVC2	1.6%	9.8 [10.6]	1.8 [1.9]	3.7 [4.0]	272 [178]
SPVC3	5.3%	9.8 [8.0]	1.8 [1.7]	3.7 [3.3]	237 [117]
SPVC4	10.0%	9.5	1.5	3.2	180

to the study of aggregates (see Pregel state below), this might not be the determining effect, but would, rather, be due to the growth of fibril cross-section.

As can be seen by inspection of Table 10.1, r_{max} seems to increase to a minor extent with increasing degree of modification while r_{min} seems to decrease slightly. The molecular density ρ as determined from the Porod regime decreases slightly, although this is probably not significant as its determination relies upon r_n and ultimately upon r_{max} and r_{min}.

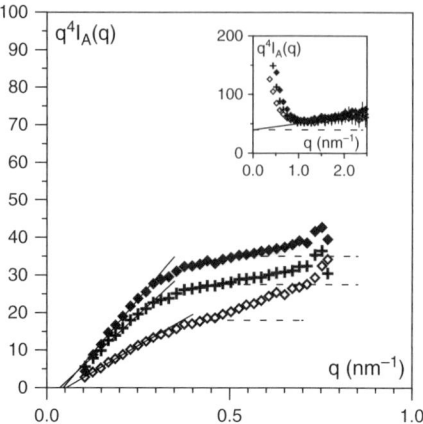

Fig. 10.13. Scattering curves $q^4 I_A(q)$ vs q for PVC (♦), EPCV2 (+) and EPVC3 (◊) at $T = 70°C$. The solid lines correspond to the best fit with $\lambda = 1$, while the dotted lines stand for the possible location of the plateau regime in the Porod range (see Table 10.1 for details). Inset: results obtained in the asymptotic range plotted by means of a Kratky representation. (After Lòpez et al. 1994.)

In the *asymptotic range*, the three samples display the same $1/q^2$ behaviour. According to Lòpez et al. the appearance of polyethylene-like sequences are defects randomly dispersed amidst sheet-like arrangement of the complex. As a result, the distance between sheets should be globally preserved, hence the absence of any noticeable alteration of the behaviour in this q-range.

The intensities scattered by the same samples after heating to 70°C, i.e. above the melting of the complex, are shown in Figure 10.13. In all cases the $1/q^6$ terms have vanished, indicating a very low degree of interaction between fibrils. As expected, this corresponds to a drop in elastic modulus. Conversely, r_{max} and r_{min} are little affected. This is consistent with the assumption that fibril formation is chiefly controlled by the crystallization of the highly syndiotactic sequences: at 70°C the syndiotactic crystals are still present and sustain the global molecular morphology of the fibrils.

Significant departure from the $1/q^2$ behaviour occurs in the asymptotic range. The scattered intensity is now better described by equation (10.14), which implies chains to be locally spaced apart (except in the 'syndiotactic' crystals). This clearly indicates the disappearance of the complex, and is in line with the conclusions drawn from the behaviour in the Porod range. The spacing apart of the chains is further borne out by the large decrease of the molecular density ρ of the fibrils, which hints at a swelling of the fibril.

SPVC samples

The substitution of chlorine by a benzenethiolate group entails still more dramatic consequences as is made apparent in Figure 10.14 (Lòpez et al. 1994). The $1/q^6$ terms

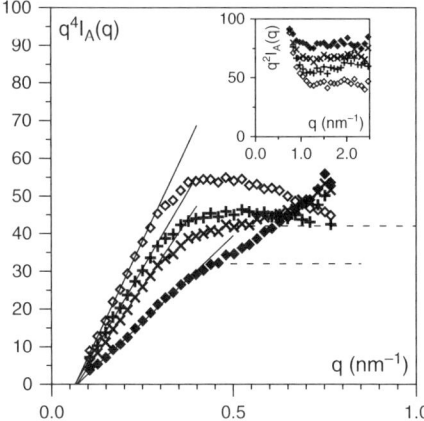

Fig. 10.14. Scattering curves plotted by means of a $q^4 I_A(q)$ vs q representation for PVC (◇), SPVC1 (+), SPVC2 (×) and SPVC3 (◆). **Inset:** results obtained in the asymptotic range plotted by means of a Kratky plot ($q^2 I_A(q)$ vs q). $C_{PVC} = 0.05\,g/cm^3$. (After Lòpez et al. 1994.)

vanish for a degree of modification as low as 1.6% (against 11% for the EPVC samples). For higher degrees of modification the scattering curves in the transitional and the Porod ranges are reminiscent of those obtained at 70°C for EPVC samples, which suggests that the complex formation is strongly impeded. It is worth underlining that this is correlated with a strong decay of the elastic modulus (Lòpez et al. 1996), which does show that the $1/q^6$ terms are related to the density of junctions between fibrils.

In all cases, the value of λ can be still taken as λ = 1. Values of r_{max} shown in Table 10.1 show a slight decrease of this parameter as opposed to a slight increase for EPVC samples. Values of r_{min} are also slightly lower. The fibril molecular density decreases more markedly than would be accounted for by experimental uncertainties.

In the asymptotic range all the samples scatter as $1/q^2$ yet the intensity level increases with increasing degree of modification. Lòpez et al. have provided no clear-cut explanation as to what occurs at the short-range molecular level. In fact, in terms of sheet, the increase of intensity would mean an increase in mass per unit area. Replacing a chlorine by benzenethiolate would lead to a maximum increase of about 12%, against 70% presently found. Another effect may arise from the fact that the contrast is calculated on the basis of a random copolymer which is but an approximation. Still, that the $1/q^2$ behaviour is reached at smaller q suggests the receding of adjacent sheets from one another. In the framework of the interdigitation process, a larger spacing between sheets corresponds to a loosening of the grip between fibrils, which results in a drop of physical junction density.

At 70°C the situation worsens in that SPVC3 no longer forms any gel, and its scattering response is that of individual chains. For the other two samples, fibrillar aggregates are still present whose global dimensions have not been dramatically altered.

The experiments carried out on modified PVCs suggest the following two comments.

1. The decay of complex formation is concomitant to the removal of chlorine atoms from potential sites where diesters could establish electrostatic interaction. This essentially affects the propensity of establishing pair-wise junctions between chains, while the integrity of the fibrils is virtually preserved. This emphasizes that crystallization of the syndiotactic sequences is the driving process in PVC gelation. The ***building-up of the complex structure allows the increase of the intrinsic fibril stiffness together with the formation of additional junctions between fibrils, and thereby plays a role in the ageing process***.

2. The nature of the group replacing chlorine atoms highlights the subtle balance needed for forming an organized PVC/solvent complex structure. Complex formation is impeded in the absence of electrostatic interactions, but too strong an affinity also turns out to be highly unfavourable.

The pregel state

Molecular structure of PVC aggregates

The study of finite aggregates instead of the infinite network has provided new information that has proven to be extremely relevant to the understanding of the gel state. As will be discussed, experimental techniques such as light scattering and viscometry enabled the discovery of new aspects of the gelation phenomenon and gave access to structural parameters such as the fractal dimension of the fibrils.

PVC aggregates

Below the critical gel concentration, only aggregates of finite size can be formed through a clustering process of the PVC chains. In diesters this critical concentration is about $C_{gel} \approx 0.022 \, \text{g/cm}^3$ as was determined by Candau et al. (1987) using inelastic light scattering. Although the experimental value of C_{gel} is close to the overlap concentration C^* (de Gennes 1979), it is merely fortuitous. To associate C_{gel} and C^* would only be relevant if below this concentration aggregation did not take place. What determines C_{gel} is related to the fibrillar morphology of the aggregates, and consequently their cross-sections (see Chapter 4).

Former studies on PVC aggregates by light scattering were first carried out by Kratochvil et al. (1967), who observed the classic downturn due to cluster formation only in a Zimm representation ($I^{-1}(q)$ vs q^2). The same type of study on diesters by Mutin and Guenet (1986) has revealed an unexpected maximum in the scattering pattern, as shown in Figure 10.15.

The origin of this maximum was not correctly accounted for until Dahmani et al. (1994) performed some decisive experiments by investigating aggregate suspensions at very high dilutions. These authors have observed that the maximum q_m shifts towards lower q-values and eventually vanishes at dilutions of about $C/25$ (see Figure 10.15).

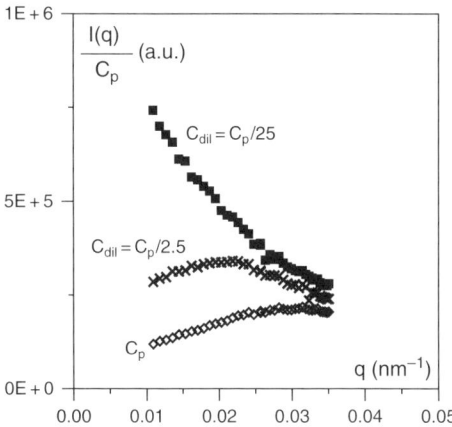

Fig. 10.15. Light scattering on aggregates prepared at $C_p = 0.01$ g/cm^3 in DEO and diluted to C_{dil} as indicated. (After Dahmani et al. 1994.)

They therefore suspected that the occurrence of a maximum arises from a liquid-like order between the aggregates: the larger the dilution the smaller the q_m. Once the maximum has vanished under appropriate dilution, the aggregate parameters such as the *z-average* radius of gyration, R_z and the weight-average molecular weight M_w can be determined (see Part I). These parameters may describe the system created at the preparation concentration provided that dilution does not break up the individual aggregates. As emphasized by Dahmani et al., the fact that gels immersed in an excess of DEO do not swell suggests that the only effect of dilution is to separate clusters from one another without breaking them up.

Dahmani et al. have established the following relation:

$$q_m \propto (C/M_w)^{1/3} \qquad (10.15)$$

where C is the actual polymer concentration (i.e. taking into account the dilution). This relation with an exponent 1/3 simply shows that q_m is proportional to the reciprocal of the average distance between aggregates as expected for a liquid-like order.

Equation (10.15) retrospectively implies that R_z and M_w do describe the aggregates. The size of the aggregates increases with the increase of the preparation concentration C_{prep} and eventually diverges at C_{gel}. A plot of $R_z(C_{prep})$ vs $M_w(C_{prep})$ has given the following relation:

$$M_w(C_{prep}) \propto R_z^{1.5 \pm 0.1}(C_{prep}) \qquad (10.16)$$

An additional investigation of aggregates produced close to the critical gel concentration, and for which $qR_z > 1$, has given the following behaviour for the intensity:

$$I(q) \propto q^{-1.5} \qquad (10.17)$$

The occurrence of power laws in both cases with nearly the same exponent suggests some sort of *fractal dimension*. According to Dahmani et al. (1997) this may well correspond to the *longitudinal fractal dimension of the fibrils*. The longitudinal fractal dimension D_f is the fractal dimension of the axis of an object possessing circular symmetry, and is defined by means of the classic relation:

$$<S_F^2> \propto L_F^{2/D_f} \tag{10.18}$$

where $<S_F^2>$ and L_F are the mean-square end-to-end distance and the contour length of the long axis, respectively (see Chapter 4).

This analysis can only be valid if the contour length of the long axis is far larger than the fibril cross-section. For an object displaying cylindrical symmetry the radius of gyration is written:

$$R_z^2 = <R_L^2>_z + <r_\sigma^2>_z \tag{10.19}$$

R_L^2 is the squared radius of gyration of the long axis (see equation (4.2)), and r^2 is the quadratic radius of gyration of the cross-section. The brackets denote z-averaging for polydispersity, assuming that polydispersity in length and in cross-section are not correlated (see Chapter 4 for details).

The cross-sectional radii as derived from neutron scattering (see Table 10.1) compared with the values of $R_z(C_{prep})$ as determined by light scattering (i.e. from 30 nm to 160 nm) entail that presently the dominant term in equation (10.19) is $<R_L^2>$. As a result, the exponent 1.5 stands most likely for the fibril longitudinal dimension. *This therefore implies that PVC fibrils cannot be regarded as straight cylinders*. This statement is in agreement with the morphological observations made (Yang and Geil 1983; Cho and Park 2001), and also with the rheological behaviour discussed in the next section.

That the same value is found in equation (10.17) for aggregates formed close to C_{gel}, i.e. under the condition $qR_z > 1$, gives further support to identifying this exponent with the fractal dimension of the fibril long axis.

Note that the interpretation of neutron data by models involving straight cylinders still holds provided l_p, the persistence length of the fibril long axis, is larger than about 10 nm ($l_p > 1/q_{min}^{neutrons}$). On the other hand, light scattering data require l_p to be smaller than about 35 nm ($l_p < 1/q_{max}^{light}$). Therefore, 10 nm $< l_p <$ 35 nm.

Light scattering measurements carried out with other esters (dibutyl oxalate, ethyl pelargonate and ethyl valerate) in such a way that $qR_z > 1$, show the same behaviour as in equation (10.17) with the same exponent (1.5 ± 0.1).

Viscometry measurements also differentiate diesters from monoesters, as shown in Figure 10.16. The intrinsic viscosity of aggregates first increases linearly with increasing preparation concentration independent of ester type. For diesters, it eventually levels off beyond $C_{prep}^* \approx 0.016$ g/cm^3.

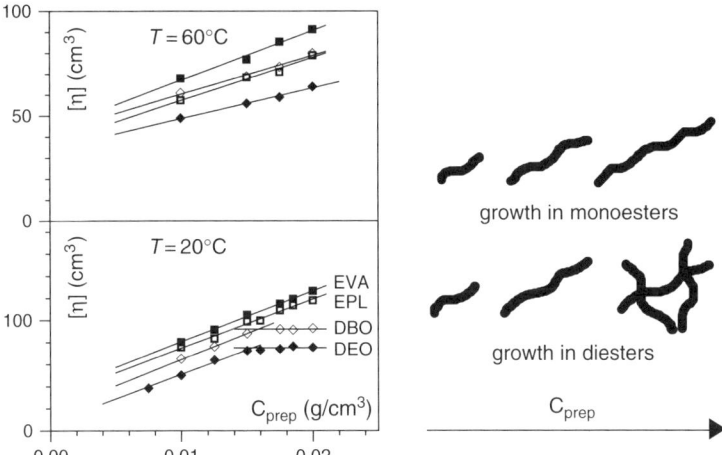

Fig. 10.16. **Left:** Intrinsic viscosity [η] as a function of the preparation concentration for PVC aggregates in diesters and monoesters ((♦) = diethyl oxalate (DEO); (◊) = dibutyl oxalate (DBO); (□) = ethyl pelargonate (EPL); (■) = ethyl valerate (EVA)). **Lower diagram:** $T = 20°C$; upper diagram $T = 60°C$. **Right:** schematic representation of the growth of aggregates in monoesters (upper) and diesters (lower). (After Dahmani et al. 1997.)

These differing types of behaviour arise from the fact that monoesters cannot interact with two chains simultaneously, unlike diesters. They are therefore unable to form the same PVC–solvent complexes. This dissimilarity was first pointed out by Najeh et al. (1992) by means of a study of the gel modulus as a function of solvent type, which is detailed below in the section on the mechanical properties of gels (see Gel rheology).

Dahmani et al. have therefore interpreted the cross-over in the viscosity behaviour of the aggregates in diesters as due to the formation of those '*weak*' *links* between fibrils through the PVC–diester complex. The levelling-off of the intrinsic viscosity can be accounted for qualitatively by considering the following relation derived by Weill and Des Cloizeaux (1979):

$$[\eta] \propto \frac{R_H R_G^2}{M} \propto \frac{R^3}{M} \tag{10.20}$$

where R_H is the hydrodynamic radius of the particle, R_G^2 its quadratic radius of gyration and M its molecular weight. R represents typical particle size and is used here for pedagogic purposes. The increase of viscosity as a function of preparation concentration occurring in the first regime for both types of esters implies that the typical volume R^3 increases more rapidly than the molecular weight. As neutron scattering findings point to the absence of cross-sectional growth, this suggests only a longitudinal growth of the fibrils. Since equation (10.16) yields $R^3 \sim M^2$ it follows that $[\eta] \sim M$, which implies that $M \sim C_{prep}$. This is again consistent with a longitudinal growth.

Now, if at higher preparation concentration the particle consists of a cluster of n fibrils (see Figure 10.16, right), its size varies approximately as $R_c^3 \propto n^\varepsilon R^3$ where $1 < \varepsilon < 3/2$, while its molecular weight varies as nM. Some levelling off of the intrinsic viscosity $[\varepsilon] \propto n^{\varepsilon-1}R^3/M$ is expected, in particular when $\varepsilon = 1$. The occurrence of superaggregation of fibrils in diesters is given support by the viscosity measured at 60°C, a temperature located above the melting temperature of the *'weak' links* (see Figure 10.16). PVC aggregates in monoesters and diesters are no longer differentiated at the highest concentrations. This is in agreement with the fact that above 50°C the capability of diesters to form a complex with PVC vanishes, which makes monoesters and diesters indistinguishable. Above this temperature PVC aggregation can only proceed through the crystallization of the highly syndiotactic sequences.

Incidentally, equation (10.20) together with equation 10.19 qualitatively suggests that fibrils in DBO possess smaller cross-sections than those produced in DEO, since their viscosity is higher. This agrees with the conclusions drawn from neutron scattering findings.

Dahmani et al. have also established the relation between intrinsic viscosity and molecular weight in DEO for preparation concentrations below the cross-over concentration $C^*_{prep} \approx 0.016\,\mathrm{g/cm^3}$:

$$[\eta] \propto M_w^{1.12 \pm 0.03} \tag{10.21}$$

Here it is worth recalling that, for polydisperse systems, the weight-average momentum of the intrinsic viscosity $[\eta]_w$ is experimentally accessible. The average molecular weight M_v associated with $[\eta]_w$ is then written:

$$M_v = \left[\sum_i C_i M_i^\alpha \bigg/ \sum_i C_i \right]^{1/\alpha} \tag{10.22}$$

in which α occurs in $[\eta]_w \propto M_v^\alpha$ and is written:

$$\alpha = (3 - D_f)/D_f \tag{10.23}$$

If $\alpha = 1$, then $[\eta]_w$ is associated with the weight-average molecular weight M_w. Interestingly, as the exponent determined experimentally equals 1.12, the average molecular weight associated with the viscosity is therefore close to the weight-average molecular weight M_w. Under these conditions, one derives $D_f = 1.42 \pm 0.02$ by means of equation (10.23). This figure is slightly lower than, but still in good agreement with, that derived from the light scattering study.

Modified PVC aggregates

Reinecke et al. (1997b, 2000) have recently studied aggregate formation with modified EPVC samples and SPVC samples in diethyl oxalate.

For EPVC samples, they have observed that the maximum in the scattering curve for as-prepared aggregates is shifted towards lower q-values when increasing the degree of chemical modification. With regard to equation (10.15) the observed shift suggests a nearly twofold increase of the cluster molecular weight for the 6%-modified EPVC.

After subsequent dilution, the radius of gyration R_z of the aggregates has been found to vary very little with increasing degree of modification. Conversely, the molecular weight M_w is seen to increase by about twofold for the 6%-modified EPVC, as expected from the shift of the scattering maximum. Nearly the same exponent as that of equation (10.16) is obtained (1.45), which indicates conservation of the fractal dimension of the modified PVC fibrils.

The invariance of both the radius of gyration and the fractal dimension against the twofold increase of the molecular weight has been interpreted by Reinecke et al. in terms of fibril cross-sectional growth. As the cross-sectional radius contributes very little to the global radius of gyration of the fibrils (see discussion about this question in relation to equation (10.19)), lateral growth does not noticeably alter this parameter. Conversely, the molecular weight increases substantially, since it varies as the square of the fibril cross-section. These experiments therefore reveal that *chemical replacement of chlorine atoms by hydrogens essentially promotes a cross-sectional growth of the fibrils without alteration of the longitudinal fractal dimension.* This statement is in line with the neutron findings obtained on gels: no significant change of fibril local conformation, with some hint at cross-sectional growth.

The conclusions drawn from this investigation weaken the case for the bottleneck effect contemplated by Lòpez et al. (1994) for interpreting the decay of the number of interactions between fibrils. While this bottleneck effect probably takes place within the fibrils, it might not be the main effect accounting for the decrease of the interactions between fibrils since fibril cross-sectional growth inevitably entails a reduction of the number of junctions per unit volume.

Why do fibrils grow sideways? This question may be tentatively answered by keeping in mind that impeding the formation of a PVC–diester complex results most probably in expulsion of solvent from the fibril. This process is liable to release the constraints upon the interfaces of the different domains of the fibril, and thereby permits cross-sectional growth (see Chapter 4).

In the case of SPVC aggregates, both the molecular weight and the radius of gyration decrease with the degree of modification (Reinecke et al. 2000). Reinecke and co-workers have given arguments to show that this result implies, chiefly, a decrease of fibril length. They also give a series of arguments for highlighting the dramatic effect fibril shortening has on the gel modulus, far more dramatic than the cross-sectional growth observed for EPVC samples. As will be discussed below, this agrees with the variation of the gel elastic modulus as a function of the degree of modification.

Gel rheology

PVC gels

The mechanical properties of PVC thermoreversible gels have been, and still are, a matter of interest (e.g. see Walters 1954; te Nijenhuis and Dijkstra 1975; Li and Aoki 1997, 1998).

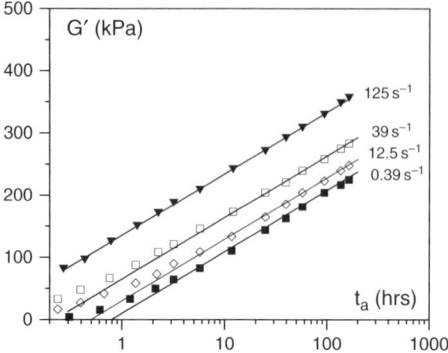

Fig. 10.17. Storage modulus G' plotted as a function of the ageing time for different angular frequencies. PVC/dioctyl phthalate gels $C_{pvc} = 9.9\%$ (w/w). (After te Nijenhuis and Dijkstra 1975.)

As highlighted in Figure 10.2 and reported some 25 years ago by te Nijenhuis and Dijkstra (1975), PVC gels show considerable ageing effect; this ageing does not depend upon the frequency in oscillatory mechanical measurements (Figure 10.17). The results of these researchers are in line with other experimental evidence such as the appearance of a low-melting endotherm. The ageing process is related to the formation of the 'weak' links through the occurrence of a PVC–diester complex. Mutin and Guenet have effectively shown that a gel can be 'rejuvenated' by heating at 60°C, as its elastic modulus drops to the value measured in the nascent state (Mutin and Guenet 1989).

Dynamic modulus measurement by Lòpez et al. (1996) of G' as a function of temperature shows a drop of modulus at about 50–60°C for dioctyl phthalate, and a similar result has been obtained by Cho and Park (2001) for dibutyl phthalate, both these chemicals being diesters (Figure 10.18). The abruptness of the modulus decrease is

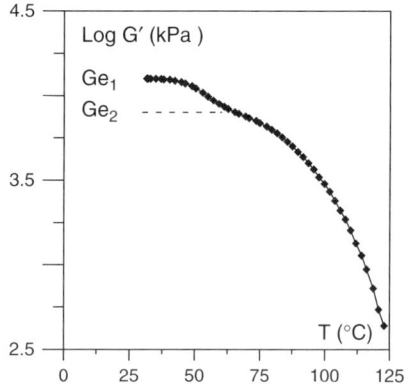

Fig. 10.18. Variation of G' as a function of temperature for a gel in Dioctyl phthalate ($C_{pvc} = 20\%$ w/w). At the transition temperature $T \approx 50°C$ the storage modulus G_{e2} drops to about 1/3 of its value at room temperature G_{e1}. (After Lòpez et al. 1996.)

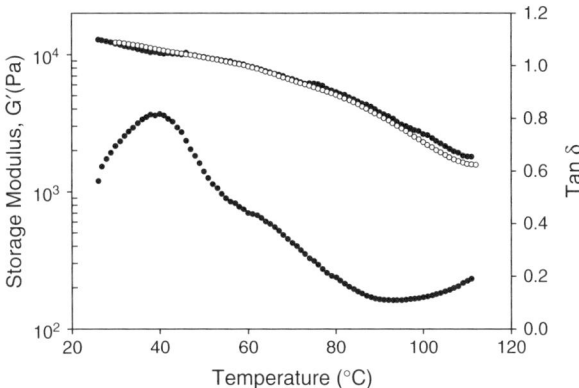

Fig. 10.19. Storage modulus G' and tan δ for an aged gel (●) and for a fresh gel (○), PVC/butyl benzoate. (From Cho and Park 2001. Reprinted with permission from Wiley Interscience.)

consistent with the relatively sharp DSC transition observed by Mutin and Guenet (1989).

Cho and Park (2001) have recently reported that gels prepared from butyl benzoate, a monoester, display neither ageing effect nor any sharp transition of G'. This again highlights the different types of gel behaviour that occur depending on whether a monoester or a diester is used (Figure 10.19).

As a rule, PVC gels do not display much relaxation when subjected to constant deformation. The rate dLog σ/dt is about 0.1% (Mutin and Guenet 1989). Oscillatory experiments carried out as a function of frequency show that G' is always about one order of magnitude higher than G'', as expected for a true gel. Clearly, PVC gels can be characterized by a modulus at zero frequency which is consistent with the existence of crystalline junctions as opposed to 'nematic-like' junctions such as those reported for isotactic polystyrene (see Chapter 14).

With the aim of throwing some light on the origin of the *'weak'* links, and particularly of testing the model of Abied et al. (1990), Najeh and co-workers performed mechanical experiments in which they compared the evolution of the elastic modulus of gels produced in diesters and in monoesters (Najeh et al. 1992). The study relied upon the following rationale: if the *'weak'* links result from electrostatic chain pairing by diesters, then monoesters, which are only 'one-handed', cannot possibly achieve the same effect. The use of a series of diesters and monoesters was intended for comparing moduli at equivalent solvent quality. All the gels were prepared at the same starting concentration and were allowed to age in an excess of preparation solvent so as to reach swelling equilibrium. Solvent achieving the same equilibrium swelling ratio ($G_\infty = V_\infty/V_o$) can be regarded as having the same quality with respect to PVC. The results are given in Figure 10.20, where the equilibrium elastic modulus is plotted as a function of the equilibrium swelling ratio (50 days of ageing are usually needed for equilibrium to be reached).

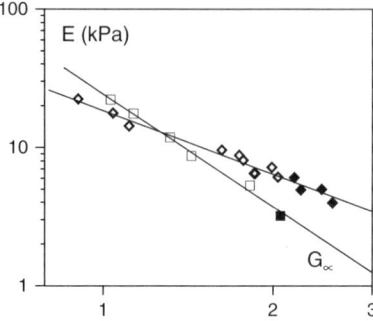

Fig. 10.20. Variation of the elastic modulus as a function of the swelling ratio in a series of diethyl esters (◊), of dibutyl esters (♦), of ethyl monoesters (□) and isoamyl acetate (■). (After Najeh et al. 1992.)

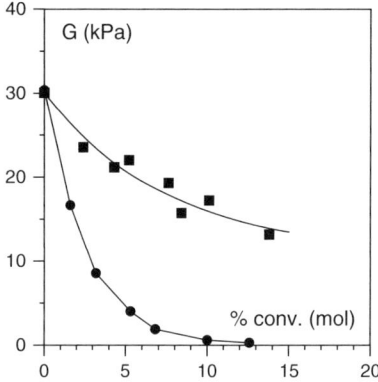

Fig. 10.21. Elastic modulus vs the degree of conversion for EPVC samples (■) and SPVC samples (●). (After Lòpez et al. 1996.)

Power laws are obtained whose exponents depend only upon whether diesters or monoesters are used, but not at all upon the alcohol substituent:

$$E_{diesters} = 18.6 \times G_\infty^{-1.54 \pm 0.1} \, kPa \tag{10.24}$$

$$E_{monoesters} = 24.5 \times G_\infty^{-2.72 \pm 0.1} \, kPa \tag{10.25}$$

These results do show that monoesters and diesters behave differently yet not as expected. In fact, no complex formation can occur in monoesters and yet at low swelling ratios the gels produced from them possess moduli higher than those produced in their diester counterparts. Najeh et al. have shown by DSC that, in monoesters giving low swelling ratios, the chain sequences that are not engaged in the syndiotactic crystals (disordered domains) stand below their glass transition temperature allowing fibril rigidity to be maintained. Unlike the case with diesters, chain rigidity is not due to complex formation but to chain vitrification. As a result, increasing the solvent quality of the monoester produces nothing but a plasticization effect, hence the high exponent of the power law (2.72). Conversely, in diesters, complex formation is thought to take place in all circumstances, and hence there is a significantly lower exponent (1.5). In this case, using a theoretical approach based on Jones and Marquès' theory for systems displaying enthalpic elasticity (Jones and Marquès 1990) (see Chapter 4), Najeh et al. have derived the relation:

$$E \propto \frac{er^4}{G_\infty} \qquad (10.26)$$

where e is the intrinsic Young's modulus of the fibril and r its cross-sectional radius. The assumptions for deriving equation (10.26) are: (1) a constant fibril longitudinal fractal dimension, i.e. independent of the swelling ratio, and (2) the fibril persistence length increasing with increasing the swelling ratio (i.e. the fibrils are straight over larger distances when increasing the swelling ratio). The exponent obtained from this theoretical approach is -1 against -1.54 for the experimental value. Yet, in equation (10.26) r is treated as a constant. As emphasized by Najeh et al. it suffices that r varies slightly with the swelling ratio (such as $r \propto G_\infty^{-0.135}$) for obtaining the right exponent. Such a behaviour of r is suggested by neutron scattering results obtained under near-equilibrium conditions (for 2% samples for which r is lower in DBO than in DEO).

These results prompted Dahmani et al. to re-examine the variation of the equilibrium elastic modulus as a function of polymer concentration in the light of the theory developed by Jones and Marquès (1990) (see Part I). The modulus concentration values of the relations differ and are expressed as a function of the fractal dimension of the object (in the case presented here, the longitudinal fractal dimension of the fibrils):

$$\text{entropic elasticity } E \propto \varphi^{\frac{3}{3-D_f}} \qquad (10.27)$$

$$\text{enthalpic elasticity } E \propto \varphi^{\frac{3+D_f}{3-D_f}} \qquad (10.28)$$

φ_N is the network volume fraction, which is not necessarily equal to the polymer volume fraction φ_p. If free/pendant chains are present, then $\varphi_N < \varphi_p$, a situation occurring in agarose gels.

The exponents β for the modulus–concentration relation ($E \sim C^\beta$) found experimentally by Dahmani et al. with gels aged in an excess of solvent are given in Table 10.2. As is apparent from this table, the values derived from the modulus–concentration relation by considering *enthalpic elasticity*, i.e. a rigid network, are in very good agreement with those determined by light scattering on aggregates for DEO, DBO and EPL. Conversely there is no agreement for EVA. In the first three cases, the less stereoregular domains of the fibrils are stiffened either by complex formation (DEO and DBO) or by vitrification (EPL). None of these phenomena occurs in EVA, so a fresh theoretical approach is required. The light scattering data on PVC aggregates suggesting that the value 1.5 corresponds to the longitudinal fractal dimension of the fibril is borne out by the conclusions drawn from the mechanical properties. In spite of their intrinsic

Table 10.2. Comparison between the values of the longitudinal fractal dimension as deduced from light scattering on aggregates and from the exponent β of the relation modulus-concentration ($E \sim C^\beta$) by means of equations (10.27) and (10.28). DEO = diethyl oxalate; DBO = dibutyl oxalate; EPL = ethyl pelargonate; EVA = ethyl valerate. (After Dahmani et al. 1997.)

solvent	β	D_f light	D_f enthalpic	D_f entropic
DEO	3.06±0.2	1.49±0.09	1.52±0.06	2.02±0.06
DBO	2.74±0.1	1.48±0.1	1.4±0.04	1.91±0.04
EPL	3.19±0.2	1.52±0.1	1.57±0.06	2.06±0.06
EVA	4.02±0.3	1.54±0.1	1.81±0.06	2.25±0.06

complexity, some PVC gels can therefore be described with only one parameter. Here it is worth emphasizing that the PVC–solvent complex must percolate throughout the fibril (i.e. no disordered domain completely crosses the fibril), otherwise flexibility would occur.

One may still wonder why the elastic moduli are still so high for gels prepared from solvents or heated to a temperature at which neither complex formation nor vitrification occur. In fact, just above the transition at 50°C the modulus drops to about 1/3 of its original value (see Figure 10.18) despite the total melting of the complex. Clearly, one would, in the normal run of things, have expected a larger decrease.

The melting of the complex chiefly prevents fibrils establishing interfibrillar links. But, as highlighted in equation (10.26), the elastic modulus is governed both by the number of interfibrillar links per unit volume and by the intrinsic modulus of the fibrils. The latter parameter is dependent upon the mobility of the fraction of chain sequences not engaged in 'syndiotactic' crystals. Complex formation or vitrification are two different ways for reducing chain mobility to the extent of freezing it in. Yet, a good solvent, by swelling the disordered part, will also considerably reduce the mobility, although the chains remain flexible.

Gels from chemically modified PVCs

As has been mentioned above, Lòpez et al. (1996) have shown that the elastic modulus of gels prepared from chemically modified PVCs drops markedly as a function of the degree of conversion (Figure 10.21). The decay is far more dramatic with SPVC samples than with EPVC samples, since, for a 10% conversion, the gel is very weak. These results are consistent with conclusions drawn from the light scattering data. In fact, the shortening of the fibrils is liable to have a more dramatic effect on the gel modulus than the cross-sectional growth of the fibrils.

CHAPTER 11

Polyaniline

Polyaniline (PANI) was first synthesized in 1862 but systematically investigated only at the turn of the twentieth century (Green and Woodhead 1910). This polymer can be converted into a conducting polymer by doping it with strong acids such as H_2SO_4 (Doriomedoff et al. 1971). The process is usually achieved by starting from the emeraldine base of polyaniline, as shown in Figure 11.1.

Clearly, the formation of some kind of polymer–solvent complex occurs, although chemical modification takes place. In this sense, the case of PANI should not be included in this book, which is oriented only towards those polymer–solvent complexes that form without any chemical change. However, as will be discovered in the following text, doped PANI also forms compounds with organic molecules. This is often referred to as *secondary doping* as the conducting properties of these systems are seen to be significantly improved (MacDiarmid and Epstein 1994).

Polyaniline deserves a book in itself. As was emphasized by MacDiarmid and Epstein, the study of PANI systems long remained a strictly phenomenological approach. While many studies are therefore available, the underlying phenomena, together with their understanding, have only recently received growing attention. Therefore, this section will be essentially focused on the main topic of this book, namely the formation of compounds, while leaving aside other phenomenological aspects that do not seem to be well understood. The propensity of PANI to form fibrillar structures only is the attribute that places it unquestionably in this chapter (e.g. see Yang et al. 1993).

Properties of solutions

Melt processing of PANI is impossible owing to polymer decomposition occurring below its softening temperature or its melting point. To overcome this problem, several attempts were made to process it from solution. Some were successful and therefore led to increased knowledge in relation to the solution properties of this polymer. From

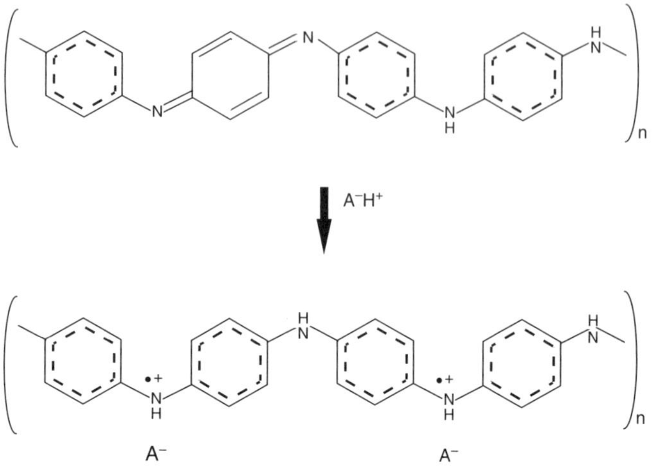

Fig. 11.1. The emeraldine form of PANI, which can be regarded as a copolymer of reduced and oxidized units, is complexed by means of a strong acid. The resulting polymer is therefore electrically conducting due to a half-filled polaron band.

these solutions, it was observed that PANI–solvent compound could be formed (e.g. see MacDiarmid and Epstein 1994; Ikkala et al. 1995).

Until the early 1980s, PANI had been considered an intractable polymer because of the poor degree of solubility observed in organic solvents. Some 'exotic' solvents such as N-methylpyrrolidinone or dimethyl sulphoxide (Angelopoulos et al. 1988), pyrrolidine or tripropylamine (Han et al. 1991), and concentrated sulphuric acid and strong acids (Andreatta et al. 1988; Cao et al. 1989) can, however, dissolve PANI to a reasonable extent. More recently, several researchers have succeeded in preparing PANI in its conducting form by using functionalized protonic acids denoted $H^+(M^-R)$, where the counter-ion contains a group R, which displays strong compatibility with non-polar or moderately polar organic solvents (Li et al. 1987; Cao et al. 1992). Two systems have been abundantly studied: p-dodecylbenzene-sulphonic acid (DBSA) and camphor-10-sulphonic acid (HCSA) (Figure 11.2).

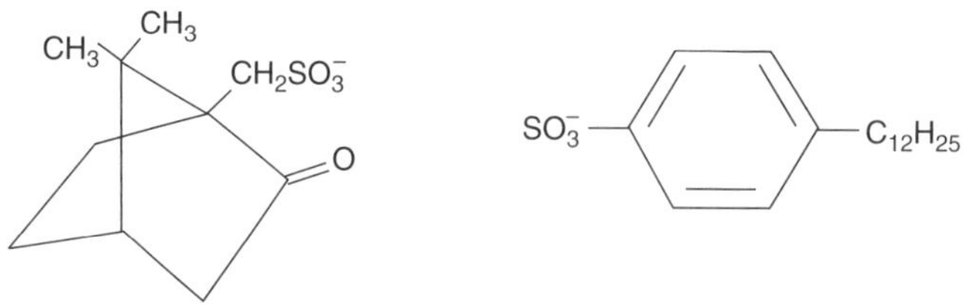

Fig. 11.2. Camphor-10-sulphonic acid (HCSA) (**left**), and p-dodecylbenzene-sulphonic acid (HDBSA) (**right**).

These solvent molecules act to some extent as surfactants, especially *p*-dodecylbenzene-sulphonic acid, and screen electrostatic effects, thus promoting solubilization in organic solvents such as *m*-cresol ($CH_3C_6H_4OH$). They also promote chain stiffening through the increase of the persistence length (see Chapter 4), although this is not known directly but is clearly apparent from circumstantial evidence: *concentrated solutions in m-cresol form liquid-crystalline phases* (Cao and Smith 1993). The increase of chain persistence length by sideways attachment of bulky molecules to polymer chains is intuitively expected. Frederickson (1993) has given a theoretical development of this effect as a function of various parameters whose details can be found in Chapter 4.

Unfortunately, little is known about the conformation in solution, i.e. to date no systematic studies of the chain dimensions in dilute solution have been carried out (either by light scattering or by neutron scattering). Cao and Smith had to rely on viscometry data obtained on other semi-flexible polymers to estimate the molecular weight of their PANI samples. They accordingly showed that the critical volume fraction V^* at which a nematic phase formed obeyed fairly well the relation derived by Flory (1956):

$$V^* \approx (8/x)[1-(2/x)] \approx 8/x \tag{11.1}$$

where x is the axial ratio (persistence length/cross-section radius).

MacDiarmid and Epstein (1994) carried out a series of experiments on these points while studying the concept of secondary doping. They showed by means of viscosity measurements that the chain conformation varies considerably on varying the proportion of solvent in mixtures of chloroform and *m*-cresol. As is displayed in Figure 11.3, the intrinsic viscosity increases dramatically when increasing the fraction of *m*-cresol, as does the conductivity and crystallinity of free-standing films cast from these solutions. As has been described for PVC, an increase of intrinsic viscosity implies an increase of the coil size, and correspondingly of the persistence length. MacDiarmid and Epstein

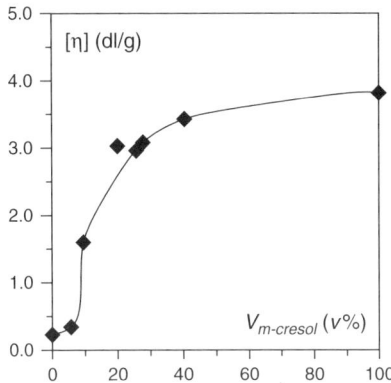

Fig. 11.3. Variation of the intrinsic viscosity of PANI doped with HCSA as a function of the volume fraction of *m*-cresol in a binary solvent *m*-cresol/chloroform. The PANI concentration is constant and amounts to 0.33 wt%. (After MacDiarmid and Epstein 1994.)

account for this effect by contemplating a strong solvation of HCSA molecules by m-cresol, which results in a similar effect observed with polyelectrolytes when no salt is added: *charges of same sign on the chain are no longer screened and have repulsive effects, hence indicating an expanded coil.* This expanded coil will facilitate chain alignment while casting films from solutions, and eventually promotes a higher degree of crystallinity. Clearly, there is formation of a polymer–solvent complex which promotes chain extension. Upon removal of m-cresol, the extended conformation is retained, so a higher crystallinity is achieved.

Two additional experiments carried out by these authors are worthy of mention. In m-cresol solutions of emeraldine base (EB), addition of HCSA entails first the increase of intrinsic viscosity, which goes through a maximum at a ratio $HCSA/EB = 2$ and then decreases (Figure 11.4). The conductivity behaviour of free-standing films cast from these solutions is the same. The experiments are performed at constant volume and constant polymer concentration. Here, addition of an excess of H^+ and CSA^- charges, in a medium with a decreasing fraction of m-cresol, will tend to create oppositely charged domains on the chain resulting in a global contraction of the conformation.

A similar study was performed on polyaniline/dodecylbenzene-sulphonic acid (HDBSA) in chloroform. As the emeraldine base is not soluble in chloroform, this study started at a ratio $HDBSA/EB = 2$. Again, polymer concentration and the total volume were kept constant. Here, as shown in Figure 11.4, the viscosity does increase but no clear maximum is observed. Similarly, the maximum conductivity of free-standing films cast from these solutions does not correspond to the intrinsic viscosity maximum. In fact, at a molar ratio $HDBSA/EB = 2$, the chain is said to be coil-like and becomes expanded

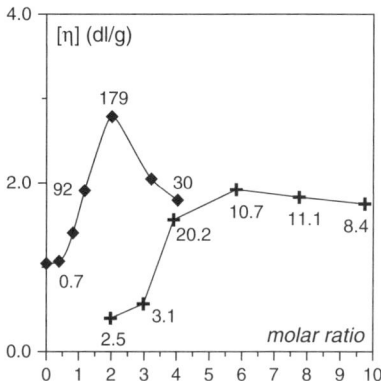

Fig. 11.4. Variation of the intrinsic viscosity as a function of the dopant/emeraldine base molar ratio. (♦) = HCSA/EB and corresponds to addition of HCSA in m-cresol solutions; (+) = HDBSA/EB and corresponds to addition of HDBSA in chloroform solutions. Figures are the conductivities expressed in S/cm of free-standing films prepared from these solutions. The PANI concentration and the total volume are kept constant so that addition of dopant occurs to the detriment of the solvent. (Data from MacDiarmid and Epstein 1994.)

at rather high HDBSA/EB ratios. As discussed by MacDiarmid and Epstein, two effects operate: (1) the screening effect owing to the excess of acid (HDBSA), which tends to contract the chain, and (2) the solvation effect, which tends to swell the coil. This solvation effect occurs here because of the dissimilarity of HDBSA, which displays a non-polar group capable of interacting with chloroform and a polar group interacting with the emeraldine base. As the part of HDBSA exposed to the solvent is the non-polar group, coil expansion is likely to occur. This effect is quite consistent with the theoretical predictions derived by Frederickson (1993) (see Chapter 4). Again, some kind of polymer–solvent compound is likely to occur. Interestingly enough, formation of organized structures, and then of a crystalline state, is favoured by the expanded chain conformation. Coil expansion can also be achieved by submitting PANI/HCSA films prepared from chloroform to *m*-cresol vapours (Xia et al. 1994).

Solid state molecular structure

Cao et al. (1995) have reported the existence of two phases, the main difference between them lying in the chain conformation. In the terminology used by these authors, *phase I* corresponds to the main chain taking on a helical conformation (with a 2_1 screw axis), while *phase II* is a trans-transoid conformation with m-mirror symmetry as proposed earlier by Langer (1987) and Shacklette et al. (1988). The characteristic features of the diffraction pattern are given in Table 11.1:

Table 11.1. Main X-ray reflections for the two different phases of PANI complexes. (After Cao and Smith 1995.)

	Phase I	Phase II
Equator	0.35 nm	0.603 nm
Meridian	0.467 nm	0.358 nm

These two phases are both obtained for PANI/HCSA complexes, the only difference being the solvent used. *Phase I* is grown from *m*-cresol while *phase II* occurs in DMSO. As shown by UV spectroscopy the conformation is the same in solutions and in films cast from these solutions. *Phase I* is the highly conducting phase (200–400 S/cm) while *phase II* is poorly conducting ($10^{-4} - 10^{-1}$ S/cm) because the π-band conjugation is significantly decreased.

Clearly, the specific interactions between the solvent and the PANI salt complex play a role in the formation of either phase.

Ikkala et al. (1995) studied these specific interactions by molecular modelling. They contemplate the existence of PANI/solvent compounds occurring through the mediation of molecular recognition, a notion already applied earlier to PEO/*para*-dihalogenobenzene compounds and to iPS/decalin compounds (see Part IV). PANI compounds are still best viewed as complexes. Indeed, as is detailed below, the main driving force is the creation of hydrogen bonds or electrostatic interactions.

The study of Ikkala et al. shows that, in sulphonic acid-doped PANI complexes, the sulphonate anion is a potential site for interaction with strong hydrogen bonding donors such as phenols. These authors have focused their investigation on two complexing agents of the emeraldine base: HCSA and HTSA (tosylene sulphonic acid). They show that HTSA and *m*-cresol essentially interact through dipolar intercalation: the polar sulphonic group orientates the OH group such that van der Waals interactions that could occur between the phenyl rings do not play any significant role.

The optimized structures calculated with HTSA and HCSA are displayed in Figures 11.5 and 11.6. Concerning HTSA doped-PANI there are two possible sites for association for *m*-cresol: either the sulphonate anion or the amine moiety of PANI. Calculation indicates that *m*-cresol is bonded to the sulphonate anion (Figure 11.5). In the case of HCSA-doped PANI, *m*-cresol can interact with three possible sites for association: the sulphonate anion, the amine moiety and the carbonyl group of HCSA. The optimized structure reveals the formation of a >C=O—HO hydrogen bond between HCSA and *m*-cresol through orientating dipolar forces. In addition, the phenyl ring of *m*-cresol turns out to be coplanar to the neighbouring PANI ring thus leading to gaining energy. This double interaction is probably why HCSA gives the unique conductivity properties to PANI. According to Ikkala et al. *m*-cresol can be intercalated on top of PANI rings due to the cavity created by the bulkiness of HCSA molecules. These authors therefore suggest that intercalation of *m*-cresol molecules twists PANI into a more planar conformation (*phase I*) which is kept on drying despite removal of the solvent (namely *m*-cresol), hence the high conducting properties.

To date the scientists involved in these studies have claimed that the ternary complexation (PANI/HCSA/solvent) only exists in the amorphous state while the solvent

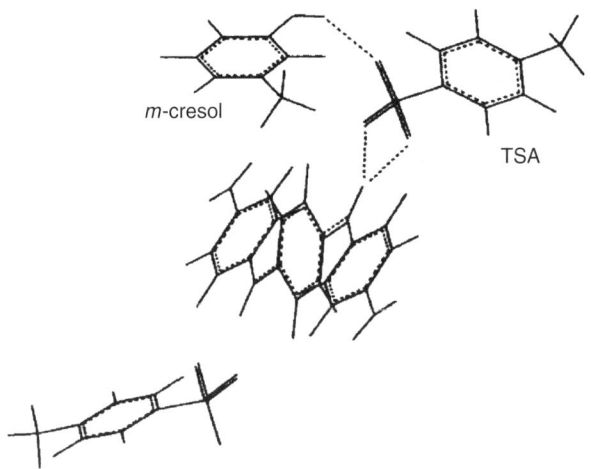

Fig. 11.5. Optimized structure of PANI doped with tosylene sulphonic acid (TSA) in contact with *m*-cresol. (From Ikkala et al. 1995. Reused with permission from Olli T. Ikkala, Lars-Olof Pietilä, Lisbeth Ahjopalo, Heidi Österholm, and Pentti J. Passiniemi, *Journal of Chemical Physics* **1995** *103* 9855. Copyright 1995, American Institute of Physics.)

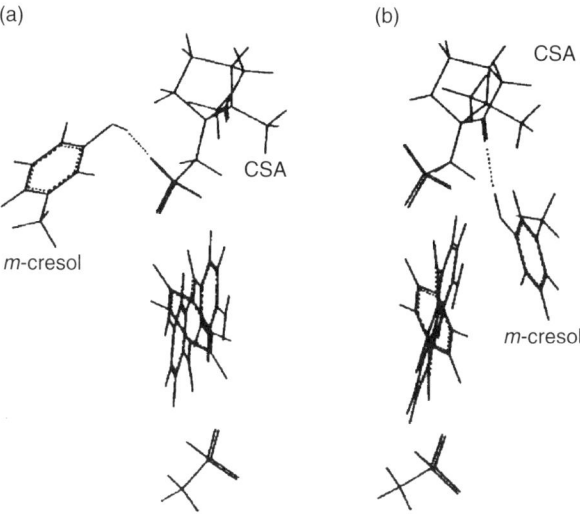

Fig. 11.6. Optimized structures of PANI doped with CSA in contact with *m*-cresol: A = *m*-cresol hydrogen-bonded to the SO$_3$ group; B = *m*-cresol hydrogen-bonded to the carbonyl group of CSA and bound by van der Waals interaction with the PANI ring. (After Ikkala et al. 1995. Reused with permission from Olli T. Ikkala, Lars-Olof Pietilä, Lisbeth Ahjopalo, Heidi Österholm, and Pentti J. Passiniemi, *Journal of Chemical Physics* **1995** *103* 9855. Copyright 1995, American Institute of Physics.)

is expelled from the PANI crystalline phase. It is not clear, however, whether some mesophases made up with the ternary complex occur or not, namely what is the actual structure of the amorphous state? As will be discovered in Part IV, the existence of mesophases has been observed with isotactic polystyrene in the gel state. Compound formation produces only some kind of liquid-crystalline order most probably because the solvated chains cannot form highly ordered crystals, and because isotactic polystyrene possesses an unusually large persistence length in these specific solvents. There is no reason why such a situation should not occur in the case of PANI, which intrinsically possesses a large persistence length, and which is known to form liquid-crystalline phases (Cao and Smith 1993). It would be of interest to determine whether, in these liquid-crystalline phases, solvent molecules are intercalated in the way described by Ikkala et al. A temperature–concentration phase diagram would be most helpful to settle this issue.

Vikki et al. (1996) have extended the study of solvent effect to resorcinol, which is solid at room temperature. At high temperature (200–220°C), PANI complexed with HCSA or HDBSA displays an exceptionally high solubility in this solvent, which allows preparation of highly homogeneous systems. An interesting result is the variation of the resorcinol melting enthalpy with resorcinol weight fraction (Figure 11.7). This may provide information on the stoichiometry of the compound (see Part I). In both cases about eight resorcinol molecules per PANI repeat unit are found (as shown in Figure 11.1, i.e. four PhN subunits).

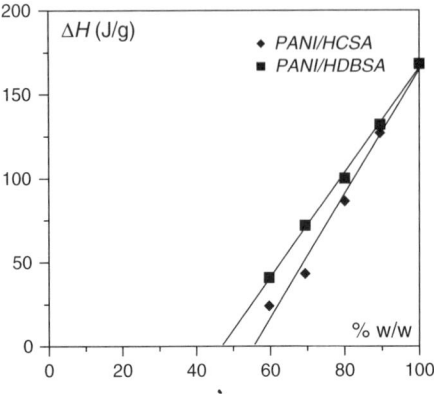

Fig. 11.7. Melting enthalpy of resorcinol as a function of resorcinol weight fraction in PANI/HCSA and PANI/HDBSA mixtures. (After Vikki et al. 1996.)

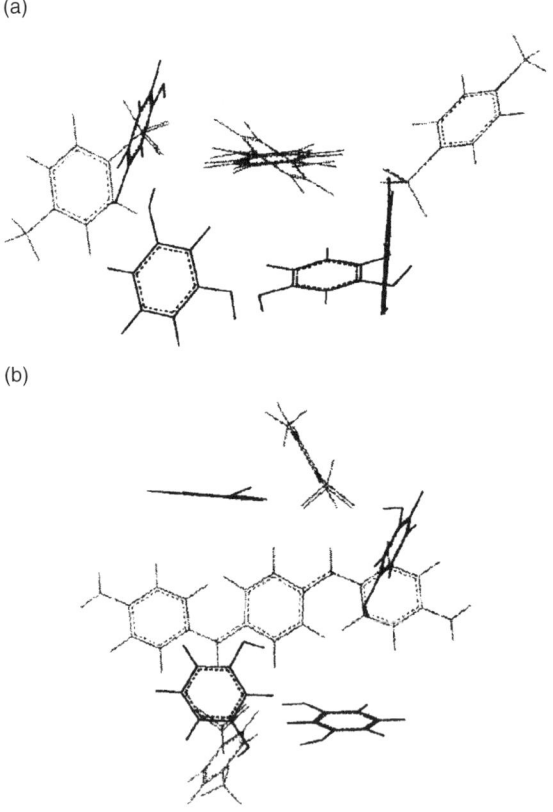

Fig. 11.8. Optimized structures for PANI/TSA model compound with 4 resorcinol molecules viewed: (a) along the chain and (b) from the top of the chain. (From Vikki et al. 1996. Reprinted with permission from ACS.)

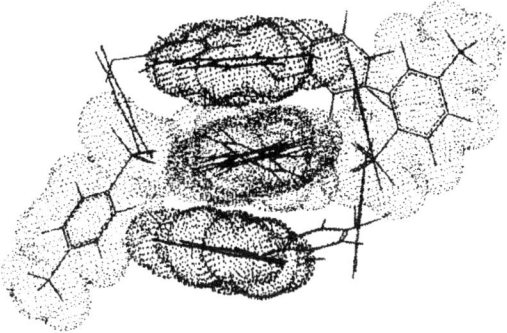

Fig. 11.9. Optimized structure for PANI/TSA model compound with eight resorcinol molecules viewed along the chain. The PANI/TSA model compound and the resorcinol molecules on the top of the chain are shown with their van der Waals surfaces. (After Vikki et al. 1996. Reprinted with permission from ACS.)

Here again, computational modelling shows the formation of specific and oriented interactions with resorcinol (for the sake of simplicity the authors consider HTSA as the complexing agent). A typical optimized structure is shown in Figure 11.8. The association is monitored by the strong orientational effect of the sulphonate group on one of the hydroxyl groups of resorcinol. Four resorcinol molecules can undergo hydrogen bonding directly with the two sulphonate groups. The association of eight resorcinol molecules, as derived by the solvent melting enthalpy method, is depicted in Figure 11.9. In this case the additional resorcinol molecules are bonded both by hydrogen bonds and by phenyl–phenyl interactions.

Gels

Recently, Vikki et al. (1997) have reported the formation of thermoreversible gels from PANI/DBSA systems provided that the solutions are first made in formic acid, which is eventually removed by vacuum drying. The gel is then made up of a randomly dispersed system of PANI/DBSA complex in DBSA as dispersing medium.

The structure of dodecylbenzene-sulphonic acid (DBSA) is portrayed in Figure 11.10, where $R_1 = C_mH_{2m+1}$ and $R_2 = C_nH_{2n+1}$, the dominant length of the alkyl chain being such that $m + n + 1 = 12$.

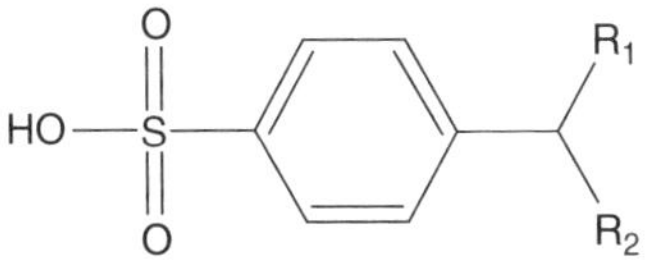

Fig. 11.10. Dodecylbenzene sulphonic acid.

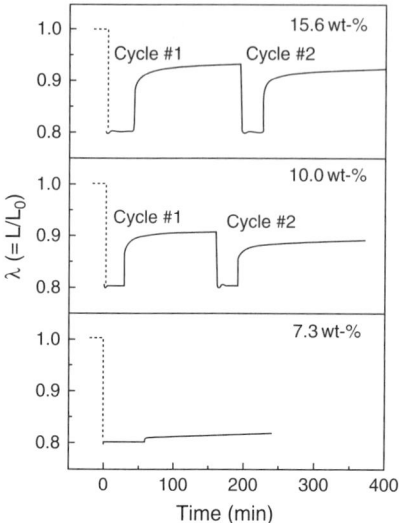

Fig. 11.11. Relaxation experiments carried out on PANI/DBSA gels. Cycles consisting in applying and releasing deformation are used to test the recovery behaviour of these systems. (From Vikki et al. 1997. Reprinted with permission from ACS.)

The gel character of the system has been tested by using a modification of the method devised by Daniel et al. (1994) and described in Chapter 4. Here, Vikki et al. observe the behaviour of a cylindrical sample after applying a deformation of $\lambda = 0.8$ followed by a complete release of the deformation. Two cycles are used for the most highly concentrated systems. As can be seen in Figure 11.11, recovery is partial yet quite important. These outcomes give strong support to the network status of these systems.

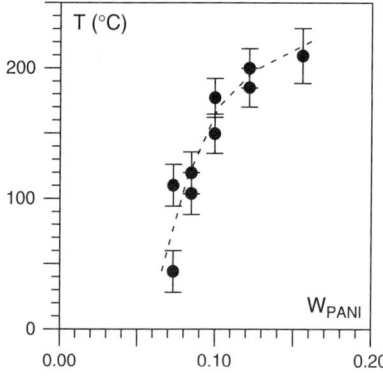

Fig. 11.12. Temperature–concentration phase diagram of the system PANI/DBSA as obtained from rheological measurements. The temperatures correspond to the disappearance of the network structure. (After Vikki et al. 1997.)

A temperature–concentration phase diagram of this system has been mapped out by determining the gel melting point as a function of DBSA concentration (Figure 11.12). This diagram has been established through dynamic rheological experiments involving the use of vibrating plates, which therefore provides temperatures wherein the 3D character of the network is lost, but does not necessarily indicate the final melting point, namely the temperature at which all organized structures vanish. At the lowest concentration the gel melting point is subject to large variations, possibly due to the 'destructive' character of the measuring procedure. Note that the gels are highly concentrated, bearing in mind that the stoichiometric composition of the PANI/DBSA complex is about $W_{PANI} \approx 0.22$ if one considers that part of the DBSA molecules protonate iminic nitrogen atoms while others interact with aminic nitrogen atoms through hydrogen bonding.

The conducting properties of these gels are quite interesting although they are not as great as those observed in the solid state. Figure 11.13 shows the variation of conductivity with PANI weight fraction. Conductivity is highly sensitive to PANI concentration, as it varies approximately as W_{PANI}^6. For a concentration the network status of which has been clearly ascertained, the conductivity is of the electronic type, suggesting direct chain-to-chain hopping. At the melting of the gel the conductivity drops quite suddenly by one to three orders of magnitude, and becomes essentially of the ionic type, reaching the level of conductivity of pure DBSA.

Jana and Nandi (2000, 2001) and Jana et al. (2002) have further studied gels systems in various surfactants, the molecular structure of which is shown in Figure 11.14 (in all cases a typical network with fibrillar morphology is observed (Figure 11.15). Also, these authors report the occurrence of two melting peaks, one probably due to the gel melting and the other to the melting of the fibrils. Note that the fibril structure is such that the surfactants interact with one another so as to form bilayered structures.

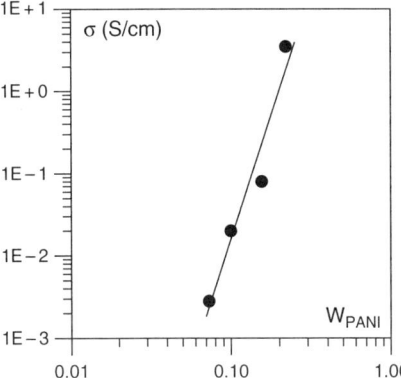

Fig. 11.13. Electrical conductivity of PANI/DBSA systems as a function of the PANI weight fraction at room temperature. (Data from Vikki et al. 1997.)

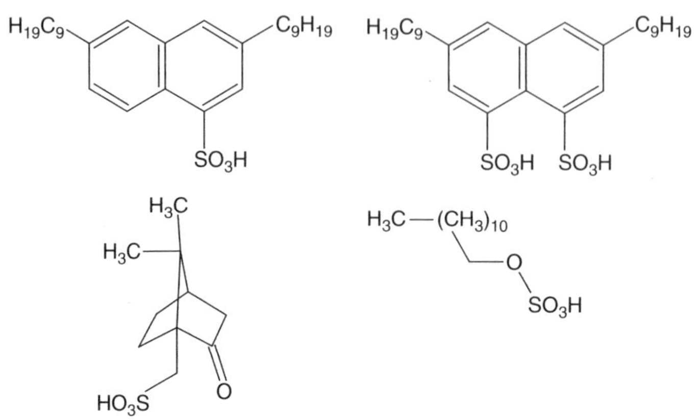

148   Polymer-Solvent Molecular Compounds

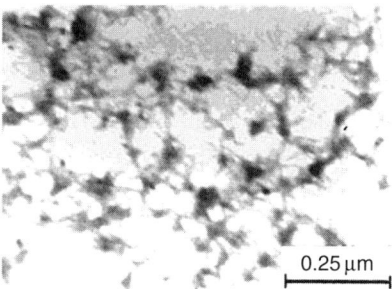

Fig. 11.14. Top left: dinonylnaphthalene sulphonic acid (DNNSA); **top right:** dinonyl disulphonic acid DNNDSA; **bottom left:** (±) camphor-(10) sulphonic acid (CSA); **bottom right:** n-dodecyloxo sulphonic acid (DOSA).

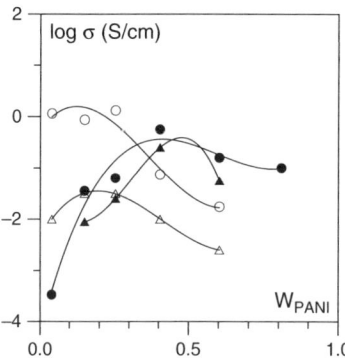

Fig. 11.15. Typical morphology as seen by TEM for PANI/CSA gels. (From Jana and Nandi 2000. Reprinted with permission from ACS.)

Fig. 11.16. Conductivity of different PANI/Surfactant gels as a function of the PANI weight fraction. (○) = DOSA; (●) = CSA; (▲) = DNNDSA; (△) = DNNSA. Lines are guide for the eyes. (Data from Jana and Nandi 2001.)

The conducting properties of the gels depend significantly on the nature of the surfactant. Results obtained by Jana and colleagues are summarized in Figure 11.16. As can be seen, the most efficient system is PANI/DOSA, for which high conductivity is obtained at low PANI concentrations. This may suggest that a high degree of connectedness of the gels is already achieved at low PANI concentrations, which would entail that PANI/DOSA fibrils possess cross-sections smaller than those of fibrils in other PANI/surfactant gels. The values of conductivity of their gels is quite similar to those reported by Vikki et al.

CHAPTER 12

Poly[vinylidene fluoride]

Poly[vinylidene fluoride] (PVF$_2$), whose chemical structure is $(-CH_2-CF_2-)_n$, has been known for decades, yet received greater attention when Kawai discovered its piezoelectric properties (Kawai 1969). Many investigations have been carried out on the solid state owing to the potential applications of this polymer. An excellent review has been authored by Lovinger (1981a).

The existence of PVF$_2$/solvent compounds has been reported only recently by A.K. Nandi and his group from IACS Kolkata (India). These researchers observed that compounds can be formed in diesters such as the diethyl series (Mal et al. 1995; Dikshit and Nandi 2000). Later on, other compounds were discovered in camphor and in ethyl acetoacetate (Dasgupta et al. 2005, 2006).

To date, no crystalline lattice has been derived for these molecular compounds. The known forms of PVF$_2$ are therefore presented for the sake of completeness. The nomenclature suggested by Lovinger in his review will be used. Four forms have been observed and described: the α form, β form, γ form and δ form. The α form is the commonest crystalline form obtained from melt-crystallization (Figure 12.1). The unit cell is orthorhombic with $a = 0.496$ nm, $b = 0.964$ nm, $c = 0.462$ nm and space group $P2$ cm (Hasegawa et al. 1972a, b; Bachmann and Lando 1981). The chain conformation is a slightly distorted $tgt\bar{g}$ with $t = 179°$ and $g = 45°$, with the lowest potential energy for all the PVF$_2$ polymorphs (Hasegawa et al. 1972a; Görlitz et al. 1973).

The β form is obtained by mechanical deformation of melt-crystallized films, and turns out to be the form with piezoelectric properties. The unit cell is also orthorhombic (Figure 12.2) with $a = 0.858$ nm, $b = 0.491$ nm, $c = 0.256$ nm and space group $Cm2m$ (Hasegawa et al. 1972a). The chain conformation is almost all-*trans* although some statistical distortion occurs owing to steric hindrance between fluorine atoms (Gal'perin et al. 1965).

The γ form (Figure 12.3) is obtained from films crystallized from solution (Cortili and Zerbi 1967; Gal'perin et al. 1970), or from crystallization under high pressure (Doll and

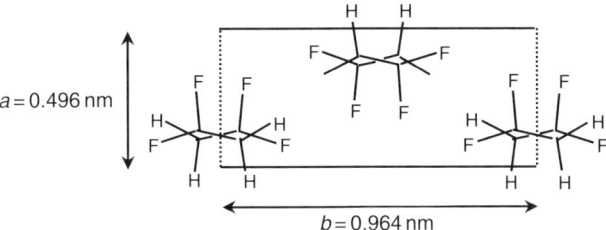

Fig. 12.1. The unit cell of the α form of PVF$_2$ as seen parallel to the chain axis. (After Bachmann and Lando 1981.)

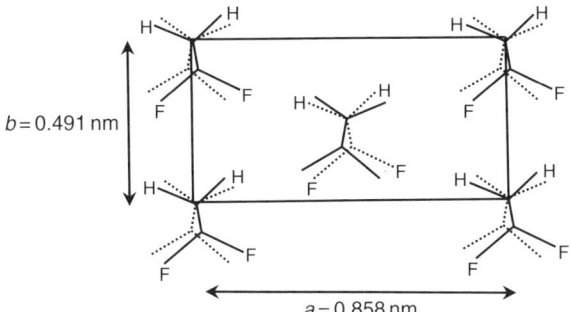

Fig. 12.2. The unit cell of the β form of PVF$_2$ as seen parallel to the chain axis. (After Hasegawa et al. 1972b.)

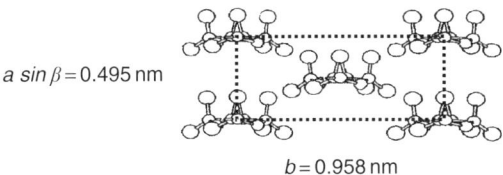

Fig. 12.3. The unit cell of the γ form of PVF$_2$ as seen parallel to the chain axis. (After Takahashi and Tadokoro 1980. Reprinted with permission from ACS.)

Lando 1968) or at high temperature from the melt (Lovinger and Keith 1979). In the latter case, the growth rates of the α form and of the γ form have been studied as a function of the annealing temperature (Prest and Luca 1975; Lovinger 1980). These authors have observed that at low annealing temperatures the α form grows up to 7 times faster than the γ form, but that at higher annealing temperatures this trends is reversed and the γ form grows faster (Figure 12.4). There is some consensus as to the unit cell, which is monoclinic with $a = 0.496$ nm, $b = 0.958$ nm, $c = 0.923$ nm and $\beta = 93°$, and which belongs to the space group Cc (Takahashi and Tadokoro 1980; Lovinger 1981b).

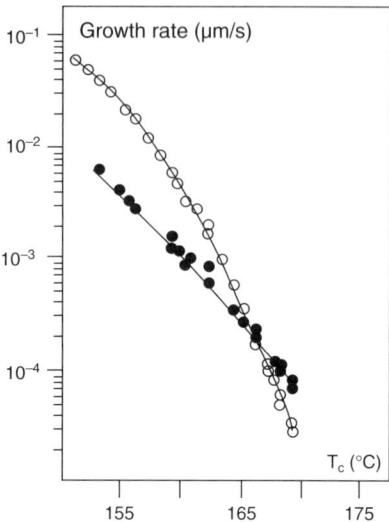

Fig. 12.4. Growth rate of the α form (○) and of the γ form (●) as a function of the crystallization temperature T_c. (Data from Lovinger 1980.)

The chain conformation is $t_3gt_3\bar{g}$. It is worth mentioning that melt-crystallized PVF_2 often displays two endotherms in DSC experiments. The first endotherm may arise from a solid transformation α → γ while the second endotherm might be the melting of the γ form (Prest and Lucas 1975).

Finally, there exists a fourth modification: the δ form. This form is obtained by poling the α form using high electric fields (Davis et al. 1978; Naegele et al. 1978). The X-ray reflections occur at the same positions as those of the α form but exhibit different intensities. This therefore indicates that the α form and the δ form have the same unit cell but different chain packing. Indeed, the δ form unit cell shown in Figure 12.5 was found to be orthorhombic with $a = 0.496$ nm, $b = 0.964$ nm, $c = 0.462$ nm, and space group $P2_1cn$ (Bachmann et al. 1980).

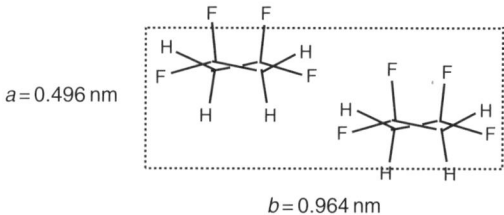

Fig. 12.5. The unit cell of the δ form of PVF_2 as seen parallel to the chain axis. (After Bachmann et al. 1980.)

Dikshit and Nandi have observed that molecular compounds are formed between PVF_2 and a series of diethyl esters (diethyl malonate, diethyl succinate, diethyl glutamate, diethyl pimelate and diethyl azelate). Their conclusion relies essentially on a DSC study in which the variation of the melting enthalpies of those systems forming compounds plotted as a function of the PVF_2 weight fraction is seen to depart conspicuously from linearity (Figure 12.6). Conversely, the couple PVF_2/diethyl oxalate, which does not form a molecular compound, does obey linearity. The non-linear variation is probably due to the existence of two endotherms that cannot be resolved even at low heating rates.

This means that the excess enthalpy with respect to the linear variation stands for the enthalpy associated with the transformation of the complex (congruent, incongruent or singular melting). The maximum observed for the excess enthalpy should therefore correspond to the stoichiometric composition of the compound. This composition is about $X_{PVF_2} \approx 0.47 \pm 0.03$, which eventually gives about *1 solvent molecule/4 monomer units* for PVF_2/diethyl azelate and *1 solvent molecule/3 monomer units* for PVF_2/diethyl succinate.

The temperature–concentration phase diagrams have also been mapped out by these authors. Their shapes, together with the variation of the gel melting enthalpies, are consistent with those of incongruently melting compounds or singular melting compounds (Figure 12.7).

Interestingly, a fibrillar morphology is observed in some systems, which suggests that chain folding is impeded as with stereoregular polystyrenes (Figure 12.8). The origin of this unusual morphology is certainly due to the PVF_2/diester molecular compounds.

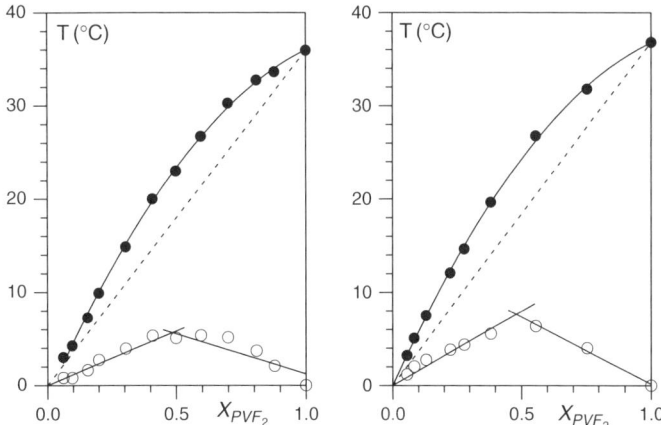

Fig. 12.6. Variation of the melting enthalpies as a function of the PVF_2 weight fraction. The dotted line highlights the linear variation expected in the absence of compound. (●) = measured enthalpies; (○) = excess of enthalpy with respect to linear variation. **left:** PVF_2/diethyl succinate; **right:** PVF_2/diethyl azelate. (Data from Dikshit and Nandi 2000.)

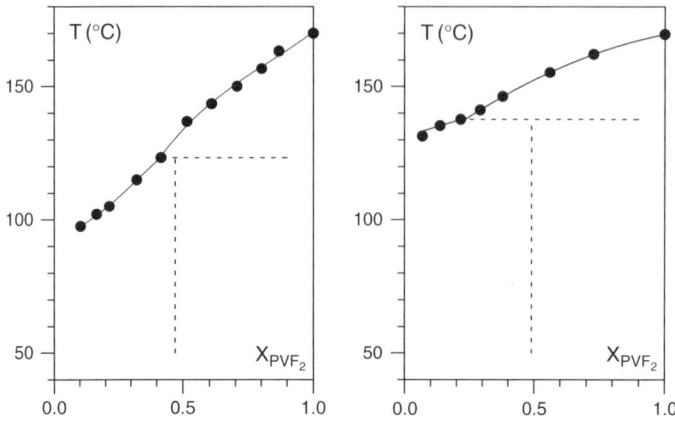

Fig. 12.7. Temperature–concentration phase diagrams for **left**, PVF_2/diethyl succinate, and **right**, PVF_2/diethyl azelate. (Data from Dikshit and Nandi 2000.)

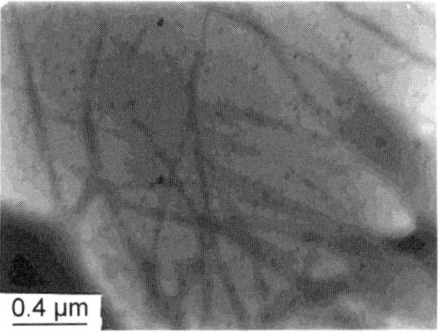

Fig. 12.8. Transmission electron micrograph of 0.3% w/w solution-deposited PVF_2/diethyl azelate systems showing a fibrillar morphology. (From Dikshit and Nandi 2000. Reprinted with permission from ACS.)

Nandi and co-workers discovered that other compounds formed while studying the PVF_2/camphor system (Dasgupta et al. 2005). Unlike the diesters, camphor is a solvent solid at room temperature with a very high melting point ($\approx 178°C$). Also, endotherms with multiple meltings are easily resolved for the PVF_2/camphor system, so all the thermal events, particularly the non-variant events, can be identified beyond doubt. These authors have mapped out the temperature–concentration phase diagram, which is shown in Figure 12.9. Owing to the thermal behaviour of PVF_2 in the solid state, namely the occurrence of two melting endotherms that do not arise from a recrystallization effect, the temperature–concentration phase diagram is rather complex. In addition to the occurrence of molecular compounds, it also presents a *metatectic* transition. As was discussed above, the existence of these two endotherms in the bulk state might indicate an $\alpha \rightarrow \gamma$ transformation, something which is certainly supposed to produce a

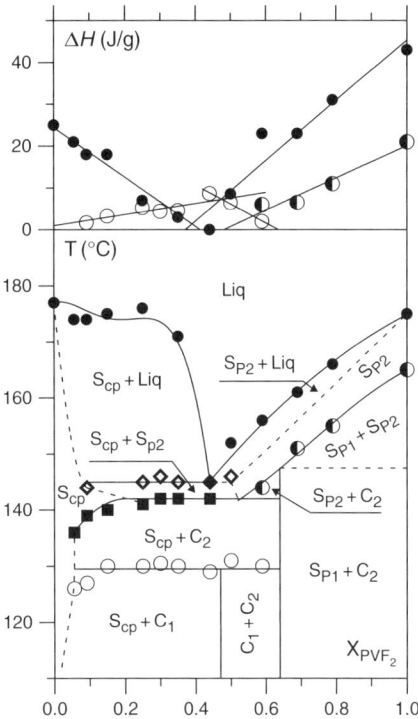

Fig. 12.9. Temperature–concentration phase diagram and Tamman's diagram for PVF_2/camphor systems. C_1 and C_2 are molecular compounds, while S_{P1} and S_{P2} are PVF_2 solid phases, and S_{cp} is a camphor solid phase. There is a direct correspondence between the symbols in either diagram. (Data from Dasgupta et al. 2005.)

metatectic transition in binary polymer/solvent systems (see Chapter 1). In their paper, Dasgupta et al. only mention the existence of two types of lamellae. If these lamellae differ only in size – hence the differing melting points – then the transformation should rather be designated as *pseudo-metatectic* since a true metatectic transformation should involve two structurally different phases, one of which transforms into the other on heating. Note that this might be the first example of such a *metatectic* transformation in polymers.

Three non-variant events are observed at $T = 130°C$, $142°C$ and $145°C$ respectively. From the phase diagram, and by applying Gibbs phase rules, the existence of the following phases can be inferred: *two molecular compounds* whose stoichiometries are of *1 solvent molecule/2 monomer units* for compound C_1 (incongruently melting compound), and *1 solvent molecule/4 monomer units* for compound C_2 (compound with singular melting); *two solid phases* S_{P1} and S_{P2} for PVF_2 for concentrations $X_{PVF_2} \geq 0.5$, hence the metatectic transition at $T = 142°C$; and *one solid phase* S_{cp} for camphor for PVF_2 for concentrations $X_{PVF_2} \leq 0.5$. The solid phases from PVF_2 probably incorporate solvent molecules into the PVF_2 amorphous domains, while the solid phase from camphor remains to be characterized as to the placement of the PVF_2 chains in this phase.

Optical microscopy investigations have revealed well-defined textures for the various phases, as is shown in Figure 12.10 (left). At room temperature for $X_{PVF_2} = 0.15$, the texture is reminiscent of eutectic systems (e.g. see Wittmann and St John Manley 1977) although in this domain the camphor solid phase S_{cp} and compound C_1 coexist. From optical microscopy it is difficult to distinguish either phase. At $T = 140°C$ the phase diagram indicates that compound C_1 has transformed into compound C_2, and indeed a striking change of texture is seen. Finally, at $T = 160°C$ the only remaining solid phase is the solid camphor phase S_{cp}, which again displays a conspicuously different texture.

SEM micrographs obtained on samples freed from camphor through sublimation at room temperature are shown in Figure 12.10 (right) for different PVF_2 fractions.

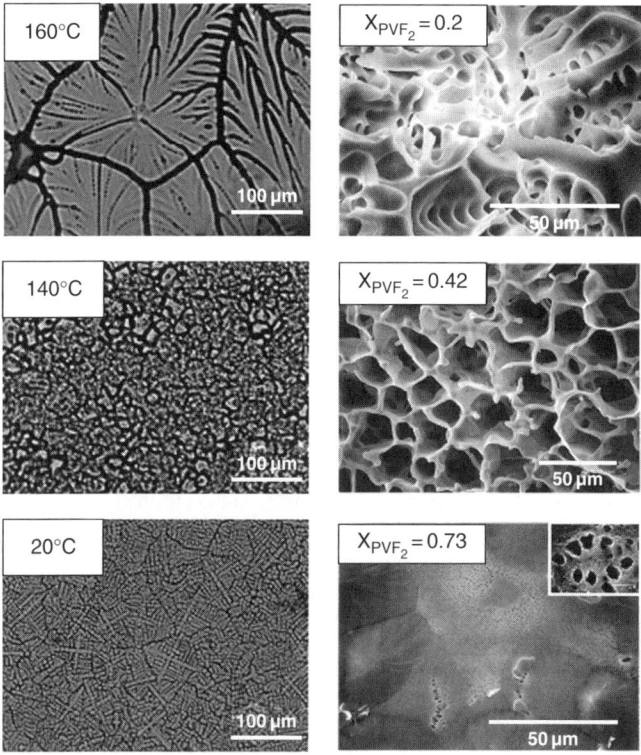

Fig. 12.10. Left: optical micrographs obtained for $X_{PVF_2} = 0.15$ at different temperatures (as indicated) corresponding to the domains of $S_{cp} + C_1$ (bottom); $S_{cp} + C_2$ (middle); $S_{cp} + $Liq (top). **Right:** Scanning electron micrographs obtained after camphor sublimation at room temperature for samples prepared at different X_{PVF_2} (as indicated). $S_{cp} + C_1$ (top); C_1 (middle); $S_{Pl} + C_2$ (bottom). (From Dasgupta et al. 2005. Reprinted with permission from ACS.)

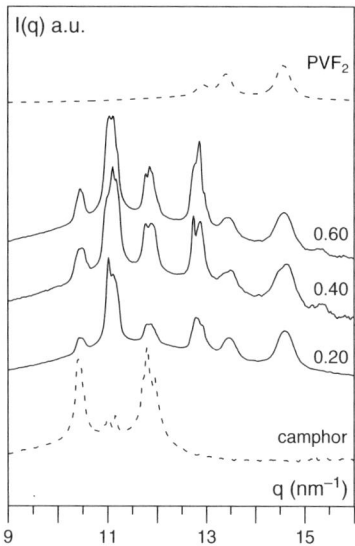

Fig. 12.11. X-ray diffraction patterns obtained at room temperature for different PVF_2 weight fractions, and for the pure components. The reflection at $q = 11\,nm^{-1}$ is typical of molecular compounds C_1 and C_2. (Data from Dasgupta et al. 2005.)

Again, these micrographs span several domains: at $X_{PVF_2} = 0.2$ the phases are $S_{cp} + C_1$, at $X_{PVF_2} = 0.42$ virtually only compound C_1, and at $X_{PVF_2} = 0.73$, $S_{P1} + C_2$. In the latter case spherulites can be seen. The different morphologies do show that the domains identified thought the phase diagrams contain contrasting phases.

As was stressed above, to date no crystalline lattice has been described for these compounds. X-ray diffraction patterns show the appearance of new reflections that do not belong either to camphor or to bulk crystallized PVF_2 (Figure 12.11), in particular the reflection at $q = 11\,nm^{-1}$. These results therefore confirm the outcome of the phase diagram, namely the existence of PVF_2/camphor molecular compounds.

Nandi and co-workers have reported on another system producing molecular compounds, i.e. PVF_2/ethyl acetoacetate (Dasgupta et al. 2006). They have mapped out the T–C phase diagram by DSC investigations (Figure 12.12). Note that ethyl acetoacetate is a liquid at room temperature, and therefore differs from camphor. As with PVF_2/camphor, a *metatectic* or *pseudo-metatectic* transition is seen.

Two compounds, designated as C^a and C^b, are also identified, yet the stoichiometry is the same while their structure is not. Whether these are different crystal structures or simply due to lamellae of differing thicknesses remains to be determined. The stoichiometry as derived from Tamman's plot is about *1 solvent molecule/monomer unit*. Two solid phases from PVF_2 are observed owing to the existence of the metatectic transition. The existence of molecular compounds has received further support from FTIR investigations through the appearance of a new band at $592\,cm^{-1}$. X-ray diffraction patterns also reveal

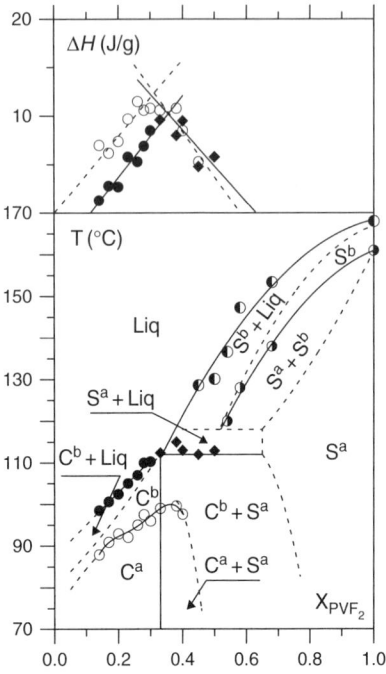

Fig. 12.12. Temperature–concentration phase diagram and Tamman's diagram for PVF$_2$/ethyl acetoacetate systems. C^a and C^b are compounds of the same stoichiometry but different structure. S^a and S^b are solid phase from PVF$_2$. The metatectic transition is supposed to be located in the vicinity of $T = 117°C$. (Data from Dasgupta et al. 2006.)

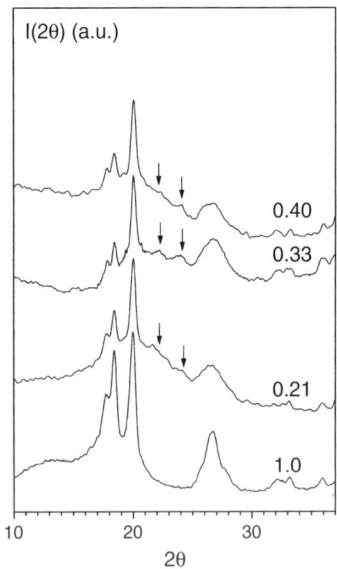

Fig. 12.13. Diffraction patterns of PVF$_2$/ethyl acetoacetate systems (composition indicated). Arrows highlight the new reflections. (From Dasgupta et al. 2006.)

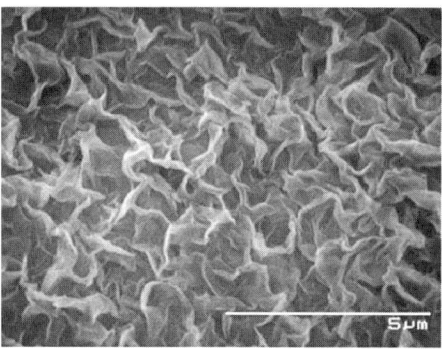

Fig. 12.14. Scanning electron micrograph of PVF$_2$/ethyl acetoacetate systems for a composition $X_{PVF_2} = 0.17$ w/w. (From Dasgupta et al. 2006. Reprinted with permission from ACS.)

two new reflections (Figure 12.13) at $2\theta = 22.2°$ and $24°$ ($q = 15.6\,\text{nm}^{-1}$, $d = 0.4\,\text{nm}$ and $q = 17\,\text{nm}^{-1}$, $d = 0.369\,\text{nm}$) whose intensity is at a maximum for $X_{PVF_2} = 0.33$, which corresponds to the stoichiometric composition.

The morphology of these systems resembles a crumpled cloth when studied by SEM (Figure 12.14). It is not clear whether this morphology arises from the existence of fibrils in the wet sample. The initial morphology can be dramatically altered during the drying process. Interestingly, this morphology is very similar to that observed for sPS/benzophenone systems (see Solution-cast compounds, Chapter 16).

CHAPTER 13

Liquid-crystalline polymers

Some liquid-crystalline polymers can form crystallo-solvates (Iovleva and Papkov 1982; Papkov 1983; Cohen et al. 1991; Cohen and Dagan 1995; Cohen 1996; Cohen and Adams 1996).

Cohen and co-workers have studied the case of poly(p-phenylene benzobisthiazole) (PBZT) and poly(p-phenylene benzobisoxaazole) in polyphoshoric acid and water (PBZO) (see Figure 13.1 for chemical structure).

These polymers are very rigid and form liquid-crystalline mesophases. They are used for preparing materials with high-tensile strength (Wolfe et al. 1981).

In the case of PBZT Cohen et al. (1991) have observed two molecular compounds designated as form I and form II. Form II is observed when slow coagulation in 85% phosphoric acid is carried out. Form I is obtained by exposure to vacuum for several hours. Form I is more ordered than form II. Cohen et al. have derived a 2D lattice due to the low number of reflections with $a = 0.815$ nm, $b = 2.449$ nm and $\gamma = 94.5°$. Their estimated stoichiometry is 8 PPA/3 PBZT but no hint is given as to the water content. Form II, which contains more water molecules, displays many fewer reflections, so no information on the crystal unit cell is available and neither is any value for the stoichiometry.

Cohen and Cohen also studied PBZT in methanesulphonic acid (MAS). They identified in addition two crystallo-solvate phases designated form I and form II (Cohen and Cohen 1995). They suggest that in form I, four acid anions are complexed to one protonated PBZT repeat unit with two additional acid molecules in the unit cell. Form II is obtained after deprotonation of the polymer when the molar concentration of water equals the molar concentration of the free acid.

Studies on PBZO/PPA/water systems have also revealed the existence of two solvated forms, again designated as form I and form II. Again, form I contains a lower number of

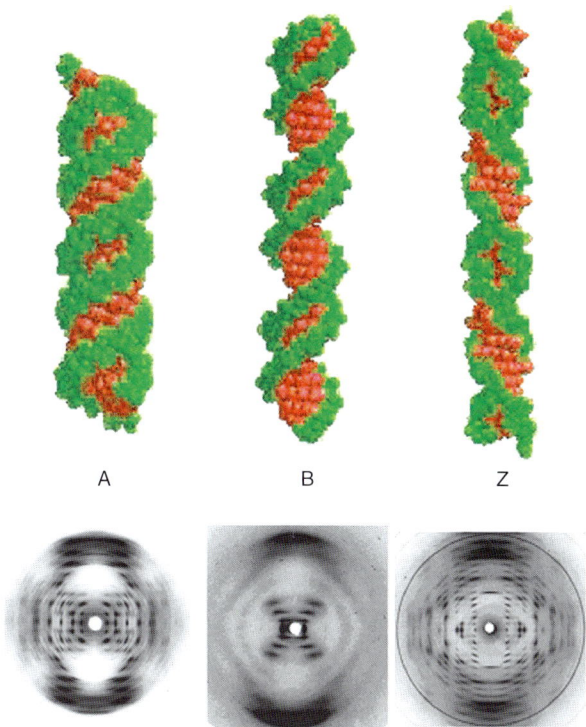

Plate 1. Molecular representations of the helical forms of A-DNA, B-DNA and Z-DNA together with the corresponding fibre diffraction patterns. All forms are double-stranded helices. (Diffraction patterns from Fuller, W., Forsyth, T., Mahendrasingam, A. *Phil. Trans. R. Soc. B* **2004** *359* 1237. Reprinted with permission from The Royal Society.) (See Fig. 9.2 on page 101)

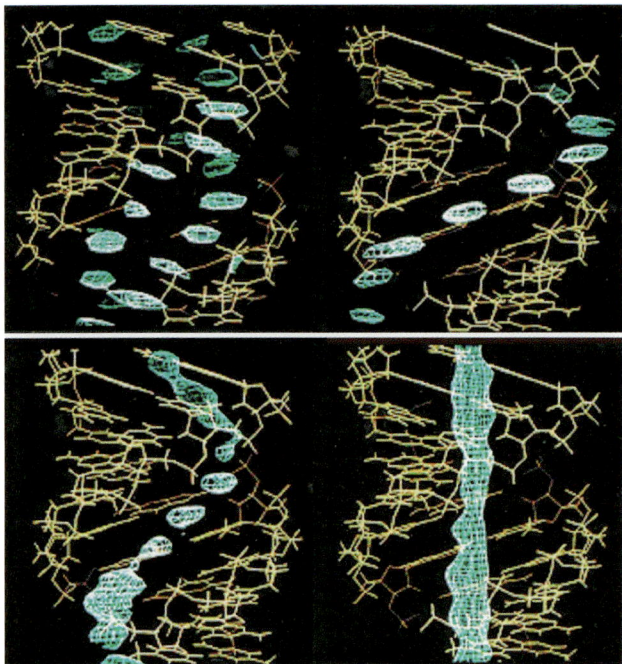

Plate 2. Fourier difference synthesis map showing the four ordered water sites in the A-DNA double helix (locations of water molecules are in blue, while the DNA is in yellow). **Top left**, site 1; **top right**, site 2; **bottom left**, site 3; **bottom right**, site 4. (From Shotton et al. 1997. Reprinted with permission from Elsevier.) (See Fig. 9.5 on page 103)

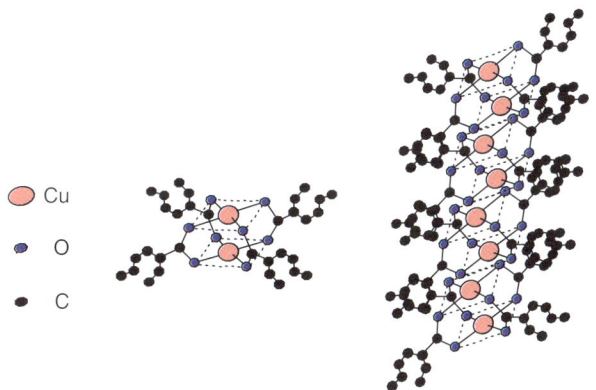

Plate 3. Structure of the bicopper complex copper (II) tetra-2-ethyl-hexanoate, and showing the way it piles up to form filaments. (Maldivi 1989.) (See Fig. 15.18 on page 204)

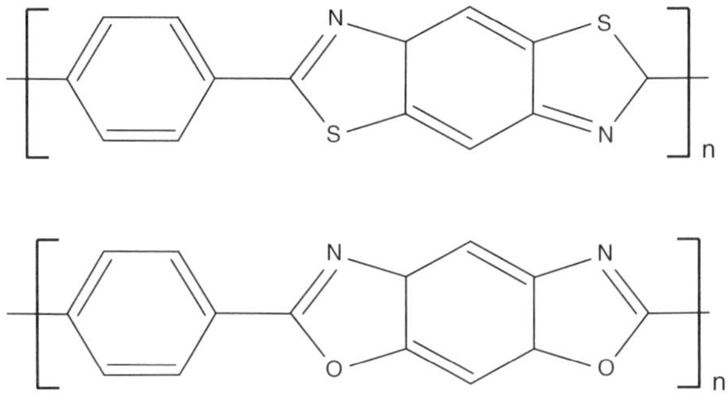

Fig. 13.1. Top: chemical structure of poly(*p*-phenylene benzobisthiazole), **bottom:** poly (*p*-phenylene benzobisoxaazole).

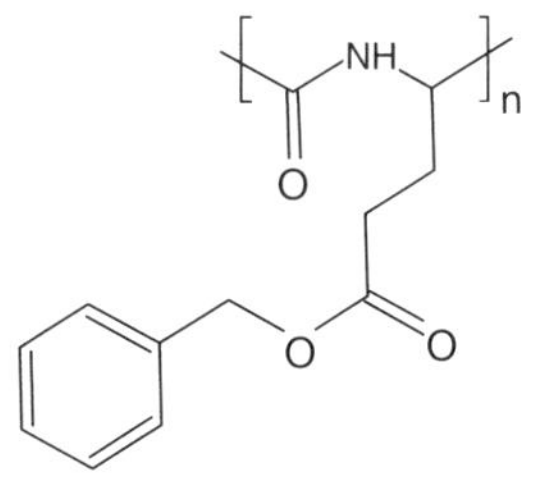

Fig. 13.2. Chemical structure of poly[γ benzyl L-glutamate) (PBLG).

water molecules than found in form II. Here both forms are relatively highly organized, so crystalline lattices were able to be proposed by Cohen and Adams (1996). For form I, Cohen and Adams have suggested a monoclinic unit cell with $a = 1.26$ nm, $b = 1.16$ nm, $c = 1.22$ nm (chain axis) and $\gamma = 98°$. A monoclinic unit cell is also proposed for form II, with $a = 0.7$ nm, $b = 0.58$ nm, $c = 1.20$ nm (chain axis) and $\gamma = 99°$.

Cohen and co-workers have further examined the case of poly[γ-benzyl L-glutamate) in benzyl alcohol (Figure 13.2). This system produces thermoreversible gels with a fibrillar morphology (Hikata et al. 1977; Sasaki et al. 1982). According to Cohen and Dagan, reflections observed at 1.9 and 3.8 nm^{-1} in concentrated systems do not correspond to the known crystalline forms of PBLG (Cohen and Dagan 1995).

These authors conclude that the formation of a crystallo-solvate is highly probable, and suggest the possible schematic phase diagram drawn in Figure 13.3 (note that this type of phase diagram was also proposed for PBZT systems).

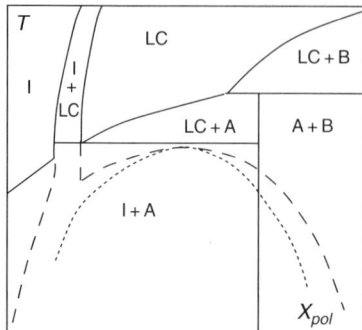

Fig. 13.3. Hypothetical T–C phase diagram for explaining the behaviour of PBLG/benzyl alcohol systems. I = isotropic solution; LC = liquid crystalline phase (nematic); A = molecular compound phase; B = solid phase. The dashed line defines the binodal (miscibility gap), while the dotted line stands for the spinodal line. (After Cohen and Dagan 1995.)

I believe that Science must be understood as a social phenomenon, a gutsy, human enterprise, not the work of robots programmed to collect pure information.

Stephen Jay Gould in *The Mismeasure of Man* (1981)

PART IV

Intercalates and clathrates

These systems are characteristic of synthetic polymers that form molecular compounds with organic solvents. The term **clathrate** was coined by Powell in 1948 to describe a particular form of molecular compound in which one component, the *host*, forms a cage structure imprisoning the other, designated as the *guest*. This term is borrowed from the Latin word *clathratus* (meaning closed or protected by cross bars or trellis), which in turn comes from the classical Greek word *kleithron*, or *klathron*. This word means a 'lock' or a 'bar' (earlier origins are found in Indo-European languages, but this is another story!).

As regards **intercalates**, they are seen as two-dimensional, open, standing layer-type or sandwich-type inclusion compounds. As will be discovered in Chapters 14–19, which comprise Part IV, some systems can be designated as intercalates on account of the sandwich-type structure but also as clathrates because the solvent molecules are not in contact with one another, and can thus be considered imprisoned by polymer chains (for details see Weber 1987).

The driving force involved in the growth of these compounds is chiefly a mechanism of 'molecular recognition'. The solvent tends to be housed in cavities created by the microstructure of the polymer chains without any special interactions such as hydrogen bonding or electrostatic interactions involved in biological systems. These microstructures can be the groove created by the helical structure, or cavities arising from the regular stacking of the lateral groups of the chain, or again empty spaces generated by assemblies of helices. To some extent, the formation mechanism of this type of molecular compound can be regarded as being essentially an *entropic* process, although in the case of some poly[ethylene oxide] compounds hydrogen bonds also come into play.

These systems can produce both spherulitic morphologies and fibrillar thermoreversible gels. In the latter case, the gel state is the direct consequence of the formation of a molecular compound with intrinsically flexible polymers such as polystyrenes. Compound formation therefore plays a major role in the formation of unusual structures.

In Part IV, the molecular compounds produced from four synthetic polymers will be described: poly[ethylene oxide] (Chapter 14), polystyrene (isotactic (Chapter 15), syndiotactic (Chapter 16) and atactic (Chapter 17)), syndiotactic poly[*para*-methyl styrene] (Chapter 18), and poly[methyl methacrylate] (isotactic and syndiotactic) (Chapter 19).

CHAPTER 14

Poly ethylene oxide

The synthesis of poly[ethylene oxide] (PEO) dates back to the early 1940s, first being produced at a factory based in Germany (Anorgana).

This polymer crystallizes spontaneously both from the molten state and from solution. This is due to the absence of asymmetric carbons, as can be seen from its chemical structure (Figure 14.1). Although one would think at first that PEO would crystallize with a planar zig-zag conformation, it in fact takes on a 7_2 helical structure, as has been shown by Tadokoro et al. (1963, 1964a, 1964b). This situation arises from the orbitals of the oxygen atom that produce the same effect as if side-groups were present (the sequence is then *trans-trans-gauche* (*ttg*)). The crystalline lattice these authors derived from the X-ray diffraction pattern is monoclinic with $a = 0.816$ nm, $b = 1.299$ nm, $c = 1.93$ and $\beta = 126°5$.

Owing to the possibility of synthesizing low molecular weight samples, the number of folds in single crystals can be monitored, and PEO can therefore crystallize in an all-extended form, namely without chain-folding. There exists an abundant literature on this, which is, however, outwith the scope of this book.

The propensity of PEO to form molecular compounds was first observed with urea (Parrod et al. 1958, 1964; Tadokoro et al. 1964b). These systems form *clathrates* due to the special structure of urea. Later, Point and co-workers discovered other compounds of the *intercalate* type with a series of benzene derivatives, such as *para*-dihalogenobenzene (Point and Coutelier 1985; Point et al. 1986a, b). Compound formation arises chiefly from molecular recognition between the shape of the solvent and the microstructure of the PEO helix. Finally, a third category has been observed where PEO chains are threaded through the cyclodextrin molecules. These compounds are usually designated as inclusion compounds. These three categories will be detailed in what follows. Note that compounds produced between PEO and salts will not be considered because they essentially concern polyelectrolyte aspects.

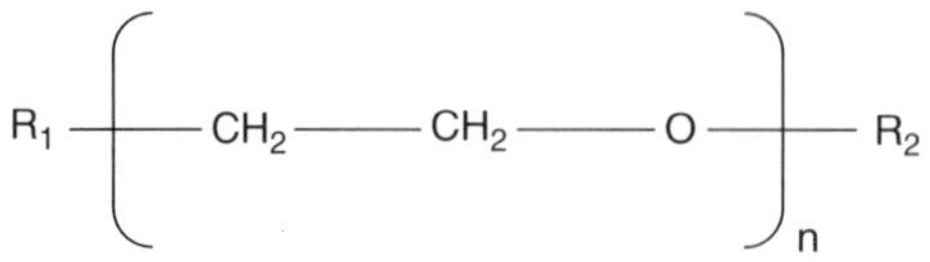

R₁ ─[─ CH₂ ─── CH₂ ─── O ─]ₙ─ R₂

Fig. 14.1. Chemical structure of poly[ethylene oxide]. R_1 and R_2 stand for terminal groups that can be varied from H atoms to very large groups.

Clathrates and inclusion compounds

The formation of PEO/urea compounds was first observed from the crystallization of PEO in a saturated methanolic solution of urea by Parrod et al. (1958, 1964). Tadokoro et al. (1964b) have obtained oriented samples by immersing oriented PEO crystalline film in a methanolic solution of urea. These compounds can also be prepared by quenching homogeneous solutions prepared from PEO/urea mixtures above the melting point of urea.

The crystal structure of urea under atmospheric pressure consists of ribbons of molecules linked in a head-to-tail fashion along the tetragonal c-axis. Each ribbon is surrounded by four identical, orthogonally orientated ribbons. Such an arrangement creates channels with square cross-section 0.394 nm × 0.394 nm. The existence of channels is probably a key factor in the formation of clathrates of PEO with urea. As has been shown by Tadokoro et al. (1964a) and later confirmed by Chenite and Brisse (1991) on pseudo-hexagonal single crystals, the polymer chains are inserted inside channels formed by the surrounding urea molecules (Figure 14.2).

The resulting crystalline structure consists of a trigonal unit cell of dimensions $a = b = 1.054$ nm, $c = 0.91$ nm and $\gamma \approx 120°$ of space group $P3_121$. The PEO chains take on a nearly 4_1 helical structure, which is slightly more squashed compared with the 7_2 helix in pure PEO, and therefore with the 7_2 helical structure first proposed by Tadokoro et al. (1964a) for PEO in these clathrates. The stoichiometry derived from crystallographic data is 9/4 (9 urea molecules for 4 PEO units), which corresponds to the earlier stoichiometry, namely 2/1 (Parrod et al. 1958; Bailey and France 1961). In the crystal, two thirds of the urea molecules form the walls of the channels while one third are linked to the PEO chains through hydrogen bonds.

Vasanthan et al. (1996) have given, by ^{13}C NMR, circumstantial evidence for the existence of these urea molecules bonded to PEO chains. Indeed, the PEO mobility is seen to be close to that for PEO in the bulk crystalline state, as opposed to what is usually observed with other inclusion compounds. Vasanthan et al. therefore conclude that these bonded urea molecules slow down the motion of PEO chains within the urea channels.

Wagner et al. (2005) established the temperature–concentration phase diagram shown in Figure 14.3 by means of calorimetry investigations (DSC). The shape of the phase diagram indicates that the compound is of the *congruently melting* type. The stoichiometry derived from the phase diagram is in excellent agreement with that deduced from the crystallographic data, namely 9/4.

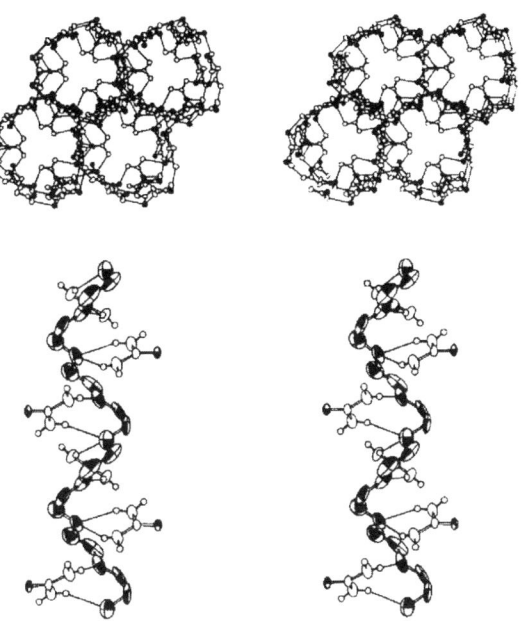

Fig. 14.2. Upper: stereo pair displaying the channels formed by hydrogen-bonded urea molecules and the extra urea molecules present in the channels. **Lower:** stereo pair displaying the association between PEO chains and urea molecules that are located within the urea channels. (After Chenite and Brisse 1991. Reprinted with permission from ACS.)

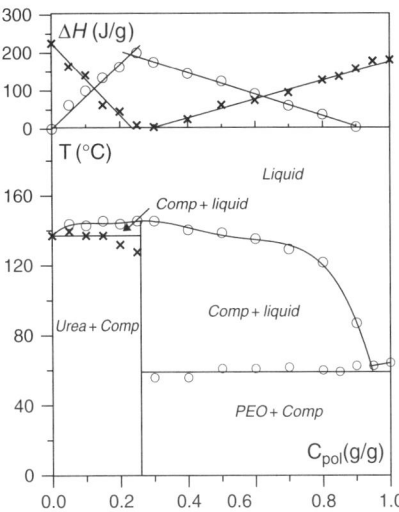

Fig. 14.3. Temperature–concentration phase diagram (**lower**) and Tamman's diagram (**upper**) of the PEO/urea system. The samples were obtained by quenching homogeneous PEO/urea solutions prepared at a temperature above the melting point of urea. (From Wagner et al. 2005.)

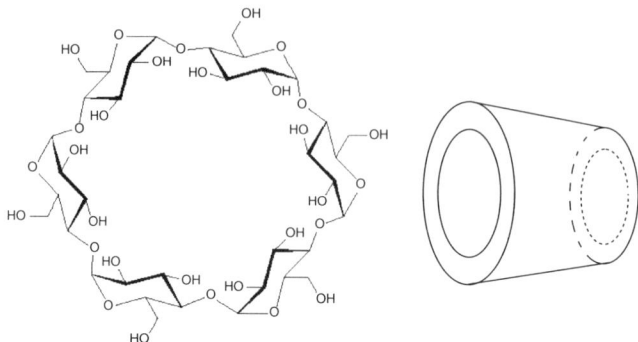

Fig. 14.4. Left: Chemical structure of α-cyclodextrin. **Right:** schematic drawing highlighting the flower-pot-like shape of these molecules.

Harada et al. (1992) have reported that PEO also forms inclusion compounds with cyclodextrins, particularly α-cyclodextrin and γ-cyclodextrin (Figure 14.4). Cyclodextrins are cyclic sugars, α-cyclodextrin containing six glucose residues and γ-cyclodextrin, eight. These molecules have a 'flower-pot-like' molecular structure.

Compound formation occurs through the threading of the PEO chains through the centre of the cyclodextrins, as shown in Figure 14.5. Only one chain is involved in α-cyclodextrin (Harada et al. 1992), while two PEO chains are threaded through γ-cyclodextrin (Harada et al. 1994). Harada et al. designated the latter case as a double-stranded inclusion complex, although there is no indication of PEO taking on a double helical conformation.

For *PEO/α-cyclodextrin* compounds the stoichiometry is 2/1 (2 PEO monomer units/1 cyclodextrin) and for *PEO/γ-cyclodextrin* compounds the stoichiometry has been found to be 4/1 by Harada and colleagues.

Note that no compound forms with the β-cyclodextrin, which is made up of 7 units. The inner hole of this molecule is too large for housing one chain but not large enough for accommodating two chains.

As reported by Huang et al. (1998), DSC thermograms reveal no melting peak due to pure crystalline PEO, which suggests the involvement of PEO chains in the compound. Also, no melting peak of the compound is observed as it decomposes before melting at

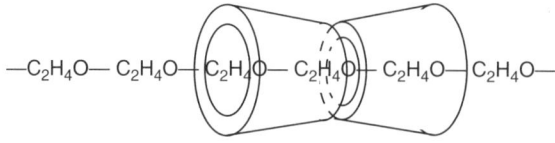

Fig. 14.5. Structure of PEO/α-cyclodextrin compounds. (After Harada et al. 1992.)

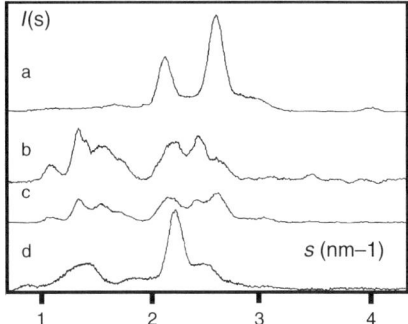

Fig. 14.6. Diffraction patterns of: (a) pure crystallized PEO, (b) cyclodextrin, (c) a mixture of PEO and cyclodextrin, (d) the PEO/α-cyclodextrin compound. (From Huang et al. 1998. Reprinted with permission from Elsevier.)

about 334°C. Note that the compound displays a higher decomposition temperature than that of pure α-cyclodextrin ($T_{decomp} = 315°C$). The occurrence of thermal decomposition before melting therefore prevents the establishment of temperature–concentration phase diagrams.

The diffraction pattern of these compounds conspicuously differs from that of either component or a mixture of PEO and cyclodextrin (Figure 14.6). This is a clear indication of complex formation (Huang et al. 1998).

Intercalates

In the mid-1980s other types of PEO compounds were discovered by Point and co-workers (Point and Coutelier 1985; Point et al. 1986a; Point and Demaret 1987). This discovery was fortuitous and due to an attempt to prepare single crystals from *ortho*-dichlorobenzene, a liquid solvent at room temperature. Surprisingly, the resulting crystals would melt at about 90°C instead of 62°C, which is the usual melting temperature of PEO crystals. It turned out that PEO/*para*-dichlorobenzene compounds would form, simply because *para*-dichlorobenzene was present as an impurity in the *ortho*-dichlorobenzene used.

Similar types of compound also form with *dihydroxy*benzene, such as hydroquinone and the like. While in the first case molecular recognition between the solvent molecule and the chain microstructure is clearly the driving force for the genesis of the compounds, in the second case other types of interaction are in play (the chemical structures are given in Figure 14.7).

Intercalates through molecular recognition

In their first paper of a series on these systems, Point and colleagues both mapped out the temperature–concentration phase diagram and determined the crystal structure of

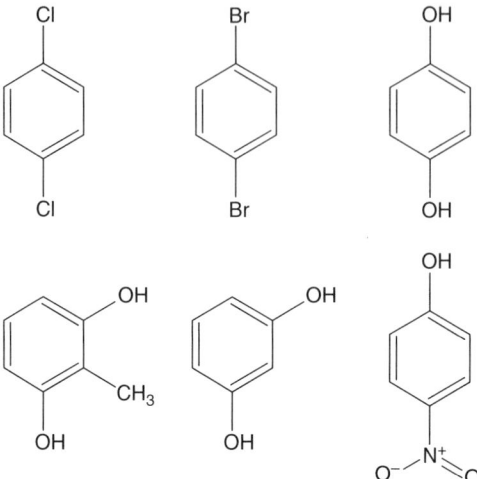

Fig. 14.7. Chemical structures of the solvent used for preparing PEO intercalates: **top**, from left to right: *para*-dichlorobenzene, *para*-dibromobenzene, hydroquinone; **bottom**, from left to right: 2-methyl resorcinol, resorcinol, *para*-nitrophenol.

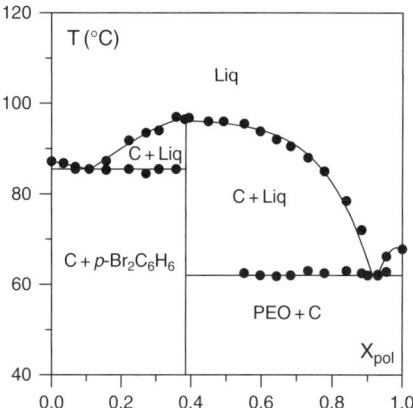

Fig. 14.8. Temperature–concentration phase diagram for PEO/*para*-dibromobenzene compounds (C). The compound possesses a stoichiometry of 2/3 (2 solvent molecules for every 3 monomer units). (Data from Point and Coutelier 1985.)

PEO/*para*-dibromobenzene compounds. The phase diagram reveals a congruently melting compound, as shown in Figure 14.8. The stoichiometry is 2/3 (2 solvent molecules for every 3 monomer units).

Point et al. further determined the crystal structure by X-ray diffraction, and later confirmed the position of the solvent molecules by FTIR and neutron diffraction (Point

Table 14.1. Unit cells for PEO/p-C_6H_4XY intercalates. All belong to space group $Cmc2$. ρ is the crystal density. (After Point et al. 1986a.)

X	X	a	b	c	ρ (g/cm^2)
Cl	Cl	1.648	0.951	2.786	1.341
Br	Br	1.674	0.968	2.798	1.681
Cl	Br	1.658	0.958	2.810	1.510
Br	F	1.638	0.946	2.856	1.449
Cl	I	1.678	0.973	2.824	1.665

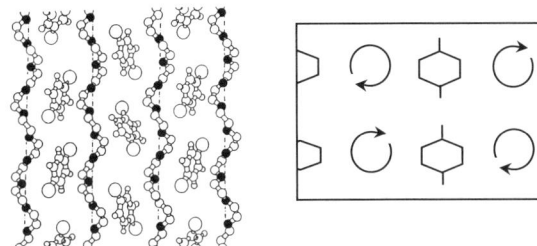

Fig. 14.9. Left: the PEO/*para*-dibromobenzene crystal unit cell as seen perpendicular to the PEO helical axis. Right: the unit cell as seen parallel to the helical axis. The circular arrows mimic the chains and their helicity. (After Point and Coutelier 1985; Point et al. 1986a, b. Reprinted with permission from Wiley Interscience.)

and Coutelier 1985; Point et al. 1986a, b; 1991a, b). Typically, in the systems shown in Table 14.1 the unit cell is basically orthorhombic of space group $Cmc2_1$ wherein the PEO chains take on a 10_3 helical structure (Figure 14.9). Note that this helix is actually very close to the 7_2 helix in the pure PEO.

The crystalline organization is therefore typically that of an *intercalate* as rows of polymer stems alternate with 'sheets' of solvent molecules.

Note that Point et al. further discuss both the existence of disorder, such as domains with non-stoichiometric composition, and departure from the perfect 10_3 helix.

Other *para*-dihalogenobenzenes of the type p-C_6H_4XY, where X and Y are various halogens, have been systematically studied by Point's group (Point et al. 1986a). In Figure 14.10 is shown the temperature–concentration phase diagram of PEO/*para*-dichlorobenzene, which is essentially similar to that established for PEO/*para*-dibromobenzene (Paternostre et al. 1998). In Table 14.1 the unit cell parameters obtained with these different solvents are brought together. In all cases the unit cell is orthorhombic, the helical structure of PEO is close to a 10_3 helix, and the stoichiometry is 2/3. Energy minimization also highlights the fact that the polymer layers form voids that can accommodate the *para*-dihalogenobenzene guest molecules extremely well.

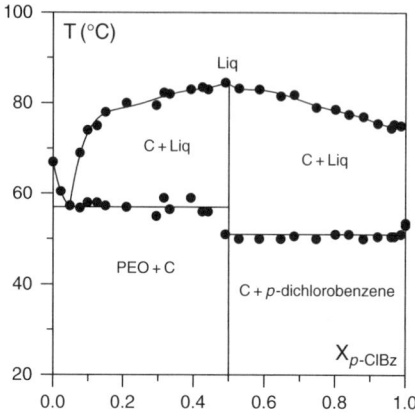

Fig. 14.10. Temperature–concentration phase diagram for PEO/*para*-dichlorobenzene compounds (C). The compound possesses a stoichiometry of 2/3 (2 solvent molecules for every 3 monomer units). Note that the x-axis stands for the weight fraction of *para*-dichlorobenzene. (Data from Paternostre et al. 1998.)

PEO/hydroxybenzene compounds

Myasnikova and co-workers were the first to report on the formation of *PEO/resorcinol* compounds (resorcinol = *meta*-dihydroxybenzene) in the mid-1970s (Myasnikova 1976; Myasnikova et al. 1980). These authors suggested a monoclinic unit cell, yet Delaite et al. (1992) determined that the unit cell was in fact orthorhombic, the parameters later being refined by Ianelli et al. (1999). These parameters are $a = 1.054$ nm, $b = 1.018$ nm, $c = 0.989$ nm, and the space group $Pna2_1$. The PEO chains take on a conformation close to the usual *ttgttg* arrangement, which gives a structure close to a 4_1 helix.

Interestingly, as shown in Figure 14.11, no direct PEO chain-to-chain interaction can be established as the chains are surrounded by resorcinol molecules as opposed to what is seen with *para*-dihalogenobenzene. So, *PEO/resorcinol* compounds are not strictly speaking intercalates, but rather resemble clathrates, although no resorcinol channels are present as in the case with urea. This situation probably arises from the hydrogen bonds between the OH of resorcinol and the oxygens of PEO (Ianelli et al. 1999).

From the crystalline lattice a stoichiometry 1/2 is determined (1 resorcinol molecule for every two PEO monomer units). This is confirmed by the temperature–concentration phase diagram, which is shown in Figure 14.12 (Myasnikova et al. 1980; Delaite et al. 1992). This diagram shows that the *PEO/resorcinol* compound is of the congruently melting type with the existence of two eutectic systems. Delaite et al. have further discovered that there exists a *metastable* form of the PEO/resorcinol compound, designated as the β *form*, which transforms into the stable α form by simply ageing a few minutes at room temperature. The β *form* spherulites are observed with the *metastable* α form (Figure 14.13).

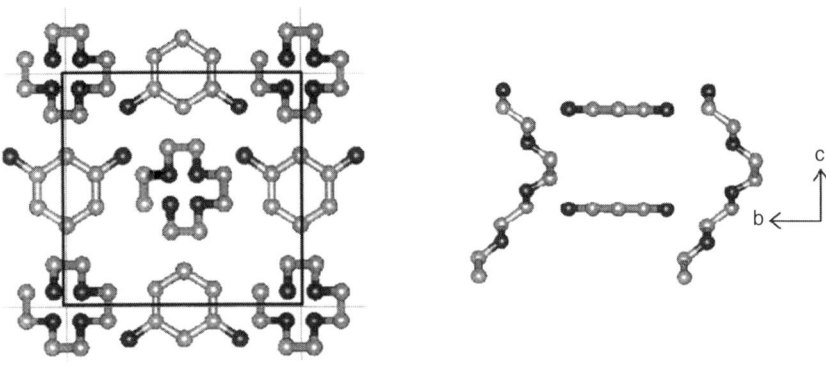

Fig. 14.11. Schematic representation of the *PEO/resorcinol* unit cell as seen parallel to the chain axis (**left**), and perpendicular to the chain axis (**right**). (After Delaite et al. 1992; Paternostre et al. 1999b. Reprinted with permission from ACS.)

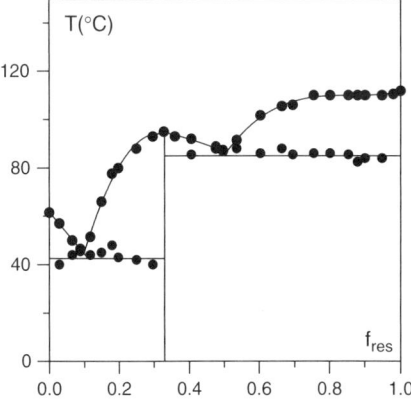

Fig. 14.12. Temperature–concentration phase diagram for the stable α phase of the *PEO/resorcinol* compound (here the x-axis stands for the molar fraction of resorcinol). (After Myasnikova et al. 1980; Delaite et al. 1992.)

Investigations have been further carried out with a resorcinol derivative, namely 2-methyl resorcinol. In fact, Belfiore and Ueda (1992) have argued that 2-methyl resorcinol introduces an additional steric effect as the methyl group, being *ortho* to both hydroxyls, lies therefore very close to the PEO chain when the latter preferentially establishes hydrogen bonds with both hydroxyl protons of the same molecule.

It turns out that *PEO/2-methyl resorcinol* system is unique in the series of PEO compounds, as it produces two complexes of distinct stoichiometries and melting temperatures. This is illustrated by the temperature–concentration phase diagram shown in Figure 14.14 (Belfiore and Ueda 1992; Paternostre et al. 1999a). The more solvated

174 Polymer-Solvent Molecular Compounds

Fig. 14.13. Optical micrograph of a mixture of the α form (non-banded spherulites) and the β form (banded spherulites) of the PEO/resorcinol compound. (After Delaite et al. 1992. Reprinted with permission from ACS.)

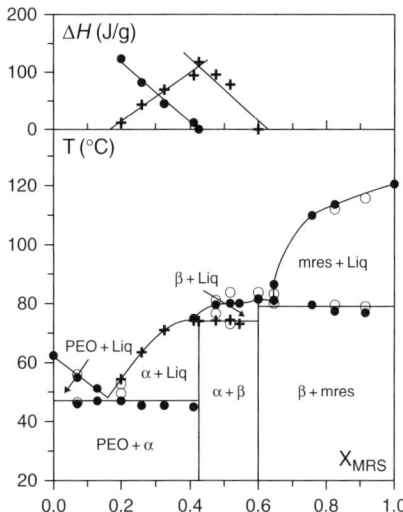

Fig. 14.14. Temperature–concentration phase diagram, and Tamman's diagram, for PEO/2-methyl resorcinol compound (here the x-axis stands for the weight fraction of 2-methyl resorcinol). Open symbols stand for temperatures as determined by X-ray experiments, the others for temperature as measured by DSC experiments. (Data from Belfiore and Ueda 1992; Paternostre et al. 1999a.)

compound, designated as *PEO/α-2MR*, melts at about $T_{\alpha-2MR} \approx 74°C$ and possesses a stoichiometry of 1/2 (one 2-methyl resorcinol molecule for every two PEO monomer units), while the less solvated compound, designated as *PEO/β-2MR*, melts at $T_{\beta-2MR} \approx 84°C$ and has a stoichiometry of 2/7 (two 2-methyl resorcinol molecules for every seven PEO monomer units). *PEO/α-2MR* appears to be of the singular-melting type while *PEO/β-2MR* is of the congruently melting type. Two eutectics are obtained at compositions $X_{MRS} \approx 0.16$ and $X_{MRS} \approx 0.65$. The unfortunate choice of α and β for designating two stable compounds can lead one to confusion, the same symbols being used for PEO/resorcinol which designate phases of different stabilities.

As can be seen from the diffraction patterns shown in Figure 14.15, these two complexes possess different crystalline structures. The crystalline structure of these two complexes was first derived by Paternostre et al. (1999a), and later refined by Ianelli et al. (Ianelli et al. 1999). An orthorhombic unit cell accounts for the diffraction pattern

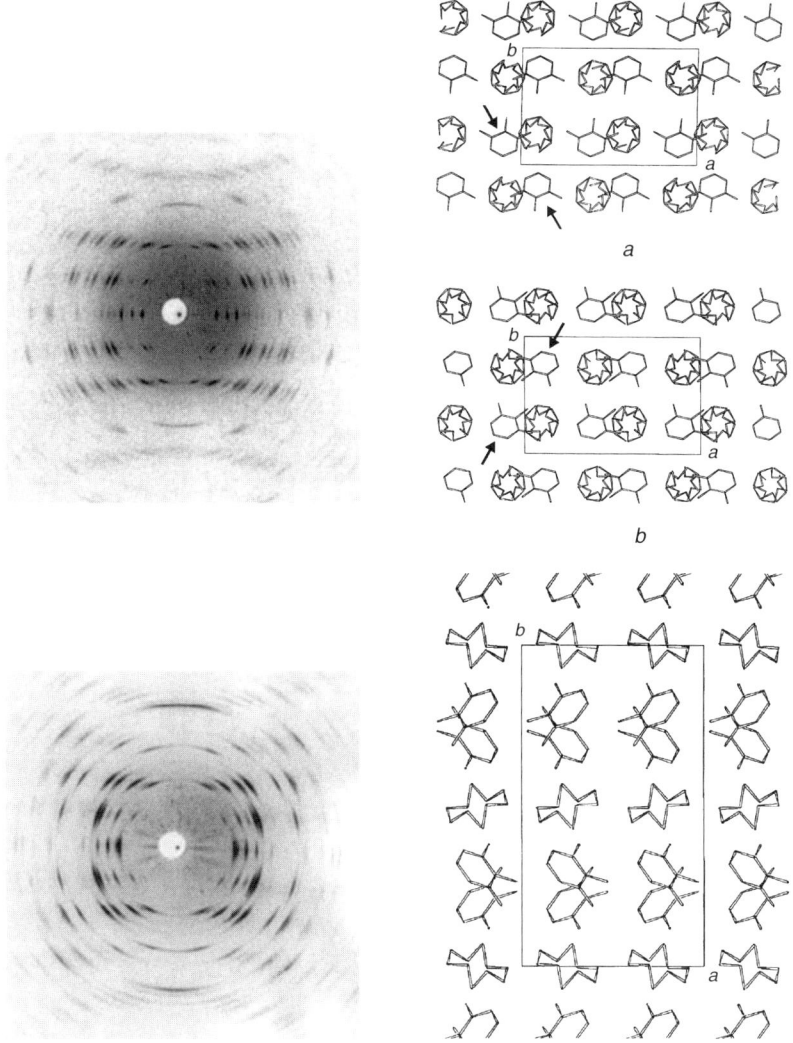

Fig. 14.15. Diffraction patterns and corresponding crystal unit cell for the different complexes of PEO/2-methyl resorcinol. **Upper:** PEO/α-2MR and **lower:** PEO/β-2MR. For PEO/α-2MR the two possible packings for the 2-methyl resorcinol molecules are shown by arrows. (After Ianelli et al. 1999; Paternostre et al. 1999a.) Reprinted with permission from ACS.

of *PEO/α-2MR* with $a = 1.624$ nm, $b = 1.057$ nm, $c = 1.879$ nm, $\gamma = 89.2°$, and space group $P2_1/a$, with unique axis c (Figure 14.15). Here the PEO chain take on a *ttg* conformation, which entails a structure close to a 7_2 helix. For the *PEO/β-2MR* complex, an orthorhombic unit cell also accounts for the results, but with different parameters, namely $a = 1.107$ nm, $b = 1.860$ nm, $c = 1.074$ nm, and space group *Pbca*. Here, the chain conformation departs slightly from the strict *ttg* conformation, so the structure is rather close to a 4_1 helix.

The occurrence of a complex between PEO and *hydroquinone* was first reported by Myasnikova in the mid-1970s (Myasnikova 1976). The phase diagram was first established by Belfiore and Ueda (1992) and later completed by Paternostre et al. (1999b), as shown in Figure 14.16. The latter derived a stoichiometry of 1/2 (1 hydroquinone molecule for every 2 PEO monomer units) both from the phase diagram and from the crystal structure.

Paternostre et al. studied the crystalline structure by X-ray diffraction and FTIR, in the latter case using hydrogenous and deuterated PEO. They have thereof suggested that the unit cell of this complex is a triclinic with the parameters: $a = 1.17$ nm, $b = 1.2$ nm, $c = 1.06$ nm, $\alpha = 78°$, $\beta = 64°$, $\gamma = 115°$ (Figure 14.17). The PEO chains take on a 4_1 helix, which, as emphasized by these authors, is quite close to the usual 7_2 helix.

Studies using ^{13}C NMR, carried out by Spěváček and co-workers, have shown that PEO chain mobility is definitely slower in complexes of resorcinol, hydroquinone and

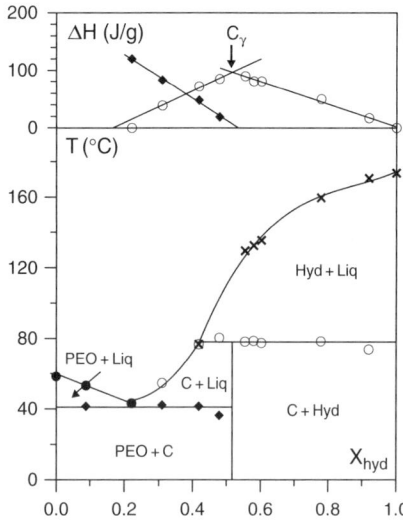

Fig. 14.16. Temperature–concentration phase diagram, and Tamman's diagram for PEO/hydroquinone (here the x-axis stands for the weight fraction of hydroquinone). (Data from Belfiore and Ueda 1992; Paternostre et al. 1999b.)

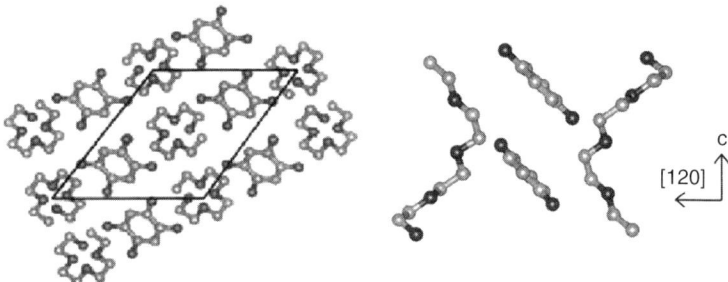

Fig. 14.17. PEO/hydroquinone compound. Projection of the crystal unit cell along the chain axis (**left**) and perpendicular to the chain axis and plane [120]. (After Paternostre et al. 1999b. Reprinted with permission from ACS.)

para-nitrophenol than in pure PEO crystals (Spěváček et al. 1998). This shows the existence of strong hydrogen bonds between the PEO chains and the solvent.

The strength of these hydrogen bonds is greatest with *para*-nitrophenol. Compounds in *para*-nitrophenol are certainly the most exotic among all the PEO/solvent complexes, as the PEO chains take on a very unusual conformation.

The phase diagram mapped out by Point and Damman (Figure 14.18) reveals the existence of one complex of the *singular-melting* type, together with one eutectic (Point and Damman 1992). The stoichiometry is 2/3, namely 2 *para*-nitrophenol molecules for every 3 PEO monomer units. The eutectic is seen for $X_{pnp} = 0.25$.

Point and Damman have also determined the crystallographic structure of the complex from X-ray diffraction patterns on rolled film as well as on spherulites. They have derived

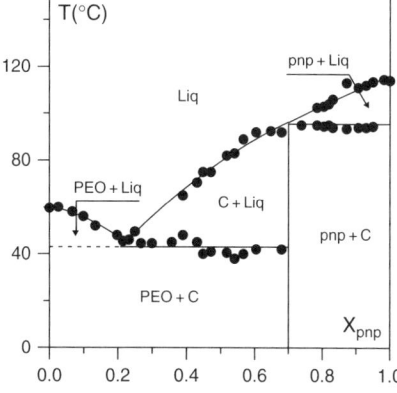

Fig. 14.18. Temperature–concentration phase diagram PEO/*p*-nitrophenol (the x-axis stands for the weight fraction of *p*-nitrophenol). (After Point and Damman 1992.)

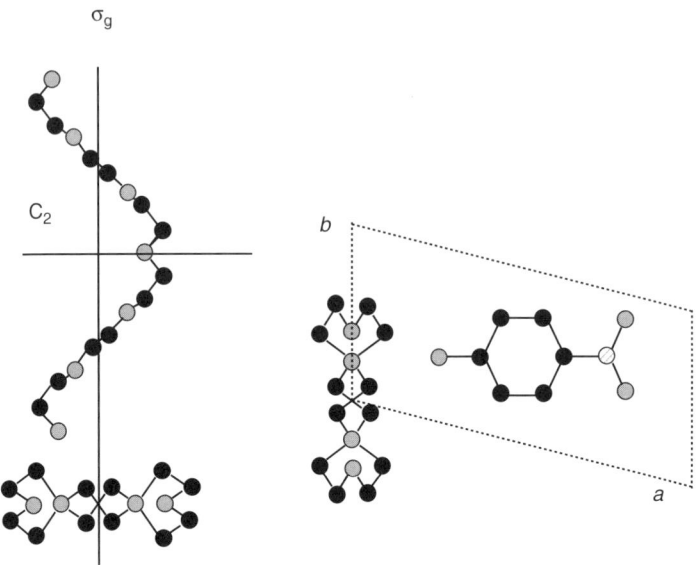

Fig. 14.19. Right: Crystal unit cell of the stable α form of the PEO/*para*-nitrophenol complex as seen parallel to the chain axis; **left:** chain conformation. (After Damman and Point 1994.)

a triclinic unit cell with parameters $a = 1.172$ nm, $b = 0.555$ nm, $c = 1.557$ nm, $\alpha = 90.7°$, $\beta = 87.1$, and $\gamma = 104.0°$ (Figure 14.19). From FTIR investigations, these authors have observed that the characteristic bands for PEO under the 7_2 helical conformation are conspicuously absent, while new bands appear. This therefore suggests that PEO chains take on an unusual conformation. This conformation was worked out by the same authors in a more recent paper wherein they assumed that the factor group of the PEO chains is C_{2h}. This led them to consider only four torsion angles to determine completely the chain conformation, which is then $ttgttgttttg'ttg'ttt$ (also noted, a $t_2gt_2gt_3t_2g't_2g't_3$ arrangement) (see Figure 14.19), the torsion angles corresponding to 180° for t (trans), 60° for g (gauche) and 300° for g' *(gauche)* (Damman and Point 1994).

It is worth emphasizing that both *ttg* and *ttt* arrangement are stable conformations for PEO, as shown by Smith et al. (1994). This conformation of PEO chains in the PEO/*para*-nitrophenol complex was further confirmed by a sophisticated NMR study carried out by Harris et al. (2000) with slight corrections of the torsion angles from the exact *trans*.

Damman and Point have also observed, as with the *PEO/resorcinol* complex, that a metastable form, designated as the β *form*, can be grown at low temperature (Damman and Point 1995). This form melts at a temperature some 30°C lower than the stable α *form*, yet the stoichiometry remains unchanged, namely 2 *para*-nitrophenol molecules for every 3 PEO monomer units. These authors have measured the spherulite growth rate as a function of the crystallization temperature (Figure 14.20). At low temperature

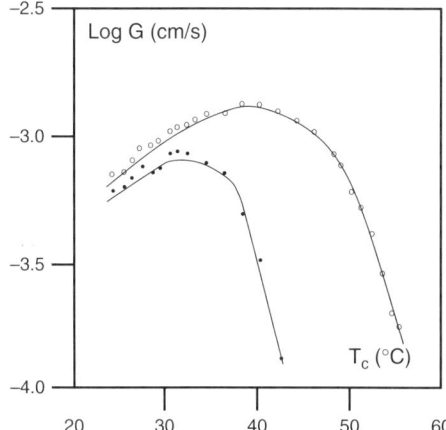

Fig. 14.20. Growth rate of the stable α form of PEO/*para*-nitrophenol complex (o) and the metastable β form of PEO/*para*-nitrophenol complex (●). (After Damman and Point 1995.)

both forms grow at approximately the same pace, although the *β form* transforms rapidly into the *α form*.

From their FTIR investigations Damman and Point have concluded that the PEO conformation differs from that taken on in the stable *α form*. Indeed, PEO vibrations that characterize the $t_2gt_2gt_3$ arrangement, such as 457 cm^{-1}, totally disappear, while new modes at unusual frequencies are observed, in particular at 667 cm^{-1}. The crystal structure remains unknown to date, owing to the high metastability of the *β form*.

CHAPTER 15

Isotactic polystyrene

Isotactic polystyrene (*iPS*) was first synthesized by Natta's group in the late fifties by using heterogeneous catalysts known as Ziegler-Natta catalysts (Natta et al. 1960). Although the raw material contains a significant amount of atactic polystyrene (about 20 to 30%), which probably arises from the thermal polymerization of styrene as the reaction is commonly carried out at 80°C, it shows a very high isotacticity once properly purified (usually above 98% isotactic content). Natta was awarded the 1963 Nobel Prize in Chemistry for the synthesis of a series of isotactic polymers, of which polystyrene is one example.

Isotactic polystyrene is a slowly crystallizing polymer with a low degree of crystallinity (50% is the usual upper limit in the bulk state) with a complex annealing behaviour, although only one helical conformation, the 3_1 helix and one crystalline lattice, is known thus far in the solid state (hexagonal lattice with $a = b = 2.19$ nm, $c = 0.665$ nm, space group *R3c*). It can be obtained very easily in the amorphous state by a rapid quench to below the glass transition temperature. Samples a few millimetres thick can be routinely processed. The thickness of chain-folded crystals, and the corresponding melting temperature, depends strongly upon the annealing temperature (Overbergh et al. 1977a, b). The extrapolated melting point is found to be about 250°C (Berghmans et al. 1979). Unlike other polymers, such as poly[methyl methacrylate], the glass transition temperature is little dependent upon the tacticity ($T_G = 95 \pm 5°C$).

The occurrence of iPS–solvent molecular compounds was first noticed in fibrillar thermoreversible gels, and in fact constitute the very essence of these gels.

Thermoreversible gels

Most of the studies on isotactic polystyrene have primarily been focused on the crystallization behaviour from the bulk state (e.g. see Berghmans et al. 1979). The interest shown in the physical gelation of this polymer was triggered by a study by Keller and co-workers that revealed the existence of an unknown molecular structure (Girolamo et al. 1976). Although, as will be seen in what follows, the issue was rather controversial, the paper by Girolamo et al. should be regarded as the starting point of the new development in the field of thermoreversible gelation. Some 20 years after its publication, much work has been done and any unusual behaviour or obscure results, of which there have been many, have now received coherent explanations. Also, Sundararajan and co-workers first showed the existence of iPS–solvent compounds, a notion that turned out to be decisive for the understanding of thermoreversible gels (Sundararajan and Tyrer 1982; Sundararajan et al. 1982). Therefore, a historical presentation in chronological order of the knowledge gathered on these gels, albeit informative, would be confusing. To gain a better insight into the overall picture, it is believed that the thermodynamics of the gelation process (phase diagrams) should be presented prior to discussing the molecular structure. In doing so the latter should appear much clearer to the reader.

Thermodynamics: temperature–concentration phase diagrams

Thermoreversible gels are formed in a domain of concentration ranging from 5 to 30%, although this statement will rely upon the way scientists are willing to define a gel. Girolamo and colleagues have reported how a gel can be produced from solution in decalin (decahydronaphthalene). If the solution is cooled very slowly or kept well above room temperature, chain-folded crystals (spherulites) will form. The resulting system is usually strongly turbid and crumbly. Conversely, if the system is cooled down more rapidly to below a certain temperature, one observes a *prise en masse* of the solution: a transparent, slightly bluish entity is formed whose aspect is quite reminiscent of biopolymer gels such as aqueous agarose gels. This entity can be extracted from the preparation vessel while retaining its as-prepared shape, which confirms its gel status.

While early studies considered the solvent as a mere diluent, it was soon realized that its role in the gelation process was more important, and more subtle, than previously thought. This was particularly noticeable for decalin, two conformers of which exist: *cis*-decalin (referred to as a *boat* conformation) and *trans*-decalin (referred to as a *chair* conformation). The transformation from one conformer into the other is physically impossible without breaking covalent bonds. Varied experimental evidence has shown that the gelation phenomenon is strongly dependent upon the conformer used (Atkins et al. 1984; Guenet and McKenna 1988).

The effect of the solvent is particularly well evidenced by the temperature–concentration phase diagrams established in either of the decalin conformers. Guenet and McKenna (1988) have established two types of diagrams: the *gel formation diagram* as determined from the formation exotherms (at a cooling rate of $-5°C/min$) and the *gel melting diagram* as obtained from the melting endotherms (extrapolated to $0°C/min$).

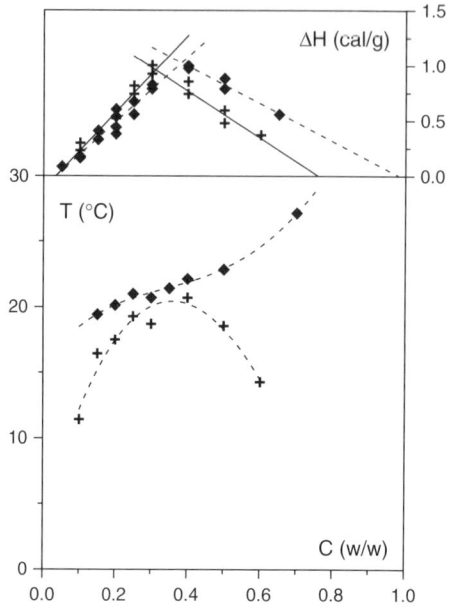

Fig. 15.1. Formation phase diagrams and the corresponding Tamman diagrams for iPS/*cis*-decalin solutions (+) and iPS/*trans*-decalin solutions (♦). (Data from Guenet and McKenna 1988.)

The formation phase diagrams together with the corresponding Tamman diagrams are given in Figure 15.1. As can be seen, they differ essentially in two aspects:

1. The formation temperatures are slightly higher in *trans*-decalin than in *cis*-decalin. Yet the main feature is the occurrence of a maximum in *cis*-decalin for $C_{pol} \approx 30\%$ (w/w), unlike what is seen in *trans*-decalin.

2. In both cases the variations of the enthalpies associated with the exotherms (Tamman's diagrams) exhibit a maximum, yet for distinct concentrations: $C_\gamma = 30\%$ (w/w) in *cis*-decalin against $C_\gamma = 40\%$ (w/w) in *trans*-decalin.

These discrepancies cannot be interpreted in the framework of Flory's theory (Flory 1953). In this theory, the polymer–solvent interaction parameter (χ_1) governs the melting point of a crystalline system wherein polymer crystals are surrounded by solvent molecules. The melting point depression due to the presence of solvent molecules is written (see also Chapter 1):

$$\frac{1}{T_m} - \frac{1}{T_m^0} = Cst \times (v_s - \chi_1 v_s^2) \quad (15.1)$$

where T_m and T_m^0 are the melting temperatures of the sample containing a volume fraction v_s of solvent and in the pure state, respectively. Equation (15.1) indicates that the better the solvent, the higher the melting point depression. It must be stressed that this relation is derived by using mean-field approximation, which means that solvent molecules are

Isotactic polystyrene

approximated to spheres and polymer chains to necklaces of spheres. The actual solvent and polymer microstructures are therefore not taken into account.

Under these conditions, as the value of χ_1 is virtually the same for *cis*-decalin and *trans*-decalin, one should not observe any significant difference, in contradistinction with the experimental facts.

Guenet and McKenna (1988) approached the question in a different way. According to these authors, the shape of the formation diagrams as well as the evolution of the formation enthalpies suggests the occurrence of polymer–solvent compounds whose thermal behaviour is strongly dependent upon the conformation of the solvent. In terms of the compound, the maxima as measured from the Tamman diagram provide two markedly different stoichiometries, i.e. 1.75 *cis*-decalin molecules/monomer unit against 1.15 *trans*-decalin molecules/monomer unit.

The melting phase diagrams together with the corresponding Tamman diagrams obtained on heating are shown in Figures 15.2 and 15.3. As can be seen, the shapes of these diagrams are quite reminiscent of the formation phase diagrams presented in Figure 15.1. Similarly, the maxima of the melting enthalpies give the same stoichiometries.

From the shape of the phase diagrams it can be stated that the iPS/*cis*-decalin system is a *congruently melting compound* while the iPS/*trans*-decalin system corresponds to *a compound with a singular point* (see Chapter 1). Interestingly, the melting points differ significantly: gels from *trans*-decalin melt some 25°C above those from *cis*-decalin. Such

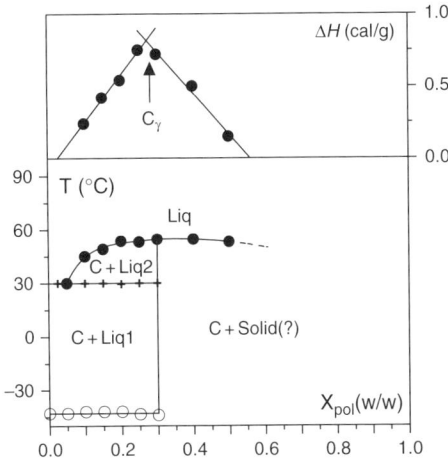

Fig. 15.2. Melting T–C phase diagram together with the corresponding Tamman diagram for iPS/*cis*-decalin gels prepared by a rapid quench to 0°C. The non-variant event at $T = 30°C$ is due to a monotectic transition indicating that the system was quenched within the miscibility gap. The non-variant event at $T = -45°C$ is the solvent melting temperature. (Data from Guenet and McKenna 1988.)

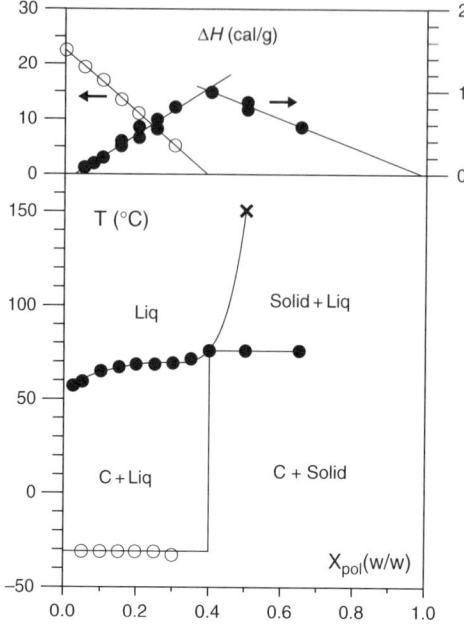

Fig. 15.3. iPS/*trans*-decalin gels: melting T–C phase diagram together with the Tamman diagram for the solvent melting (T_{solv}), which is also a way of determining compound stoichiometry. (Data from Guenet 1986; Guenet and McKenna 1988.)

a fact is exactly what is expected from the stoichiometry since the rule stipulates that, all things being equal, the higher the solvation of the compound the lower its melting point.

It is worth mentioning that the occurrence of a temperature-non-variant line can be clearly observed in the case of *cis*-decalin, which corresponds to a *monotectic transformation*. The existence of such a line implies that a liquid–liquid phase separation is involved in the gelation process. It should be stressed, however, that gelation is not due to, but has simply been interfering with, this type of phase separation (see above). Owing to the broad endotherm in *trans*-decalin, the same line is not seen although it should also exist.

These phase diagrams indicate that polymer and solvent molecules together build up molecular structures (solvated structures) that are responsible for the formation of a macroscopic gel. Also, they highlight the fact that a slight change of solvent conformation has a significant effect on the thermodynamic properties. Finally, one should notice the low values of the melting enthalpies. This may suggest that only a very small part of the chains participate in the organized structures, or it may convey something else. These comments will be of importance in elucidating the molecular structure.

The equilibrium swelling behaviour of gels as determined after immersion in an excess of preparation solvent has unexpectedly given confirmation of the stoichiometries deduced

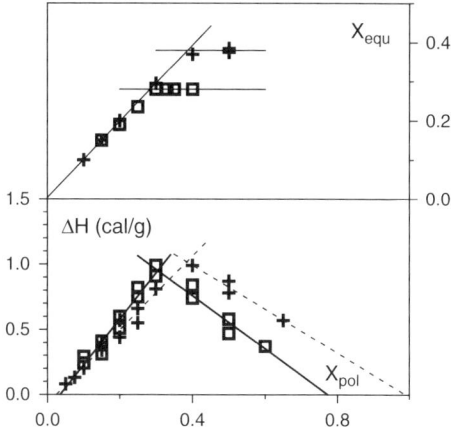

Fig. 15.4. Upper: swelling behaviour of iPS/*cis*-decalin gels (o) and iPS/trans-decalin gels (□); X_{equ} is the concentration (w/w) reached after swelling to equilibrium. **Lower:** the corresponding melting enthalpies. (Data from Guenet and Klein 1989 and Guenet and McKenna 1988.)

from the phase diagrams. As is shown in Figure 15.4, where the equilibrium concentration is plotted as a function of the starting concentration, one can distinguish two domains. Below a given starting concentration, X_s, no swelling occurs, while above X_s the system swells in such a way as to always reach the same concentration, X_∞. It turns out that $X_s \approx X_\infty$ but, more interestingly, this corresponds within experimental uncertainties to the stoichiometric concentration C_γ (see Figure 15.4). It suffices to rescale the polymer concentration by the stoichiometric concentration to obtain the same behaviour in both decalin conformers. Clearly, this type of behaviour differs markedly from that observed in chemical gels. In the latter, the swelling properties are monitored by the osmotic contribution to the free energy which obliges the flexible chains to expand (this is sometimes referred to as the C* theorem; de Gennes 1979). In the case of thermoreversible gels, where all conformations are frozen in (rigid gels), there is no such contribution. Conversely, any compound for which the stoichiometry is not fulfilled will absorb solvent so as to maximize its entropy (see Chapter 1). This can be summarized as follows: for $C < C_\gamma$ no swelling should occur as the amount of solvent exceeds the stoichiometry, while for $C > C_\gamma$ there appears a solvent deficit which will be reduced by solvent absorption until the concentration reaches C_γ. The swelling behaviour unexpectedly confirms the existence of compounds with differing stoichiometries in *cis*- and *trans*-decalin, but also indirectly highlights the rigid nature of the resulting physical network.

Other results provide additional support for the polymer–solvent compound concept. For instance, gels prepared from 1-chlorodecane (a solvent with a χ_1 parameter similar to that of decalin) exhibit a peculiar behaviour on melting (Guenet et al. 1985). The shape of the DSC thermogram is strongly dependent upon the heating rate: at a high heating rate a low-melting endotherm is predominant ($T_{low} \approx 75°C$), while a high-melting endotherm ($T_{high} \approx 150°C$) gradually appears to the detriment of the former as the heating rate is lowered. This behaviour is straightforwardly accounted for by considering an *incongruently melting* compound. At a high heating rate one can observe the metastable melting point before the incongruent melting (see Chapter 1). Conversely, lower heating rates gradually reveal the 'equilibrium' melting point, which corresponds to the solid phase (non-solvated crystals).

Morphology and molecular structure

Gels from isotactic polystyrene are fibrillar networks. A typical AFM picture is shown in Figure 15.5. The microfibrils possess cross-sectional dimensions of about 10 to 20 nm while the mesh size is about 0.1 to 1.0 μm depending upon polymer concentration (Atkins et al. 1984; Guenet et al. 1985).

The reason why such a morphology is produced in lieu of the usual suspensions of spherulites remained unclear for some years. As it was unduly thought that gels could only be produced from poor solvents and by a rapid quench to low temperature, the suggestion was made that a spinodal decomposition could well account for the experimental facts, as had already been proposed for agarose gels (Feke and Prins 1974).

Spinodal decomposition is one of the two possible mechanisms involved in a liquid–liquid phase separation that occurs when a system is quenched within a miscibility gap (see Chapter 1). The main reason for contemplating such a mechanism arose from the theory that predicted the occurrence of a network structure for the polymer-rich phase at the early stage of spinodal decomposition (Cahn and Hilliard 1959, 1965). This normally transient network, which would soon vanish under normal conditions, had therefore to be frozen in either by crystallization or by some glass transition process to produce a gel (Tan et al. 1983; Arnauts and Berghmans 1987). This view was widespread in the domain of ceramics, the latter effect being said to reveal their peculiar morphology (Zarzycky 1970). This meant that gelation should be systematically kinetically controlled (Girolamo et al. 1976; Arnauts and Berghmans 1987): high cooling rates should yield a gel, low cooling rates chain-folded crystals and intermediate rates a mixture of both.

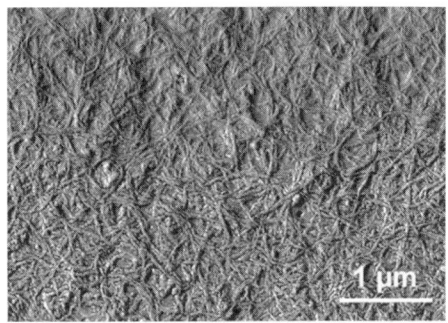

Fig. 15.5. AFM picture of a gel prepared in *cis*-decalin ($C_{pol} = 0.04\,\text{g/cm}^3$). (From Ladjyn and Guenet, unpublished.)

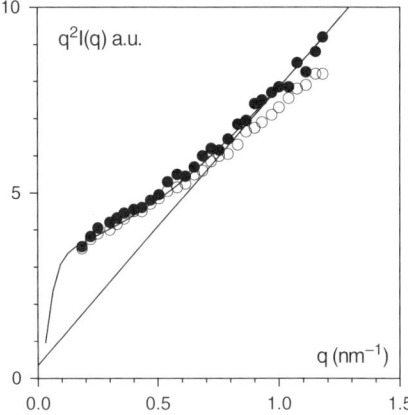

Fig. 15.6. Scattering curve for deuterated chains (C = 1% (w/w); $M_w = 2.5 \cdot 10^5$) in a 15% hydrogenous gel (w/w) in *cis*-decalin. (o) = chains in the gel state; (•) = chains in the molten state at 66°C. The full line has been calculated with the pseudo-analytical equation of Yoshisaki and Yamakawa (1980) by taking $l_p = 4$ nm and $L = 500$ nm. The straight line represents the asymptote for $ql_p > 1$ (see Chapter 2 for details). (Data from Klein et al. 1990.)

As will be discovered in the next section, gelation is a thermodynamically controlled phenomenon. Provided that the system is quenched below the so-called gelation threshold, a gel and only a gel will be produced in spite of very slow gelation kinetics.

In fact, it was suggested by Guenet (1987) that the fibrillar morphology originated essentially in the unusual chain conformation taken on by the chains in these 'special' solvents. Chain conformation determined in the gel state by small-angle neutron scattering for iPS/*cis*-decalin systems obeys worm-like statistics (Figure 15.6). Therefore, chains are locally rigid, with a persistence length about four times longer than in the usual flexible state, but are otherwise globally Brownian. For instance, the radius of gyration still varies as $M^{1/2}$ but is twice as large as that of iPS flexible chains (Guenet et al. 1979). According to Guenet (1992, 1996a), *fibrous gels are produced rather than chain-folded crystals because the chains' enhanced rigidity prevents them from folding*. Here, it is worth comparing iPS chain conformation in gel with that of agarose chains, which are intrinsically rigid as discussed in Chapter 7. Agarose does not produce any chain-folded crystals but fibrous gels only. Clearly, chain rigidity is the key factor in the making of iPS gels. Yet, what makes iPS chains more rigid than they usually are in their unperturbed state? The study of the chain conformation as a function of temperature, concentration and solvent type allows one to throw some light on this question.

The above statement has been given additional support from a study of the chain conformation that exists just after gel melting. Unlike what takes place on melting chain-folded crystals, where chains very rapidly recover their Gaussian, flexible conformation (Barham 1993), chains still retain their worm-like statistics after gel melting (Figure 15.6). The solution must be heated well above the melting point (up to 95°C; Klein et al. 1990a) to obtain the usually found flexible chains. Of further note is the invariance of chain conformation with the quenching temperature. Klein and colleagues have therefore concluded that, while spinodal decomposition may interfere, it is not responsible for gelation. Again, chain rigidity is the key factor.

The chain conformation is also dependent upon whether *cis*- or *trans*-decalin is used (Guenet 1987), and upon polymer concentration (Klein et al. 1990a). In a mixture of *cis*- and *trans*-decalin (designated as *cis/trans*-decalin), although the radius of gyration also is seen to increase by about twofold as in *cis*-decalin, the $1/q$ scattering is replaced by a $1/q^2$ behaviour instead. The loss of the $1/q$ behaviour can be explained by considering a cross-sectional effect. In fact, scattering in this domain of transfer momenta q ($q = (4\pi/\lambda) \sin \theta/2$, with λ = radiation wavelength and θ = scattering angle) is written for a cylindrical object of cross-sectional radius r for $qr < 1$ (see Chapter 2):

$$q^2 I(q) \propto 4q\pi \times \left(1 - \frac{q^2 r^2}{4}\right) \qquad (15.2)$$

If we are dealing with a hollow cross-section with inner radius r, one obtains:

$$q^2 I(q) \propto 4q\pi \times \left[1 - \frac{q^2 r^2 (1 + \gamma^2)}{4}\right] \qquad (15.3)$$

The second term in equations (15.2) and (15.3) contains the cross-section effect. For large enough radii, the $1/q$ behaviour can be smeared, and the more so in the case of a hollow cross-section. The scattering function may eventually display an apparent $1/q^2$ behaviour in this q-range which is not at all related to the behaviour of flexible, Gaussian chains.

This possibility was contemplated by Guenet, who employed the term '*distorted statistical unit*' (Guenet 1987). We shall discover in what follows that this cross-sectional effect is most probably linked to the special helical structure occurring in these gels.

The loss of the $1/q$ behaviour is also observed in *cis*-decalin for concentrations larger than 30%, i.e. above the stoichiometric concentration (Klein et al. 1990a). Decreasing the degree of solvation of the compound entails modification of the local structure. Klein and colleagues had also to take into account a cross-sectional effect to reproduce theoretically the intensity scattered in the intermediate domain and the value of the radius of gyration.

The thermodynamic study has revealed in *cis*-decalin the existence of a maximum in the melting temperature and in the melting enthalpy as a function of polymer concentration. Interestingly, the same behaviour is observed for the chain radius of gyration (Klein et al. 1990a). The largest radius is seen at a polymer concentration corresponding to the compound stoichiometry (i.e. 0.3 w/w). Fits of the scattering curves in the intermediate domain with theoretical models derived by Muroga (1988) suggest that rod-like portions of nearly 40 nm are present at the stoichiometric concentration (see Chapter 2). The radius of gyration then decreases as a solvent deficit appears beyond the stoichiometric composition. The theoretical fits suggest, as expected, the appearance of a significant number of Gaussian-like sequences on the chain together with the presence of very long rod-like structure (estimated length from the fits is about 70 nm). Provided that the fits are relevant and realistic this shows that, unlike what would be intuitively expected, desolvation is not random along the chain but seems to occur on well-defined sequences thus still allowing the existence of rigid structures. Needless to say that such a behaviour utterly differs from what is usually observed for polymer solutions (Daoud et al. 1975).

Again, these results undoubtedly emphasize the special role of the solvent, particularly in promoting formation of rigid, rod-like structures. While these local structures seem to differ slightly depending on whether *cis-* or *trans*-decalin is used, the desolvation process (i.e. increasing polymer concentration) gives similar scattering patterns, which suggests the appearance of a common molecular morphology.

Until the investigation by Girolamo et al. (1976) the only known helical structure of isotactic chains was the 3_1 form (Figure 15.7). These authors found that diffraction patterns taken from iPS gels did not reveal any feature of this helix. Here, it must be emphasized that these diffraction patterns were obtained on 'dried' and stretched gels. An unusual meridional reflection corresponding to a distance of 0.51 nm was observed instead. This led Girolamo et al. to postulate the existence of another helical form, the 12_1 helix (Girolamo et al. 1976; Atkins et al. 1984) (see Figure 15.7) which was shown to be energetically stable (Corradini et al. 1980). This 12_1 helix is generated by *tt* arrangements of the monomers instead of the *tg*-type regular sequences in the 3_1 helix. The occurrence of this helical form was questioned by Guenet (1986) on the basis of neutron diffraction data, and later by Nakaoki and Kobayashi (1991) on the basis of infra-red (IR) spectroscopy experiments. Guenet reported that the 0.51 nm reflection could not be seen in the nascent gel. Nakaoki and Kobayashi have studied

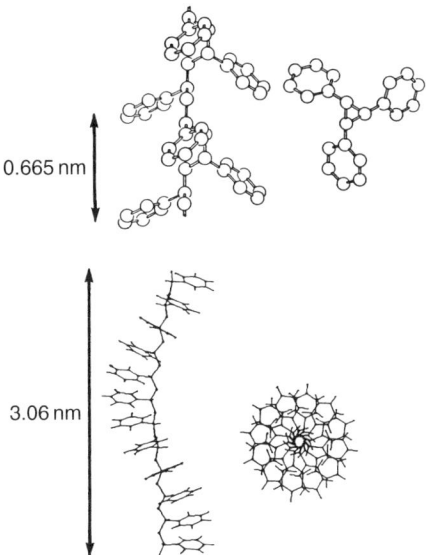

Fig. 15.7. The 3_1 helix (**upper**) and the 12_1 helix (**lower**) seen perpendicular to their axis (**left**) and parallel (**right**). Pitch as indicated. (From Sundararajan et al. 1982. Reprinted with permission from ACS.)

the conformational ordering during gelation and have observed that the IR bands whose intensity increases are those corresponding to the iPS crystalline state, i.e. those due to the tg sequences. They have concluded that the small difference in frequency for the 923 (918) and the 899 (896) cm^{-1} between the crystalline and the gel phase together with the appearance of additional bands at 1070 and 1062 cm^{-1} indicate strong polymer–solvent interactions, a conclusion in agreement with the phase diagrams established by Guenet and McKenna (1988). These authors therefore have suggested reconsidering more carefully the possible occurrence of the 3_1 helix (Guenet 1986; Nakaoki and Kobayashi 1991; Nakaoki et al. 2002). Chatani has proposed a somewhat modified version of the 3_1 form: a coiled-coil where the minor helix is the 3_1 form (Chatani 1993; Chatani and Nakamura 1993). The fibre period is 3.06 nm, contains 18 monomers and possesses an outer radius of 0.63 nm and an inner radius of about 0.2 nm. The large helix is a 6-fold helix as the basic elements are trimers taking on a 3_1 conformation. Selection rules indicate that this helical form also accounts for the 0.51 nm meridional reflection, which therefore corresponds here to the axial rise of the large helix (see Chapter 2).

The coiling of the 3_1 form turns out to be a way of explaining the existence of a cross-sectional effect as revealed by the small-angle neutron scattering experiments in cis/trans-decalin and in more concentrated samples in cis-decalin. The coiled-coil described here is equivalent to a cylinder with a hollow cross-section at the level of resolution of the experiments. In fact the '*distorted statistical unit*' reported by Guenet was nothing but the effect of coiling (Guenet 1987). This further indicates that coiling is most probably absent in nascent gels produced from cis-decalin below the stoichiometric composition but appears above it. Therefore, the drying process of the gel eventually produces the same helical structure. This highlights again that the way in which the drying procedure is achieved may give rise to different molecular structures.

Clearly, the 3_1 form and its coiled-coil version allow one to account far better for the body of results and data gathered on these gels than the 12_1 form does, particularly as regards the formation of a polymer–solvent compound. From a neutron diffraction study on the nascent gels Guenet noticed that the diffraction pattern of liquid decalin also exhibits a maximum at 0.51 nm associated with the distance between first neighbouring molecules (Guenet 1986). This distance corresponds approximately to $d \approx 0.63$ nm ($d \approx 1.23 \times 0.51$), a value which happens to be close to the pitch P of the 3_1 helix ($P = 0.665$ nm).

To put it another way, iPS chains take on a 3_1 conformation in cis-decalin because this helical form adapts better to the solvent structure. As a result this conformation can be kept, even at higher temperature. On these bases Guenet and co-workers (Klein et al. 1990a, 1991) put forward the *ladder-like model*, which describes the short-range structure of the gel (Figure 15.8). In this model the 3_1 helix is stabilized by insertion of decalin molecules within the cavities created by the phenyl rings, and these stabilized helices simply align so as to take on an arrangement reminiscent of the nematic order (i.e. order in only one direction).

Possibly, the existence of a coiled-coil in cis-trans-decalin or in cis-decalin at high concentration arises from a 'warping' effect caused by the insertion of solvent molecules within these cavities. The 'nematic-like' order in the ladder-like model is borne out by neutron scattering experiments where only the chains are deuterated (Figure 15.9)

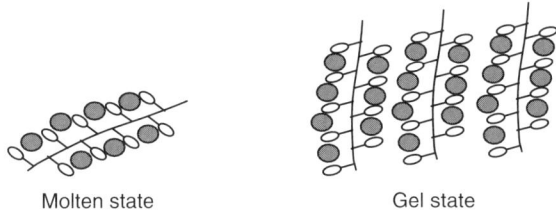

Fig. 15.8. Depiction of the ladder-like model (**left**). The chain adopts the 3_1 helical form because this conformation is in register with the solvent 'order'. Everything takes place as if the helix were stabilized by solvent molecules (black balls) housed between the phenyl groups. On cooling, these structures align (**right**) to produce a gel with a nematic-like order.

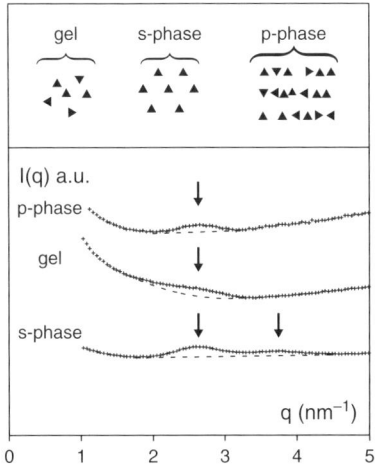

Fig. 15.9. Lower: neutron diffraction patterns for the gel ($T_{an} = 17°C$), the s-phase ($T_{an} = 25°C$ and the p-phase ($T_{an} = 35°C$). The reflections are indicated by arrows. **Upper:** models for interpreting these curves; triangles represent those helices seen parallel to their axis. (Data from Guenet et al. 1994.)

(Guenet et al. 1994). Only one maximum is detected at $q = 2.6 \text{nm}^{-1}$, which most probably reflects the distance between neighbouring chains since the reflections due to the coiled-coil 3_1 form proposed by Chatani and co-workers would occur at $q = 5$ and $q = 13.3 \text{nm}^{-1}$ (they correspond to the first layer line where $n = 1$; see Chapter 2). This distance of about 2.4 nm is quite consistent with the spacing of solvated chains under the 3_1 form (the radius of the 3_1 form is circa 0.3 nm, and the diameter of a decalin molecule is circa 0.5 nm, so a minimum distance of about 1.6 nm between helix axes is expected).

Recent experiments carried out on the nascent gel state by means of fluorescence spectroscopy are also consistent with the presence of a 3_1 helix (Itagaki and Takahashi 1995; Itagaki and Nakatani 1997; Itagaki 2001). In fact excimer formation decreases markedly when a solution in either *cis*- or *trans*-decalin undergoes gelation below the gelation threshold. If the 12_1 helix were forming, then excimer emission would be enhanced as this helix consists of near-*tt* conformers. Conversely, excimer formation is restricted very much in the case of a 3_1 or near-3_1 helix. Additional experiments with fluorescent probes (naphthalene, NP; and 1-methyl naphthalene, MN) give support to the occurrence of a polymer–solvent compound. The measure of anisotropy gives information as to the mobility of the probes. Fluorescence anisotropy reads:

$$r = (I_p - GI_v)/(I_p + 2GI_v) \qquad (15.4)$$

where I_p and I_v stand for the intensities measured when the observing polarizer is parallel and perpendicular, respectively, while G is an instrument constant.

In the case of NP, anisotropy increases with increasing iPS concentration, which suggests that the motion of this probe is strongly impeded (Figure 15.10). Conversely, in the

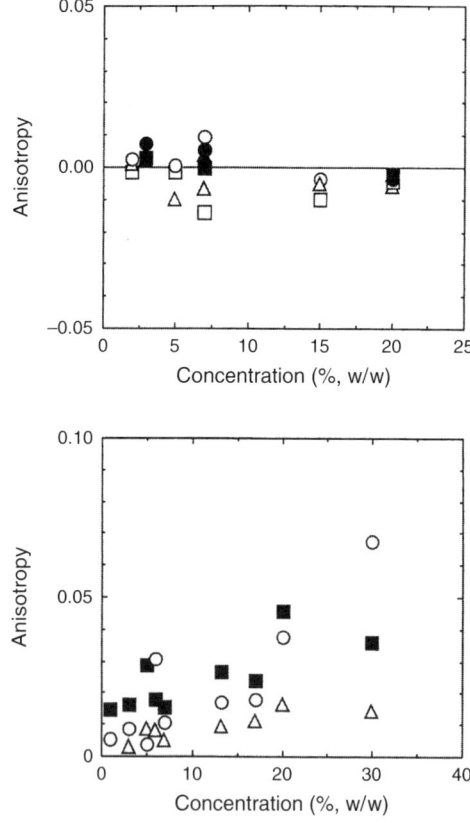

Fig. 15.10. Upper: anisotropy of 1-methyl-naphthalene in iPS/*cis*-decalin gels at 25°C. **Lower:** anisotropy of naphthalene in iPS/*cis*-decalin gels at 25°C Polymer concentration in w/w. Values obtained by exciting at 302 nm (o), 281 nm (Δ), and 257 nm (□). (After Itagaki and Takahashi 1995; Itagaki and Nakatani 1997; Itagaki 2001. Reprinted with permission from ACS.)

case of MN, no anisotropy is observed, which entails that the mobility of this probe is not affected. According to Itagaki and co-workers, the restricted mobility of NP indicates that this probe is trapped together with a part of the solvent molecules between polymer chains thanks to the existence of a polymer–decalin compound. MN molecules are free because of their size which prevents them from accessing the cavities described above. These experiments give support to the existence of a compound, but also to the model portrayed in Figure 15.8. They also emphasize that compound formation is size-dependent.

The *ladder-like model* model also accounts for most of the experimental facts observed for these gels, in particular the existence of polymer–solvent compounds whose stoichiometry and thermal behaviour is strongly dependent upon the solvent shape and size. In this respect, the 12_1 helix does not offer similar cavities, so compound formation should not be so solvent-dependent. While the 12_1 helix is energetically as probable as the 3_1 form in a vacuum (Corradini et al. 1980), this is apparently no longer so in a solvent environment.

This model is also consistent with the high solvent mobility as discovered by NMR (Pérez et al. 1988) and, as will be seen in the next section devoted to rheological properties, the high stress relaxation rates observed when a gel is submitted to deformation (Guenet and McKenna 1986).

Rheological properties

The rheological properties have been studied in detail by Guenet and McKenna 1986 and McKenna and Guenet (1988a) in *cis*-decalin, *trans*-decalin and 1-chlorododecane. These gels do not possess an equilibrium elastic modulus. Upon deformation they show considerable stress relaxation. In the investigation time range the stress relaxation can be approximated to the following form:

$$\sigma_R \propto t^{-m} \tag{15.5}$$

with $m \approx 0.1$ to 0.2.

These values are far larger than those observed for chemical gels (e.g. see Janacek and Ferry 1969a, b). Yet, such values do not come as a surprise in the light of the local molecular structure of the gels. To be sure, the limited order between chains does not favour a high degree of cohesion within and between the fibres. Gel junctions can be readily altered under deformation or stress.

The magnitude of the modulus depends markedly upon the solvent type, as well as its variation as a function of polymer concentration (Figure 15.11). In a double-logarithmic representation this variation shows rapid departure from linearity in *cis*-decalin, unlike what is seen in 1-chlorododecane. Here, the existence of polymer–solvent compounds of differing stoichiometries must be taken into account as the **relevant parameter** is no longer the polymer volume fraction **but the volume fraction of the phase responsible for the elastic properties**, i.e. the **polymer-rich phase**. Provided that the polymer concentration of the polymer-poor phase is low enough, the fraction of polymer-rich

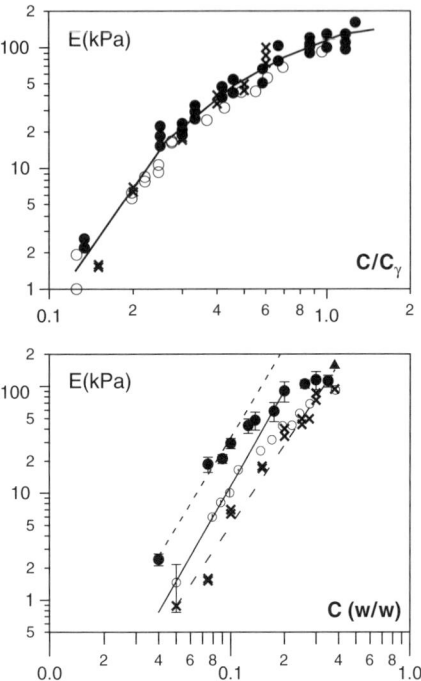

Fig. 15.11. Bottom: variation of the isochronal modulus at 120s as a function of polymer concentration for iPS/*cis*-decalin (● and ■), iPS/*trans*-decalin (○), iPS/1-chlorodecane (×). **Top:** variation of the same modulus for the three solvents now plotted as a function of C/C$_\gamma$ where C$_\gamma$ is the stoichiometric concentration (0.3 for iPS/*cis*-decalin, 0.4 for iPS/*trans*-decalin and 0.5 for iPS/1-chlorodecane. (From Guenet 1986. Modulus data after McKenna and Guenet 1988a.)

phase is simply derived by rescaling the polymer volume fraction by the polymer volume fraction under the stoichiometric conditions ($\varphi/\varphi_\gamma \sim C_p/C_\gamma$). In so doing, a master curve is obtained (Figure 15.11, upper), which shows that these gels basically possess the same structure. The observation of the master curve represents therefore a substantial confirmation of the existence of compounds, in particular accounting for the discrepancy between *cis*- and *trans*-decalin (moduli are about 4 times larger in the former than in the latter).

Similarly, the reason why the curves depart from a power variation is easily accounted for in the framework of polymer–solvent compounds. At the stoichiometric composition, the system behaves as a pure component, i.e. there is no longer free solvent. As a result, the notion of a network on which theories are derived no longer applies, and again, the relevant parameter is the rescaled concentration Cp/C_γ. Clearly, we cannot expect to observe a power-law variation up to $Cp/C_\gamma = 1$.

In the concentration domain where a power law is observed, an exponent of $\alpha \approx 2.3 \pm 0.2$ is found. If this exponent is interpreted in the framework of Jones and Marquès' theory (Jones and Marquès 1990), this leads one to consider that the fibres possess longitudinal fractal dimensions of $D_F = 1.18$ if the system is governed by *enthalpic elasticity* or

$D_F = 1.7$ if *entropic elasticity* dominates (see Chapter 4). As was discussed above, the fact that these gels do not swell at concentrations below the stoichiometric concentration rather suggests that **enthalpic elasticity** is involved. A longitudinal fractal dimension of $D_F = 1.18$ for the gel fibres implies that they cannot be straight, which is in line with the morphology as observed from electron microscopy.

Assemblies of spherulites

As was mentioned above, the gelation threshold can be located at a well-defined temperature T_{gel}: for $T < T_{gel}$ a gel and only a gel is formed while for $T > T_{gel}$ an assembly of spherulites is obtained. A striking case turns out to be iPS/*cis*-decalin, as was reported by Klein et al. (1990b). Here we have, however, something quite peculiar in the sense that the behaviour of the system as a function of the cooling rate can be very deceptive and can lead one to believe that gelation is a kinetically controlled phenomenon. If this were so, one should observe a competition between the formation of the chain-folded crystals and that of the gel. In fact, when the system is quenched just below T_{gel} a gel and only a gel is formed in spite of a very slow formation kinetics. Just above T_{gel} assemblies of spherulites are obtained. As can be seen in Figure 15.12, depending on the temperature

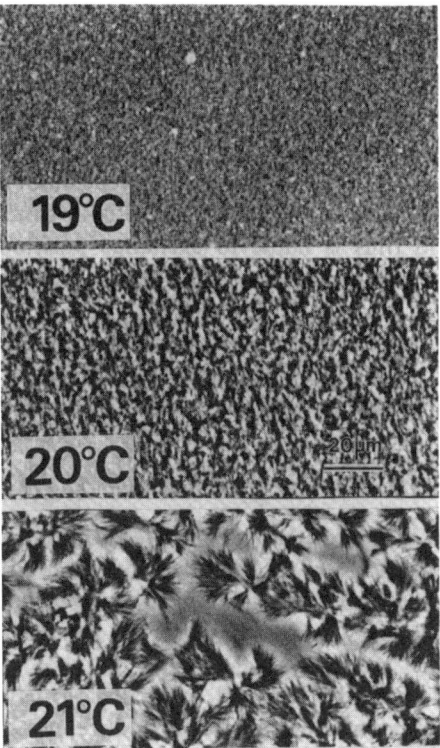

Fig. 15.12. Typical morphologies on either side of the gelation threshold ($T_{gel} = 20 \pm 1°C$, $C_{pol} = 0.3$ w/w) in *cis*-decalin. (After Klein et al. 1990b. Reprinted with permission from ACS.)

at which the system is cooled, the texture changes within 2°C from a salt-and-pepper aspect (owing to the presence of fibres) for $T = T_{gel} - 1°C$ to a spherulitic morphology for $T = T_{gel} + 1°C$.

A sharp change is also observed at the level of the molecular structure as determined from neutron diffraction (see Figure 15.9). In the gel state, at $T = T_{gel} - 1°C$, only one maximum can be seen in the diffraction curve against two maxima for those spherulites produced at $T = T_{gel} + 1°C$. While the solution must be cooled rapidly to reach T_{gel}, which may convey the impression that gelation is kinetically controlled, the occurrence of such dramatic changes at T_{gel} (different morphologies and molecular structures), implying the existence of a sharp boundary, definitely makes it a thermodynamically controlled phenomenon.

It is worth mentioning that other cases of spherulitic assemblies consisting of solvated crystals have been reported by Sundararajan and co-workers for iPS/hexahydroindane (Sundararajan et al. 1982). Yet, a thorough study as that detailed in what follows has not been performed so far on these systems.

Thermodynamics

Differential scanning calorimetry experiments carried out by Klein et al. (1990b) have shown that an iPS/*cis*-decalin 30% solution annealed at 25°C produces structures that melt at about 50°C (i.e. very close to its gel counterpart). Conversely, an equivalent iPS/*trans*-decalin solution gives structures that melt at 120°C. If the iPS/*cis*-decalin 30% solution is annealed at 60°C instead, the resulting melting point stands also in the vicinity of 120°C. The structures that melt at 120°C are nothing but chain-folded crystals in which the chains crystallize in the usual crystalline form, i.e. without solvent occluded (designated in what follows as the usual semi-crystalline state). These experiments therefore indicate that, just above the gelation threshold, the usual semi-crystalline state is produced in iPS/*trans*-decalin while a new phase is obtained in iPS/*cis*-decalin.

The thermodynamic behaviour of the iPS/*cis*-decalin system was studied in more detail by Guenet et al. (1994). Apart from the gel state and the usual semi-crystalline state, they identified two other phases: the *s-phase*, already found by Klein and co-workers, and a newly discovered phase, the *p-phase*. The *p-phase* is intermediate between the *s-phase* and the *semi-crystalline state*.

The phase diagrams established on melting are shown in Figure 15.13. Their shape depends upon the range of annealing temperatures (T_{an}). The *s-phase* is obtained typically in the range $20°C < T_{an} < 31°C$, giving below $C_p = 0.3\,(w/w)$ a single, relatively sharp endotherm whose maximum occurs at $T_\gamma = 50 \pm 2°C$. For $C_p > 0.3\,(w/w)$ a second endotherm appears at higher temperature.

The *p-phase* is produced in the range $33°C < T_{an} < 55°C$, and its melting gives rise to two endotherms which are slightly shifted to higher temperature when annealing is achieved above 45°C. The low-melting endotherm is characterized by a temperature T_π

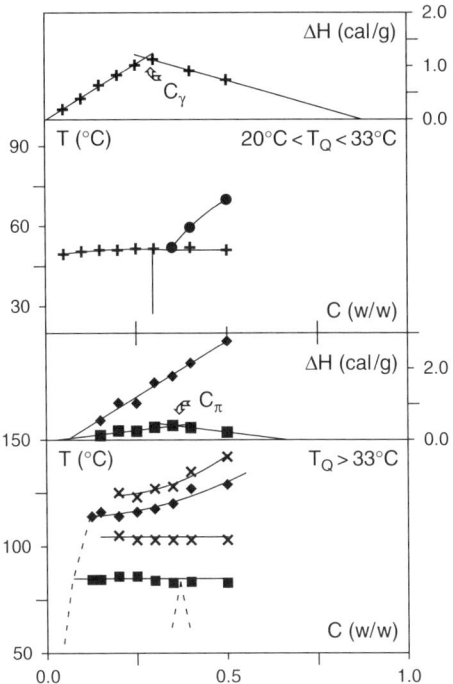

Fig. 15.13. Melting phase diagrams and their corresponding Tamman diagrams for *cis*-decalin with different domains of annealing (as indicated). For the two lower diagrams black symbols stand for $T_{an} < 45°C$ and crosses for $T_{an} > 45°C$. (From Guenet et al. 1994.)

independent of polymer concentration ($T_\pi = 85° \pm 5°C$ for $T_{an} < 45°C$ and $T_\pi = 100° \pm 1°C$ for $T_{an} > 45°C$), and the high-melting endotherm by a temperature T_σ dependent on polymer concentration.

The enthalpy associated with the sharp endotherm of the *s-phase* goes through a maximum at $C_{pol} = 0.3$ w/w. This suggests that this phase consists of a polymer–solvent compound of virtually the same stoichiometry as that of the gel state. This is not entirely surprising as the melting points of the *s-phase* and of the *gel state* are practically the same. As to the *p-phase*, the enthalpy associated with the low-melting endotherm (T_π) goes through a maximum at $C_{pol} = 0.35$ w/w while that associated with the high-melting endotherm (T_σ) increases with increasing concentration. The melting behaviour exhibited by the *p-phase* suggests the occurrence of a peritectic system. Such a system is reminiscent of an incongruently melting compound with the exception that the organization between the solvent and the polymer does not reach very high degrees of regularity (see Chapter 1).

The events occurring in the iPS/*cis*-decalin system can be summarized as follows: the higher the annealing temperature, the lower the degree of crystal solvation. The *semicrystalline* state, which is produced at high temperature, is characterized by the absence of solvent molecules in the crystalline lattice. The *p-phase* is seen to have solvent incorporated yet with a low degree of regularity, while polymer solvent compounds are obtained with the *s-phase* and the gel phase.

Morphology and molecular structure

The morphology observed for the *s-phase* is beyond doubt the most spectacular (see Figure 15.12). A few degrees above the gelation threshold very large spherulites are obtained (Klein et al. 1990b) with diameters as large as 200 μm, a size never observed for this polymer under the usual semi-crystalline state.

Diffraction experiments carried out with the intention of measuring the long-spacing L_c only reveal a shoulder which yields $L_c \approx 31.4$ nm. Such a value is larger than what is usually observed in the bulk-crystallized state of this polymer (Overbergh et al. 1977a, 1977b). The most paradoxical fact turns out to be the limited molecular order within these spherulites in spite of their unusually large size. As can be seen in Figure 15.9, only two main reflections are visible at $q = 2.65$ nm^{-1} (Bragg distance = 2.37 nm) and at $q = 3.8$ nm^{-1} (Bragg distance = 1.65 nm) (a weaker reflection can be seen at $d = 1.1$ nm but is not shown here for the sake of clarity). The first reflection is the same as that observed for the gel state, but is better resolved here. The low molecular order together with the large spherulitic structure is reminiscent of smectic arrangements, although a comparison with existing smectic types is not entirely appropriate given the present information.

According to Guenet et al. (1994) the local structure of the *s-phase* with respect to that of the gel is similar to the difference between a two-dimensional liquid and its corresponding crystal: the *s-phase* is a better ordered-form of the *gel phase* (see Figure 15.9). This explains why the stoichiometries of the compounds formed in the *s-phase* or the *gel phase* are nearly identical. This further implies that the helical forms must be the same in both cases. The three main reflections observed in the *s-phase* have been tentatively indexed by Guenet et al. as the 100, 010 and 110 of a rhombohedric crystalline lattice of parallel rods with the parameters: $a = 2.96$ nm, $b = 2.054$ and $\gamma = 125.8°$. Finally, despite the higher order obtained with the *s-phase*, there are no reflections that would support the occurrence of the 12_1 form.

A study carried out by small-angle neutron scattering shows again the similarity between the *gel* and the *s-phase* (Figure 15.14). Here, the samples have been prepared at the stoichiometric concentration and all the chains are labelled. Typically, a $1/q$ behaviour is observed in both cases, the only difference being in the value of the intensities: the intensity is lower in the case of the *s-phase* than in the *gel phase*.

This behaviour can be accounted for qualitatively by examining the intensity scattered by an assembly of parallel rods, which is written (see Chapter 2):

$$I(q) = \varphi(qr_H) \left\{ 1 - \nu \int_o^\infty 2\pi \left(1 - g(r)\right) J_o(qr) r dr \right\} \quad (15.6)$$

Isotactic polystyrene

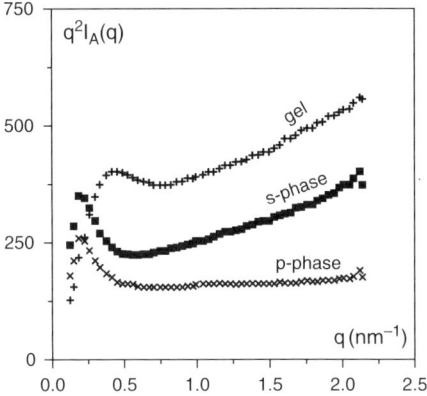

Fig. 15.14. Kratky plot for the scattering by the different phases: gel ($T_{an} = 17°C$, *s-phase* ($T_{an} = 25°C$) and *p-phase* ($T_{an} = 35°C$). All the chains are deuterated ($C_{pol} = 30\%$ (w/w) for the gel and the *s-phase*, and 35% for the *p-phase*). Solvent = hydrogenous *cis*-decalin. (From Guenet et al. 1994.)

where ν is the number of rods per unit area, $g(r)$ the radial distribution function of the rods and J_o the Bessel function of first kind and *zeroth* order. $\varphi(qr_H)$ is the form factor of an infinite rod with a given cross-sectional radius r_c.

Equation (15.6) reduces to the following expression in the explored q-range:

$$I(q) = (\pi\mu_L/q) \times \exp(-q^2 r_c^2/2) \times \{1 - 2\nu\pi q^{-0.5} f(r)\} \qquad (15.7)$$

where $f(r)$ is an oscillating function. It can be shown that equation (15.7) reduces essentially to a $1/q$ behaviour with a corrective term ($q^{-0.5} f(r)$) which depends upon the arrangement between rods. According to Guenet et al. (1994), this term must be larger in the case of the *s-phase* (better order leads to stronger interferences), which should affect the magnitude of the intensity.

Studies of the chain trajectory in the *s-phase* have also been performed for samples annealed at 25°C (Figure 15.15). The scattering curves have been interpreted by considering chain-folding yet to a lesser extent than in usual chain-folded crystals as only two folds are considered (see inset of Figure 15.15, Guenet et al. 1994).

This model is chosen because it allows one to account for the two $1/q$ regimes observed in the scattering pattern. Indeed, by neglecting the loops, the intensity scattered by such a molecular structure is written:

$$I(q) \propto (\pi M/3qL) \times [3 + 4J_o(qd) + 2J_o(2qd)] \qquad (15.8)$$

where L is the chain contour length, M its molecular weight, J_o the Bessel function of first kind and zeroth order, and d the distance between adjacent rod-like sequences. $M/L = \mu_L$, i.e. the mass per unit length.

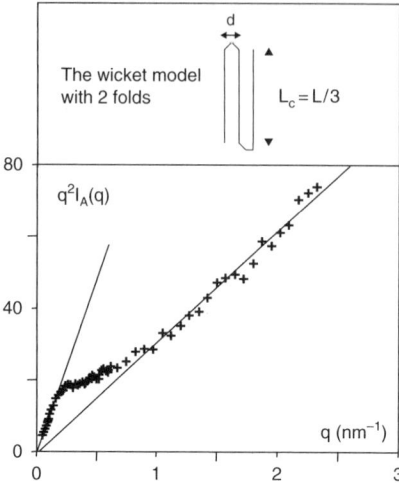

Fig. 15.15. Lower: scattering curve for a sample annealed at 25°C containing 2% (w/w) of deuterated chains and at a total polymer concentration of $C = 30\%$ (w/w). **Upper:** the wicket model used for interpreting the data. (From Guenet et al. 1994.)

In the regime where $qL > 1$ with $qd < 1$, equation (15.8) reduces to:

$$I_1(q) \propto (3\pi\mu_L/q) \times \exp(-q^2d^2/3) \tag{15.9}$$

which yields an initial slope $q^2 I_1(q) \sim 3\pi\mu_L q$.

In the regime where $qL > 1$ with $qd > 1$, the Bessel function can be asymptotically developed, which eventually gives:

$$I_2(q) \propto (\pi\mu_L/3q) \times \left(3 + \frac{2}{(\pi q d^{0.5})} \times \sigma(qd)\right) \tag{15.10}$$

where $\sigma(qd) = 2^{3/2}\cos(qd - \pi/4) + \cos(2qd - \pi/4)$.

The terminal slope is therefore $q^2 I_2(q) \sim \pi\mu_L q$. The ratio between the initial and the terminal slope is therefore equal to 3, which corresponds to the value found experimentally.

These experiments show that the main differences at the molecular level between the *gel phase* and the *s-phase* are chiefly twofold:

- a higher molecular order in the *s-phase*;
- the existence of chain-folding in the *s-phase* unlike in the *gel phase*.

This explains the different morphologies: whenever chains are allowed to fold, lamellae are formed, which eventually produces a spherulitic morphology. Conversely, if chain-folding is impaired, a fibre-like morphology is obtained instead. The gelation threshold therefore corresponds to the temperature at which the chain-folding propensity appears or vanishes. The underlying reasons for the occurrence of such a sharp transition remain to be discovered.

The morphology of the *p-phase* is also spherulitic, yet with sizes of about 1 μm, which is the usual figure for this polymer. The structure of the *p-phase* differs from that of the *s-phase* but, surprisingly, displays in the gel state only one diffraction maximum at $q = 2.65 \,\text{nm}^{-1}$ (Bragg distance = 2.37 nm) (see Figure 15.9). Admittedly, this maximum is far better resolved than in the gel state, and its occurrence at the same distance may be to some extent fortuitous. The small-angle scattering curve shows that the short-range structure of the *p-phase* certainly differs from that of the gel state as a $1/q^2$ behaviour is seen instead of the previous $1/q$ behaviour (see Figure 15.14). Such a behaviour occurs for Gaussian chains but also for sheet-like structures for which the scattered intensity is written (Porod 1951; see Chapter 2):

$$I(q) \propto \frac{2\pi\mu_s}{q^2} \qquad (15.11)$$

where μ_S is the mass per unit area of the sheet. It must be emphasized that the $1/q^2$ behaviour observed here is most probably not only due to the organized structure but to the amorphous material, which also gives off a $1/q^2$ behaviour in this range (here, 'amorphous' material corresponds to flexible, disordered chains or chain portions with the usual persistence length, i.e. $l_p \approx 1 \,\text{nm}$). The existence of amorphous material has been suggested from the thermodynamic study, and it is probably responsible for the decrease by about two orders of magnitude of the mean size of the spherulites.

This led Guenet and co-workers to consider the sheet-like structure schematized in Figure 15.9 (Guenet et al. 1994). In this model, solvent molecules are intercalated between rows of polymer stems. Indeed, this model is consistent with the conclusions drawn from the phase diagram: the degree of solvation of the chains is lower (peritectic). The *p-phase* is nothing but a partially desolvated *s-phase*, the desolvation occurring mainly in the 010 planes.

The structural investigations confirm and strengthen the conclusions drawn from the phase diagrams, i.e. the following sequence as a function of annealing temperature: *gel phase* (highly solvated) poorly organized; *s-phase* (highly solvated) better organized; *p-phase* (less solvated) relatively well organized; and finally the *semi-crystalline phase* (not solvated at all) highly organized.

Solvent-induced structures

It has been recently shown by ElHasri et al. (2004) that iPS/solvent molecular compounds can be grown by exposing solid amorphous polymers to liquid solvents. These authors have, in particular, studied the case of *cis*-decalin and *trans*-decalin. They have observed that, with amorphous iPS samples exposed to these solvents at temperatures below 10°C, fibrillar structures such as those shown in Figure 15.16 are obtained. Clearly, the morphology is reminiscent of that obtained in the gel state after preparation from

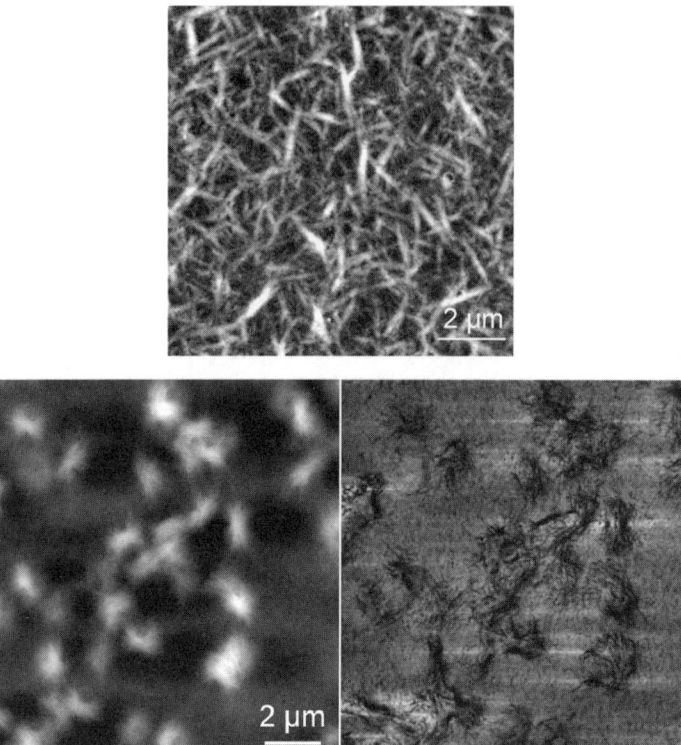

Fig. 15.16. AFM investigations. **Upper:** typical fibrillar morphology observed for iPS systems exposed either to liquid *cis*-decalin or liquid *trans*-decalin below $t = 10°C$. **Lower:** morphology observed for amorphous iPS samples exposed to *cis*-decalin. These large spherulites consist of the *s-phase*, as is obtained from solution-cast samples. Scales as indicated. (From ElHasri et al. 2004. Reprinted with permission from ACS.)

homogeneous solutions quenched down to below 10°C. Also, from the thermodynamic viewpoint, the systems display virtually the same melting as the solution-cast systems.

Increasing the exposure temperature again differentiates *cis*-decalin from *trans*-decalin. Samples exposed at 25°C show distinctive thermal behaviour. While melting occurs at about 50°C in *cis*-decalin, it takes place at 110°C in *trans*-decalin (Figure 15.17). This means that compounds are still formed in *cis*-decalin, but no longer in *trans*-decalin, a behaviour reminiscent of what has been observed when preparing samples from homogeneous solutions.

As will be further discovered in Chapter 16, which deals with syndiotactic polystyrene, it appears that, provided amorphous polymers are used, the path followed for preparing the sample does not play any role: whether the system is obtained by solution-casting, namely quenching at constant polymer concentration, or exposing to solvent, namely at

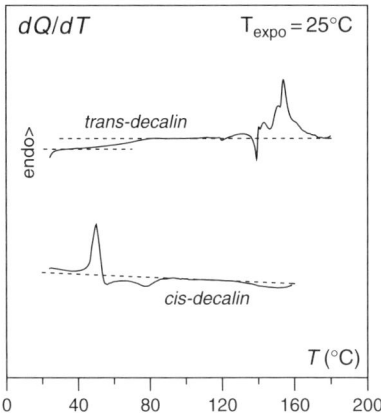

Fig. 15.17. DSC thermograms for iPS samples exposed at $T = 25°C$ to liquid *trans*-decalin (**upper**) and liquid *cis*-decalin (**lower**). (From ElHasri et al. 2004.)

constant temperature yet altering the polymer concentration, the same or very similar structures are formed.

Hybrid systems: polymer and self-assembled structures. metallogels

Self-assembling molecules that form supramolecular polymers (Lehn 1995) are being intensively synthesized and their properties studied (e.g. see Ajayaghosh and George, 2001, 2003; George and Ajayaghosh 2005; Guenet and Nandi 2006). In many cases, however, these systems cannot be used as such, and therefore require to be incorporated into a matrix. Recently, Lòpez and Guenet (1999, 2002) have shown that incorporation of these supramolecular polymers into polymer fibrils can be achieved by physical processes, such as heterogeneous nucleation, and thus nanocomposites can be prepared. The supramolecular polymer these researchers have been using is a bicopper complex (*Copper (II) tetra-2-ethyl-hexanoate*), the chemical structure of which is shown in Figure 15.18 (termed CuS8 in what follows).

This complex forms mesophases in the bulk state, but can be dissolved readily in organic solvents. In these solvents it forms long threads, as shown by the scattering curve, which can be fitted with an equation for cylinders of finite length of the type described in Chapter 2 (Terech et al. 1992):

$$q^2 I(q) \propto C_{Cu} \mu_L \times \frac{4 J_1^2(qr_c)}{q^2 r_c^2} \times \left[\pi q - \frac{2}{<L>} \right] \quad (15.12)$$

where $<L>$ is the mean length of the cylinders, r_c their cross-sectional radius and μ_L the mass per unit length. The values derived from equation (15.12), namely $r_c = 0.82 \pm 0.02$ and $\mu_L = 1350 \pm 170\,\text{g/mol}$ per nm, clearly point to the presence of one molecule in the cross-section.

204   *Polymer-Solvent Molecular Compounds*

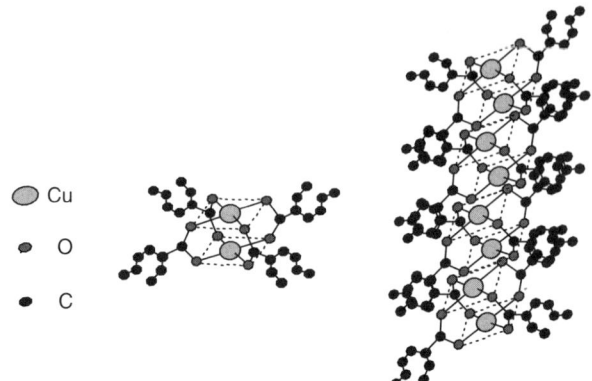

Fig. 15.18. Structure of the bicopper complex copper (II) tetra-2-ethyl-hexanoate, and showing the way it piles up to form filaments. (Maldivi 1989.) (See Plate 3)

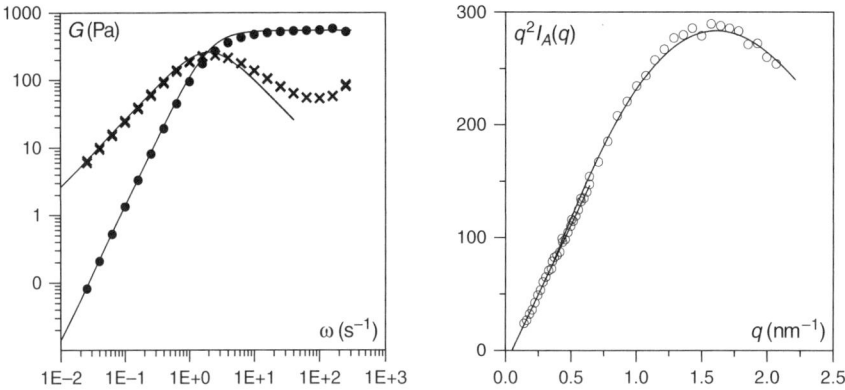

Fig. 15.19. Left, Rheological behaviour of CuS8/trans-decalin solutions; and **right,** neutron scattering curve in deuterated *trans*-decalin. Solid lines are fit as described in the text. (Dammer et al. 1995.)

The mean length $<L> = 9$ nm, is, however, small with respect to the rheological behaviour shown in Figure 15.19, left. The behaviour of G' and G", the storage and the loss moduli respectively, can be fitted with a relation developed by Maxwell:

$$G' = G_o \frac{\omega^2 \tau^2}{1 + \omega^2 \tau^2} \quad G' = G_o \frac{\omega \tau}{1 + \omega^2 \tau^2} \quad (15.13)$$

where ω is the frequency and τ the characteristic time. This behaviour indicates that the system is a dynamic supramolecular polymer in the sense that stress release essentially occurs through scission–recombination of filaments. Clearly, a filament has a finite lifetime, as opposed to the case with covalent polymers. This behaviour is generally

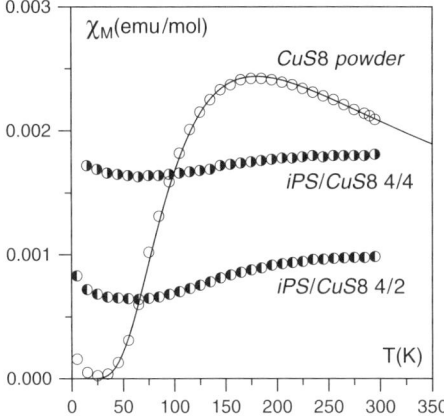

Fig. 15.20. Magnetic susceptibility χ_M as a function of temperature for CuS8 powder and encapsulated filaments ($C_{pol} = 0.04\,\text{g/cm}^3$, $C_{CuS8} = 0.04\,\text{g/cm}^3$, middle curve; and $C_{CuS8} = 0.02\,\text{g/cm}^3$, lower curve). (From Poux et al. 2003.)

accounted for by Cates' theory on these types of supramolecular polymer (Cates 1987, 1988).

Lòpez and Guenet have, therefore, suggested encapsulating these filaments for two reasons: (1) make their lifetime infinite, and (2) determine the magnetic properties in the 1D state of the bicopper complex (Lòpez and Guenet 1999). In the bulk state the bicopper complex behaves as an anti-ferromagnetic material, as shown by the behaviour of the magnetic susceptibility χ_M (Figure 15.20), which obeys the Bleaney–Bowers theory (Bleaney and Bowers 1952), namely:

$$\chi_M = \frac{N_A g^2 \mu_B^2}{kT} \frac{2\exp(2x)}{1+3\exp(2x)} \quad \text{with} \quad x = \frac{J}{kT} \tag{15.14}$$

in which N_A is Avogadro's number, g is the number of unpaired electrons in the molecule, J the magnetic exchange coupling constant and μ_B Bohr's magneton.

Lòpez and Guenet have shown that the encapsulation of the 1D filaments can be achieved by means of a heterogeneous nucleation process involving the growth of the fibrils of iPS thermoreversible gels. As the gelation threshold of iPS in *trans*-decalin is a well-defined temperature (see Thermoreversible gels above), it is expected that CuS8 filaments, which grow above this gelation threshold, should trigger the growth of fibrils in iPS/*trans*-decalin solutions. Moreover, since at high temperature CuS8 molecules are not at all assembled, ternary solutions are easily obtained. Lòpez and Guenet have therefore studied the gelation temperature as a function of the content of CuS8 filaments. Indeed, the gelation temperature increases with increasing CuS8 content, up to a critical fraction f_c^*, while above f_c^* the gelation temperature levels off (Figure 15.21).

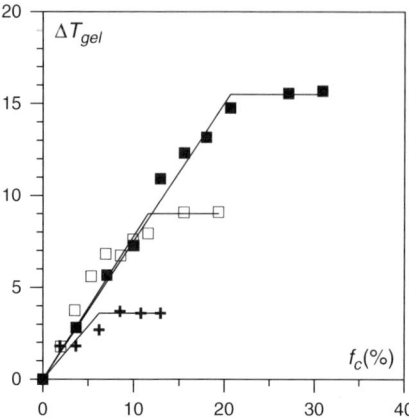

Fig. 15.21. Increase of the gelation temperature ΔT_{gel} as a function of the content of CuS8 in *trans*-decalin for different iPS concentrations: (■) $C_{pol} = 0.04\,\text{g/cm}^3$; (□) $C_{pol} = 0.08\,\text{g/cm}^3$; (✦) $C_{pol} = 0.16\,\text{g/cm}^3$. (From Lòpez and Guenet 1999.)

As seen from Figure 15.21, the extent to which the gelation threshold is increased depends upon polymer concentration: the levelling-off occurs at lower CuS8 content, the higher the polymer concentration. As stressed by Lòpez and Guenet, this may be due to an increasing incompatibility between the filaments of the supramolecular polymer and the covalent polymer. These results are certainly consistent with a heterogeneous nucleation process, the more so as the gel melting temperature together with the gel melting enthalpy remain constant independent of the CuS8 content.

These authors have further studied the structure of the ternary system by neutron scattering. By using a deuterated isotactic polystyrene with a hydrogenous CuS8 complex, they have determined the structure of each component by adjusting the scattering amplitude of the solvent. This can be easily done by using a mixture of hydrogenous *trans*-decalin and deuterated *trans*-decalin. Whether the solvent is essentially deuterated, thus matching the coherent scattering of the polymer, or essentially hydrogenous, thus matching the coherent scattering of the CuS8 complex, will reveal the molecular structure of the bicopper complex or of the polymer chains, respectively. The corresponding scattering curves are shown in Figure 15.22.

The structure of the bicopper complex is still filamentous with one molecule in the cross-section for $f_c \leq f_c^*$. This shows that the 1D structure of the bicopper complex is retained after encapsulation. For $f_c > f_c^*$ a strong upturn is seen at small q, which indicates the presence of 3D objects, as confirmed by the fit of the scattering curve, with the following relation:

$$q^2 I(q) \propto X \frac{S}{Vq^2} + (1-X)\, C_{CuS8} \mu_L \times \frac{4 J_1^2(qr_c)}{q^2 r_c^2} \times \left[\pi q - \frac{2}{<L>} \right] \qquad (15.15)$$

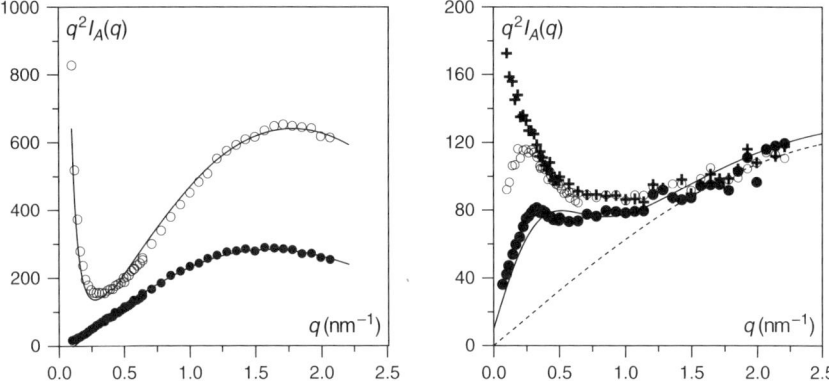

Fig. 15.22. Scattered intensities of ternary system iPS$_D$/CuS8/*trans*-decalin plotted by means of a Kratky plot (q^{2}I$_A$(q) vs q). **Left:** in an isotopic mixture of *trans*-decalin, D/H = 91/9 in v/v, thus matching the coherent intensity of the polymer and revealing the structure of the bicopper complex; the solid lines are fit with equation (15.12) for $f_c = f_c^*$ (●) and equation (15.15) for $f_c = 2f_c^*$ (○). **Right:** in an isotopic mixture of *trans*-decalin, D/H = 8/92 in v/v, thus matching the coherent intensity of the bicopper complex and revealing the structure of the polymer, (+) $f_c = 0$, (○) $f_c = f_c^*$, (●) $f_c = 2f_c^*$; the solid line is calculated using equation (15.17), while the dotted line stands for the scattering by individual chains under the 3$_1$ helix (equation (2.11)). (Data from Lòpez and Guenet 1999.)

in which the first term stands for the scattering by these 3D objects, S and V being their surface and volume respectively, and $1 - X$ their fraction. The second term in equation (15.15) corresponds to the scattering by the encapsulated filaments of CuS8.

The appearance of 3D objects, most probably the mesophase observed in the bulk state, is therefore related to the levelling-off seen above the critical CuS8 content. Clearly, the addition of CuS8 above this critical content is no longer necessary for nucleation and the excess is thus rejected into another phase.

The structure of the polymer shows a drastic change at low q. This domain is related to the fibrillar cross-section, and therefore points to a decrease of this parameter. This is consistent with the heterogeneous nucleation effect, which is expected to produce smaller organized structures. Lòpez and Guenet have been able to achieve a theoretical fit by considering four iPS chains adopting a rod-like conformation owing to their 3$_1$ helical structure, and whose position with one another fluctuates (Figure 15.23, left). The function used for expressing the positional fluctuation is:

$$w(r) = l \exp(-l^2/2l_o^2) \tag{15.16}$$

which eventually provides the following equation:

$$q^2 I(q) = \pi \mu_L C_p q \times \frac{4 J_1^2(qr_H)}{q^2 r_H^2} \times \left[1 + \exp(-q^2 l_o^2) + 2\exp(-q^2 l_o^2/2)\right] \tag{15.17}$$

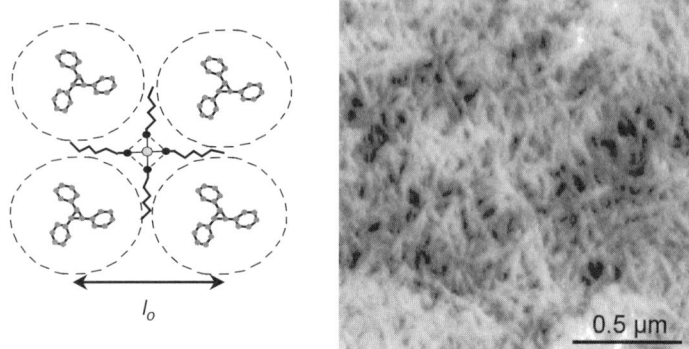

Fig. 15.23. Left: theoretical model for describing the cross-section of a fibril encapsulating one CuS8 filament (Lòpez and Guenet 1999; reprinted with permission from EDP Sciences). **Right:** AFM picture of iPS/CuS8 ($C_{iPS} = 0.04\,\mathrm{g/cm}^3$, $C_{CuS8} = 0.03\,\mathrm{g/cm}^3$, $f_c = 0.1$) after drying (Poux et al. 2001; reprinted with permission from Wiley Interscience).

where l_o is the average distance between adjacent iPS chains with a rod-like conformation. The fit yields $l_o \approx 2.4\,\mathrm{nm}$, which entails that the encapsulating fibrils possess a cross-sectional radius of about 4 nm.

It is worth noting that there is still an effect on the polymer structure beyond f_c^*, although part of the CuS8 is rejected into an additional phase.

AFM observations on dried encapsulated systems have been reported by Poux and co-workers, as shown in Figure 15.23 (right) (Poux et al. 2001). The system consists of fibrils whose cross-sectional radius is about 8 nm, a value in good agreement with the results derived by neutron analysis if one considers the relative accuracy of size determination by AFM under the present conditions.

The magnetic properties of these systems differ significantly from those in the bulk state of the bicopper complex. As can be seen in Figure 15.20, the magnetic susceptibility is no longer zero on approaching $T = 0\,\mathrm{K}$ (Poux et al. 2003). Instead, this parameter seems to remain approximately constant in the range of temperatures explored, and is clearly proportional to the copper content. As the magnetic susceptibility is rescaled by the molar content of copper, this means that this effect is not only due to the quantity of copper atoms. As the length of the filaments of the bicopper complex is proportional to the concentration, it is suggested that this effect might be due to a different piling of the bicopper complex molecules: in the encapsulated state, copper–copper interactions would replace copper–oxygen interactions as exist in the bulk (Guenet et al. 2006).

The effect of the solvent type, namely using *cis*-decalin instead of *trans*-decalin, has dramatic consequences (Lòpez and Guenet 2002). Although some alteration of the gelation temperature takes place, the behaviour is drastically different (Figure 15.24, left). Increases occur only beyond a given bicopper complex content, the latter being dependent upon the polymer concentration. Unlike systems in *trans*-decalin, the scattered

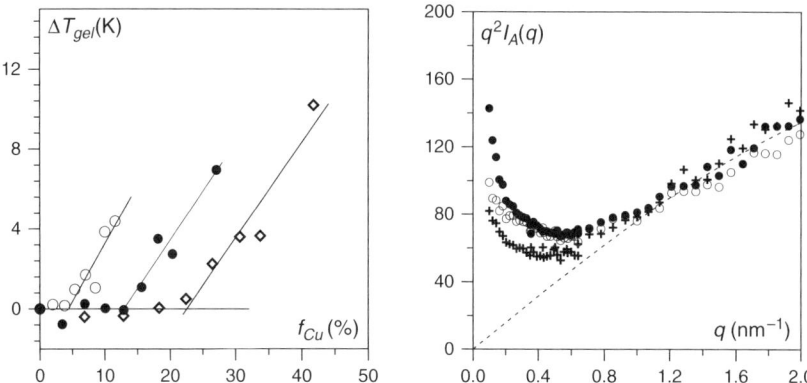

Fig. 15.24. iPS/CuS8/*cis*-decalin systems. **Left:** ΔT_{gel} vs CuS8 content; (◊) $C_{iPS} = 0.02 \, \text{g/cm}^3$; (•) $C_{iPS} = 0.04 \, \text{g/cm}^3$; (○) $C_{iPS} = 0.08 \, \text{g/cm}^3$. **Right:** intensity scattered by the polymer; $C_{iPS} = 0.04 \, \text{g/cm}^3$; (•) $f_c = 0.26$; (○) $f_c = 0.13$; (+) $f_c = 0$. (From Lòpez and Guenet 2002.)

intensity at low-q is seen to increase when increasing the bicopper complex content (Figure 15.24, right). Clearly, no heterogeneous nucleation has taken place, and the behaviour is rather reminiscent of lowering the solvent quality towards iPS. Also, the filamentous state does not display stability in this system, indicating again that no encapsulation has been achieved.

CHAPTER 16

Syndiotactic polystyrene

The success in synthesizing syndiotactic polystyrene (sPS) was most greatly welcomed by scientists interested in the effect of stereoregular configuration on molecular structure in the crystalline and the non-crystalline state (Ishihara et al. 1986; Grassi et al. 1987). This achievement has led to a wealth of papers on the topic.

Somewhat unexpectedly, the tendency of sPS to crystallize turned out to be far more complex than for its isotactic counterpart. Unlike iPS, for which only one helical form (a 3_1 helix) and one crystalline lattice have been observed to date (the 12_1 being still a subject of controversy), sPS exhibits different helical forms, namely the planar zig-zag conformation and a 2_1 helix, and also different crystalline lattices.

The two helical conformations that sPS can take on are similar to those reported on *syndiotactic polypropylene* (Natta et al. 1960). The planar zig-zag is the result of *tt* arrangements and produces an extended chain conformation. Calculations and experiments on solvent-induced samples by Immirzi et al. (1988) and Vittoria et al. (1988) have revealed the existence of a 2_1 helical form which results from *ttgg* arrangements (Figure 16.1). The pitch of the planar zig-zag is $P = 0.506$ nm with 2 monomers per repeating unit while that of the 2_1 helix is 0.77 nm with 4 monomers per turn (the basic helix unit consists therefore of 2 monomers).

It is worth stressing that the 2_1 helix can only be grown through exposure to solvent of either the amorphous state or of the α form, or from solution. Guenet has suggested that this is due to the fact that this helix with its 'fourfold' symmetry exposes phenyl rings to solvent molecules, which is favourable for polymer–solvent interactions (Guenet 2003). Conversely, the chain under the planar zig-zag conformation possesses a 'polyethylene-like face' which is likely to interact poorly with solvent molecules. Clearly, the presence of solvent molecules favours the 2_1 conformation, as will be shown in the case of sPS thermoreversible gels (Daniel et al. 1996, 1997).

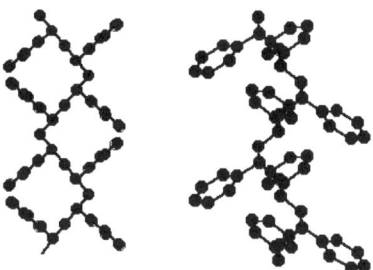

Fig. 16.1. The 2_1 helical form as seen from two different angles. (Drawing courtesy of S.J. Spells, Sheffield Hallam University.)

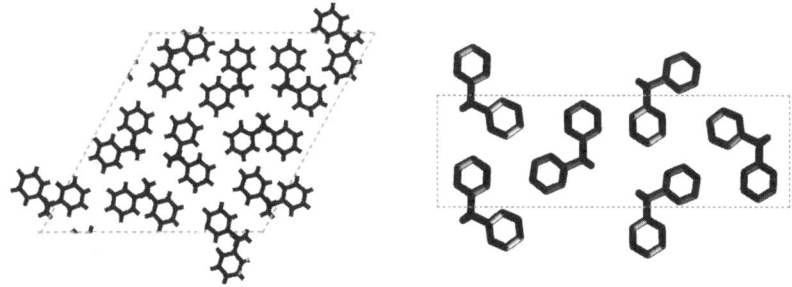

Fig. 16.2. Left: Crystal structure of the α form (hexagonal with $a = b = 2.625$ nm, $c = 0.5045$ nm). **Right:** the β form (orthorhombic with $a = 0.881$ nm, $b = 2.882$ nm, $c = 0.51$ nm). (After Greiss et al. 1988 and Chatani et al. 1988.)

For those crystalline forms where no compounds are obtained there is a consensus for designating them as α, β. The α and β forms involve the planar zig-zag conformation. What differentiates the α form from the β form is the crystalline lattice (Figure 16.2). From electron diffraction experiments Greiss et al. (1988, 1989) have assigned a hexagonal unit cell for the α form of space group $P\bar{6}2c$ with $a = b = 2.625$ nm and $c = 0.5045$ nm. The β form consists of an orthorhombic unit cell with $a = 0.881$ nm, $b = 2.882$ nm, $c = 0.51$ nm, the polymer stems are randomly distributed, which creates an orthorhombic unit cell (Chatani et al. 1988). Further modifications of these forms (designated as α', α'', β' and β'') represent minor differences of the crystalline lattices of the α form and the β form due to enantiomorphous pairs of clusters (Guerra et al. 1990).

Investigations of the crystallization habit of sPS as a function of various parameters such as the cooling rate by the same authors suggest that the β form is the stablest crystalline form when the chains are under the planar zig-zag conformation. In fact, the α form grows with rapid quench from the melt or from crystallization of the amorphous state at high undercooling. The occurrence of the α form is thus a kinetically controlled phenomenon.

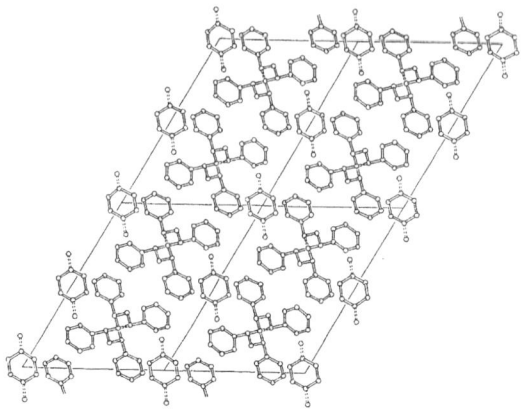

Fig. 16.3. Crystalline lattice as seen parallel to the helix axis: monoclinic cell with $a = 1.758$ nm, $b = 1.326$ nm, $c = 0.771$ nm, $\gamma = 121.2°$, and space group $P2_1/a$. sPS samples exposed to toluene vapours. (From Chatani et al. 1993; Chatani and Nakamura 1993. Reprinted with permission from Elsevier.)

For the crystalline forms consisting of sPS/solvent compounds there are now different designations: originally the first compound found was designated as the δ *form* in many papers, and is still used by some authors (Figure 16.3), although now the Italian school uses this terminology for designating the *nanoporous* crystal structure obtained after solvent removal which occurs without change of the unit cell dimensions (e.g. see Daniel et al. 2005). This must not be confused with the γ form, which is metastable and results from the elimination of the solvent together with a change of the unit cell.

Note that the original δ *form* is often designated as a *clathrate* by the Italian school although it can also be designated as an *intercalate* if one refers to the usual definition of an intercalate as a *two-dimensional open, layer-type or sandwich-type inclusion compound*. Yet, the crystalline forms discussed here could just as well be designated as *clathrates* on the basis of the solvent molecules not being in contact with one another. In this sense, solvent molecules are considered imprisoned, which is the sense conveyed by the word clathrate. It is worth noting that similarly structured PEO molecular compounds are named intercalates.

Another crystalline form involving the formation of compounds with the 2_1 helix was first discovered by Daniel et al. (1996) in *sPS/benzene* thermoreversible gels, and the unit cell (Figure 16.4) was later determined by Petraccone et al. (2005b) in the case of *sPS/norbornadiene*. This form is denoted as an intercalate by the Italian school, while Malik et al. (2006a) have proposed the designation of δ_i. These authors also propose renaming the original δ form as δ_c while designating the new form as δ_i. This would clearly underline the fact that all these crystalline forms are solvated forms with the same helical structure (2_1). Another form, which is most probably nematic-like and made up with 2_1 helices, had already been mentioned by de Candia et al. (1992), and has more recently been observed by Malik et al. in some solvents (Malik et al. 2006a).

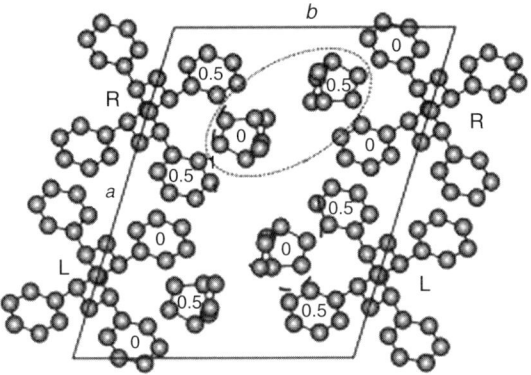

Fig. 16.4. Crystal structure of sPS/norbornadiene molecular complex (space group $P2_1/a$), monoclinic unit cell with $a = 1.75$ nm, $b = 1.45$ nm, $c = 0.78$ nm, and $\gamma = 107.8°$. The ellipses indicate couples of neighbouring guest molecules at van der Waals distance from one another. (After Petraccone et al. 2005b. Reprinted with permission from ACS.)

Table 16.1. **Left**, former designations by the Italian school of the different forms; **middle**, the current designation by the Italian school; **right**, designation by Malik and colleagues.

δ form	clathrates	δ_c form
emptied δ form	the nanoporous δ form	δ_e form
/	intercalates	δ_i form
mesophase	mesophase	δ_N form
γ form	γ form	γ form

Malik et al. have suggested naming it δ_N to maintain the existing logic. In this book both terminologies will be used although that proposed by Malik et al. is clearly preferred. Table 16.1 summarizes the different designations.

The crystalline lattice of the δ_c form (originally termed the δ form), shown in Figure 16.2, has been determined by Chatani and co-workers for sPS samples crystallized by exposure to toluene vapours (Chatani et al. 1993; Chatani and Nakamura 1993). As a rule, the diffraction pattern of the δ_c form exhibits reflections significantly broader than those from the α form or from the β form (e.g. see Guerra et al. 1990). This suggests long-range disorder which possibly originates in the accumulation of defects due to imperfect solvent placement within the crystalline lattice. As will be discovered throughout this section, the crystal lattice of the δ_c form is little dependent upon the solvent that gives rise to this form.

As aforementioned, Manfredi et al. (1995, 1997) have succeeded in preparing an emptied δ_c form, i.e. of the same crystalline lattice but without any occluded solvent. According to Malik et al. this form could be still referred to as the δ_e form, as has been the case in many papers. The γ *form* is obtained after desolvation of the δ_c *form* and a change of the crystalline lattice parameters as was first evidenced by Immirzi ct al. (1988). The transformation from the δ_c *form* to the γ *form* occurs on heating at about 120°C. Further heating yields the α *form*.

Three main types of investigation have been carried out on sPS/solvent molecular compounds:

1. On *solid samples*, either amorphous or crystallized under the α or the β *form*, subjected to solvent exposure (vapour or liquid). These samples will be designated as *solvent-induced compounds*.

2. On *solutions-cast samples* producing crystal mats (dilute solutions) or spherulites (concentrated solutions).

3. On *gels samples* prepared from homogeneous, moderately concentrated solutions that produce *fibrillar* networks.

Promising applications may be underway with regard to the δ_e form (nanoporous δ form), involving the trapping and removal of pollutants from wastewater and the storage of gases such as hydrogen (Guerra et al. 1998).

Solvent-induced compounds

Solvent sorption and crystallization

The group headed by F. de Candia and V. Vittoria at Salerno University has carried out an extensive study of the absorption properties of bulk films of syndiotactic polystyrene using various solvents (Vittoria et al. 1988, 1990, 1991a, b). The sorption phenomenon was investigated as a function of the solvent activity a, which is defined as follows.

$$a = p/p_o \quad (16.1)$$

where p is the solvent pressure at which the sample is exposed and p_o the saturation pressure of the solvent at the temperature of the experiment.

Once equilibrium is reached, the samples are dried under vacuum for several days prior to further study. This is thought to ensure that only those solvent molecules strongly bound to the polymer will remain while those trapped in the amorphous domains will be extracted. Here, it is worth noting that this drying procedure is liable to bias the determination of the stoichiometry achieved with techniques such as thermogravimetry (TGA) or NMR that are sensitive to the total fraction of mobile solvent molecules. For instance, for the same solvent (dichloromethane), Immirzi et al. (1988) have come up with a stoichiometry of *1 solvent molecule per 25 monomers*, while Gomez and Tonelli (1991) have reported a value of *1 solvent molecule per 13 monomers*. This significant discrepancy, together with the surprisingly low degree of solvation, might be due to the fact that those sPS/solvent complexes do not possess a perfect crystalline lattice. As a result, for a given crystalline structure, some supposedly equivalent solvent molecules may be less bound than others, as may arise from a small disorientation of the solvent with respect to the chain. These molecules are liable to be expelled first under vacuum. Also, Gomez and Tonelli have reported that the spin-lattice relaxation times of the 2_1 form are 2–10 times lower than those measured for the planar zig-zag form, which indicates that the former helix shows higher mobility than the latter. Under these conditions, solvent molecules may be given sufficient mobility to escape the crystalline lattice. In other words, the measured value of the stoichiometry can, then, be strongly dependent upon the way the drying procedure is achieved, and therefore be at variance with the actual value. Further, with these methods, the degree of crystallinity must be known. A stoichiometry of 1 *solvent molecule per* 7 *or* 8 *monomers* has been reported by Rapacciulo et al. (1991) after correction of the data by the degree of crystallinity.

The growth of the helical structure is then followed by infrared spectroscopy. The bands characteristic of the 2_1 helical form appear at 499, 575, 935, 1170 and 1277 cm^{-1}. Vittoria and colleagues have considered two parameters they named H_1 and H_2, which stand for the ratio between the absorbance of the helical band and the absorbance of an internal reference band:

$$H_1 = \text{absorbance } 575\,\text{cm}^{-1}/\text{absorbance } 840\,\text{cm}^{-1}$$

$$H_2 = \text{absorbance } 935\,\text{cm}^{-1}/\text{absorbance } 906\,\text{cm}^{-1}$$

In fact, these authors have shown by coupling infrared experiments with X-ray diffraction that H_1 is directly related to the appearance of the 2_1 helix while H_2 is sensitive to the growth of crystals of this helical form. The variation of H_1 and H_2 as a function of solvent activity is shown in Figure 16.5 for studies performed with dichloromethane. For solvent activities lower than $a = 0.4$, H_1 and H_2 are constant and quite low. For a larger

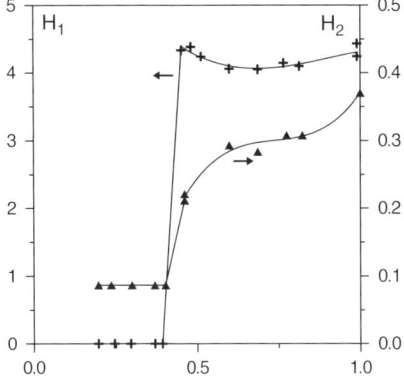

Fig. 16.5. Variation of the two parameters H_1 and H_2 (see text) as a function of the solvent activity. (Data from Vittoria et al. 1988.)

than $a = 0.4$, the value of H_1 increases abruptly and then remains virtually constant up to $a = 1$. H_2 is also seen to increase, but slowly, unlike the case with H_1.

These experiments show that a critical activity of about $a_c = 0.4$ is needed to trigger the appearance of the 2_1 helix out of totally amorphous sPS samples. However, while most of the chains assume the *ttgg* conformation (i.e. the 2_1 helix) as soon as $a = a_c$, the fraction of 'crystals' of these helices is dependent upon solvent activity, its maximum being at $a = 1$. Clearly, helix formation and crystal growth are not concomitant as is usually the case for flexible, semi-crystalline polymers. That there exists a critical activity is probably related to the glass transition of sPS. It is likely that chain rearrangements can only proceed when some plasticization has taken place in order to reduce the glass transition temperature to below the temperature at which the sorption experiment is carried out. This requires a certain solvent concentration in the vicinity of the solvent–polymer interface, which can only be achieved when a critical activity is reached.

Tashiro and co-workers have carried out a detailed study of the crystallization mechanism by FTIR, Raman spectroscopy and X-ray diffraction. They come to the conclusion that portions of helices are formed prior to any long-range organization (Tashiro et al. 2001; Gowd et al. 2003). This conclusion is in agreement with previous findings by Guenet et al. concerning the gelation mechanism of both iPS and sPS: the chain conformation is close to the 2_1 helical structure well before gelation takes place (Guenet 1992; Daniel et al. 1996).

Salerno's group has also studied the sorption of a liquid in amorphous sPS films of various thicknesses (Vittoria et al. 1990, 1991a). The evolution of the ratio C_t/C_{equ} as a function of time, where C_t is the global concentration at time t and C_{equ} the concentration reached at sorption equilibrium, shows that the sorption process is, in most cases, controlled at its early stage by Fickian diffusion (Figure 16.6). In fact, C_t/C_{equ} varies linearly as a function of $t^{1/2}/d$, where d is the sample thickness.

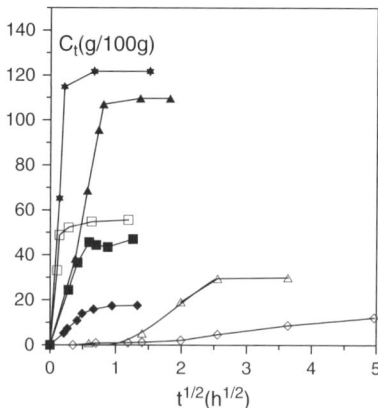

Fig. 16.6. Evolution of the ratio C_t/C_{equ} as a function of time for various solvents. (After Vittoria et al. 1991a.)

> **Note 16.1 Fickian diffusion**
>
> Consider a piece of polymer of thickness d and of infinite area where *both faces* are exposed to a liquid or a gas. At the interface but within the solid the concentration is C_{equ} (maximum concentration allowed by thermodynamics). If the diffusion of the liquid throughout the polymer sample is a Fickian process characterized by a translation diffusion coefficient D then:
>
> $$\frac{\partial C(r,t)}{\partial t} = D \frac{\partial^2 C(r,t)}{\partial r^2}$$
>
> where r is related to the axis perpendicular to the interface. The solution is:
>
> $$C(r,t) = C_{equ} Erfc\left(\frac{r}{2\sqrt{Dt}}\right)$$
>
> where *Erfc* is the complementary error function. The global concentration C_t, as defined above, is then written:
>
> $$C_T = \frac{2}{d}\int_o^\infty C(r,t)\,dr = \frac{4C_{equ}}{d}\sqrt{\frac{Dt}{\pi}}$$

Conversely, for cyclohexane and n-hexane, there is an induction time, so the diffusion process is not Fickian.

It is worth emphasizing that the concentration at a time t, i.e. C_t, is not uniform throughout the sample unlike with C_{equ}, the latter being dependent upon the solvent type (see below). C_t is determined by weighing the sample, so it depends on thickness, d.

The curves in toluene show a maximum at a later stage which, according to Vittoria et al., most probably arises from expulsion of solvent from the sample in the process of crystallizing under the δ *form*. Indeed, it seems that the sPS first swells while forming a solvated mesophase, and eventually crystallizes. The observation of birefringence in the sample, whereas no long-range order is seen by X-ray diffraction, supports such an assumption. As will be seen in the section Thermoreversible gels, below, a swollen crystalline phase can be observed.

The value of C_{equ} strongly depends upon solvent type (Figure 16.7). Vittoria et al. chose to plot this parameter as a function of the solubility parameter δ_s. A maximum is seen for chloroform but also for carbon tetrachloride. These authors came to the conclusion that adsorption, and correspondingly a higher degree of solvation of the δ *phase*, are related to solvent quality: the better the solvent, the higher the degree of solvation. It seems curious, however, that benzene and chloroform, for which δ_s is virtually the same, give markedly different values for C_{equ}. Indeed, the solubility parameter is but a rough indication of the polymer–solvent interaction. More elaborate theories (de Gennes 1979;

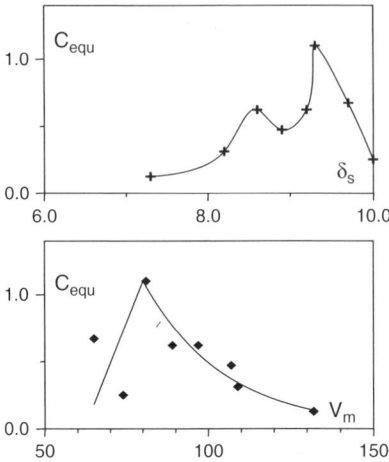

Fig. 16.7. Variation of the equilibrium concentration as a function of the solubility parameter (upper) and the solvent molar volume (lower). (From Vittoria et al.1991a.)

Des Cloiseaux and Jannink 1988) generally make use of Flory's interaction parameter χ_1. Alternatively, a plot of C_{equ} as a function of the solvent molar volume also shows a maximum for chloroform, and then a continuous decrease of this parameter as the solvent molar volume increases. As was shown by Klein and Guenet (1989), the size of the solvent molecules is an important parameter in polymer solvation effects. The investigations by Klein and Guenet have clearly highlighted that there is a connection between the degree of solvation and the features of the chain microstructure. Polymers with bulky side-groups, as are polystyrenes, possess cavities along the chain whatever the helical conformation. In these cavities solvent molecules can be housed, thus stabilizing the helical structure. Evidently, both effects, i.e. the solvent molecule size and its interaction with the polymer, must be taken into account. At equivalent polymer–solvent interaction the size of the solvent molecules is the determining factor.

Thermal stability

As has been seen in the previous section, exposure to solvent vapour or to liquid promotes the formation of the δ *phase*. Upon heating this phase transforms into the γ *phase* by expelling the solvent. This is nothing but an *incongruent-melting* wherein the chain conformation is maintained. Such a type of transformation had already been reported by Guenet et al. (1985) for isotactic polystyrene gels, yet went largely unnoticed. De Candia et al. (1995) have obtained the type of DSC thermogram shown in Figure 16.8 for samples prepared by immersing amorphous films in *methylene chloride* for 24 hours and then drying them for a further 24 hours under vacuum.

According to these authors, the first endotherm is associated with the glass transition (overshoot), while they consider the occurrence of an exotherm at about 120°C linked to the transformation of the δ *phase* into the γ *phase* (the corresponding baseline is labelled

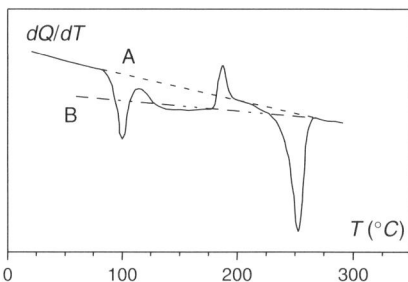

Fig. 16.8. DSC thermogram obtained for sPS sample after exposure to liquid methylene chloride. Two possible baselines (A) and (B) have been drawn (see text for details). (From de Candia et al. 1995.)

(B) in Figure 16.8). Normally, in the case of a δ *phase* more stable than the γ *phase* an endotherm should be produced instead of an exotherm. This would therefore mean that the δ *phase* is in fact unstable and tranforms spontaneously into the γ *phase* as soon as mobility permits. One may wonder, however, why this tranformation towards the stable phase did not take place sooner, during the immersion phase of the crystallization process, when the mobility was probably sufficiently high by virtue of the plasticization of the amorphous phase.

The second exotherm corresponds to the recrystallization *phase* under the α *form* of the liquid arising from the melting of the γ *phase*.

This thermogram can, however, be interpreted in several ways (see Figure 16.8). Another way consists in considering that the first endotherm observed at about 100°C is not an overshoot arising from the occurrence of the glass transition but is the transformation of the δ *phase* into the γ *phase* (the corresponding baseline is labelled (A) in Figure 16.8). As a result, at higher temperatures there is no exotherm but an endotherm extending from 110°C to 190°C which is related to the gradual melting of the γ *phase*. As the melting of the γ *phase* is completed the resulting liquid is unstable with regard to the α *phase* and therefore crystallizes under this form at about 200°C.

De Candia et al. have further argued that the δ *phase* exhibits all the characteristics of a non-equilibrium phase as annealing at temperatures lower than the transformation temperature entails loss of solvent. This argument may appear slightly dubious as, unless the solvent pressure is properly equilibrated, the sample is bound to lose solvent on heating.

In fact these conclusions rely upon the way the transformation temperature is determined. Interestingly, annealing at 100°C results in the disappearance of the δ *phase* while the diffraction pattern is consistent with the presence of a mesophase form occurring just before the appearance of the γ *phase* (Manfredi et al. 1995). Consequently, if the transformation temperature is taken at the peak of the first endotherm, i.e. 105°C, the δ *phase* does not disappear before the transition but at the DSC transition. (Note that the

DSC traces were recorded at 10°C/min so there must be some discrepancy with results obtained from annealing at constant temperature.)

Morphology

The morphologies of sPS films exposed to solvents have been investigated by de Candia et al. in dichloromethane, which first yields the δ *form*, and in acetone, a solvent which allows one to produce the γ *form*. As has been mentioned earlier in this chapter, the crystalline lattice slightly differs with the solvent type (Manfredi et al. 1995; de Candia et al. 1996).

The δ *form* texture obtained after immersion for 2 hours in dichloromethane is characterized by a network of long fibre-like structures, which embed in its cavities more irregular features (Figure 16.9). These cavities have an average diameter of 1.5 to 2 μm. A section of these films is shown in Figure 16.9. The fracture surface happens to be more regular than the polymer–solvent interface. Spherical elements are clearly visible, which might be, according to these authors, stretched craze fibres whose ends have melted back during the fracture process. Alternatively, they may be fractured fibres originally present in the sample. In this instance, Daniel et al. (1994) have obtained optical micrographs from sPS films prepared between glass slides wherein liquid toluene has been allowed to diffuse that reveal similar features. As this solvent is known to produce fibrillar thermoreversible gels, these features might well be fibres. Conversely, using *trans*-decalin instead produces large spherulites.

The transformation of the δ *form* into the γ *form* above 120°C for samples prepared from immersion in dichloromethane gives the morphology shown in Figure 16.10. It seems that the morphology has become spherulitic. Here data for the chain trajectory

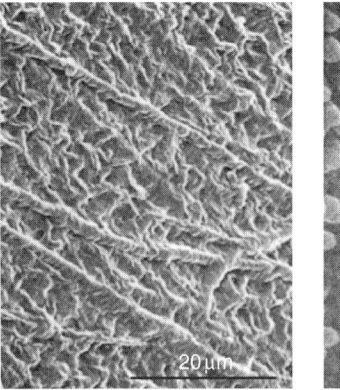 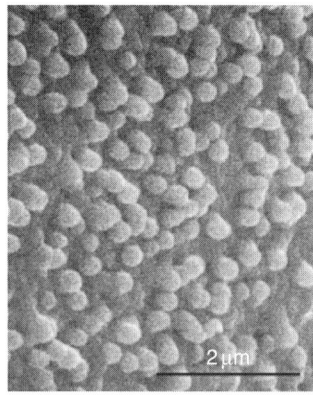

Fig. 16.9. Left: δ form obtained through immersion for 2 h in dichloromethane followed by 24 h extraction under vacuum. **Right:** section of the same film obtained by fracture in liquid nitrogen. (After de Candia et al. 1996, reprinted with permission from Taylor and Francis, www.informaworld.com.)

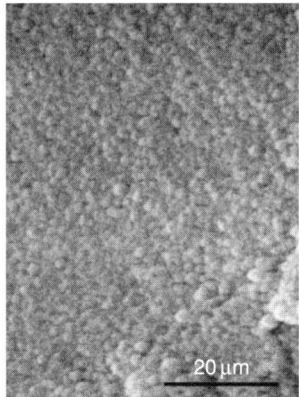 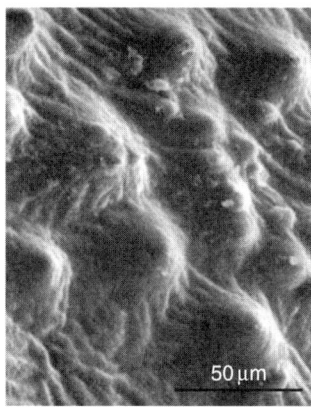

Fig. 16.10. Left: fracture surface of the γ form obtained after annealing of the δ form at 120°C. **Right:** γ form directly obtained from exposure to acetone and solvent removal under vacuum. (After de Candia et al. 1996, reprinted with permission from Taylor and Francis, www.informaworld.com.)

are still missing although they would be of great help in understanding the phenomena involved during the transformation. Indeed, a fibre-like structure is usually thought to be associated with the absence of chain-folding while the occurrence of chain-folding generates lamellae that constitute spherulites. Does the transition from the δ to the γ *form* involve a significant rearrangement of the chain trajectory? This remains a question which has been, as will be discovered below, partially addressed in the case of solution-grown crystals.

As can be seen in Figure 16.10 the morphology of the γ *form* grown directly from exposure to acetone differs markedly from that obtained after thermal processing. Sphere-like objects of about 20 μm in diameter are observed amid furrows. The fracture surface displays similar features.

It seems that the film thickness is also an important parameter. Ray et al. (2002) have obtained from very thin films (thickness below 0.15 μm) arrays of fibrils forming a network whose average mesh size depends upon the exposure temperature. Image processing reveals that the porosity of these samples obeys a log-normal distribution function (Figure 16.11):

$$X(D) = \frac{A}{\sigma_g D\sqrt{2\pi}} \exp\left[-\frac{(Ln(D/D_m))^2}{2\sigma_g^2}\right] \quad (16.2)$$

in which:

$$\sigma_g^2 = \int (LogD - LogD_m)^2 X(D)dD; \quad LogD_m = \int LogD X(D)dD \quad (16.3)$$

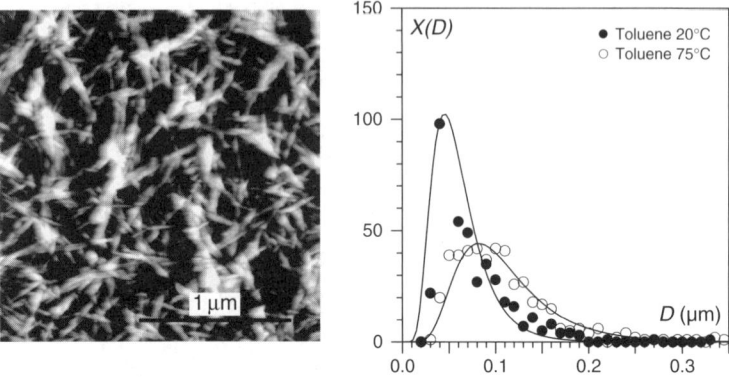

Fig. 16.11. Left: network morphology as obtained from solvent-exposure of thin sPS amorphous films. **Right:** the mesh size distribution function for two different temperatures for toluene-exposed samples. (From Ray et al. 2002. Reprinted with permission from ACS.)

where A is the amplitude and $X(D)$ is the probability density function of the diameters D. D_m is a mean diameter defined through:

$$\int_0^{D_m} X(D)dD = 1/2 \tag{16.4}$$

This procedure determines the average size of the darkest spots. Clearly, the values are not a direct measure of the mesh size but, still, are proportional to this parameter.

Crystalline structures

The crystal structure of the δ_c form (clathrate) is dependent on the solvent used. As a rule, determination of the crystalline lattice is achieved by means of solvent-induced structures that are stretched, or, in other cases, samples that are first obtained under the stretched amorphous state or the stretched crystalline α *form* and then exposed to solvent vapours.

Tarallo and Petraccone (2005) have determined the crystalline lattice of sPS/o-dichlorobenzene, which differs slightly from that derived by Chatani for sPS/toluene. The unit cell is monoclinic, of space group $P2_1/a$ with $a = 1.75$ nm, $b = 1.44$ nm, $c = 0.78$ nm, and $\gamma = 127.4°$ (Figure 16.12). Yet, the existence of two different packings with the same lattice and that are energetically equivalent leads one to describe the crystalline arrangement better with an orthorhombic unit cell of space group $Ccmb$ with $a = 1.75$ nm, $b = 2.29$ nm, and $c = 0.78$ nm which takes into account the random occurrence of each monoclinic lattice.

The same authors have also described the crystal structure of complexes obtained from CS_2 and iodine, I_2 (Tarallo and Petraccone 2004). The oriented samples of sPS/CS_2 compounds were obtained from oriented solid sPS crystallized under the α form and

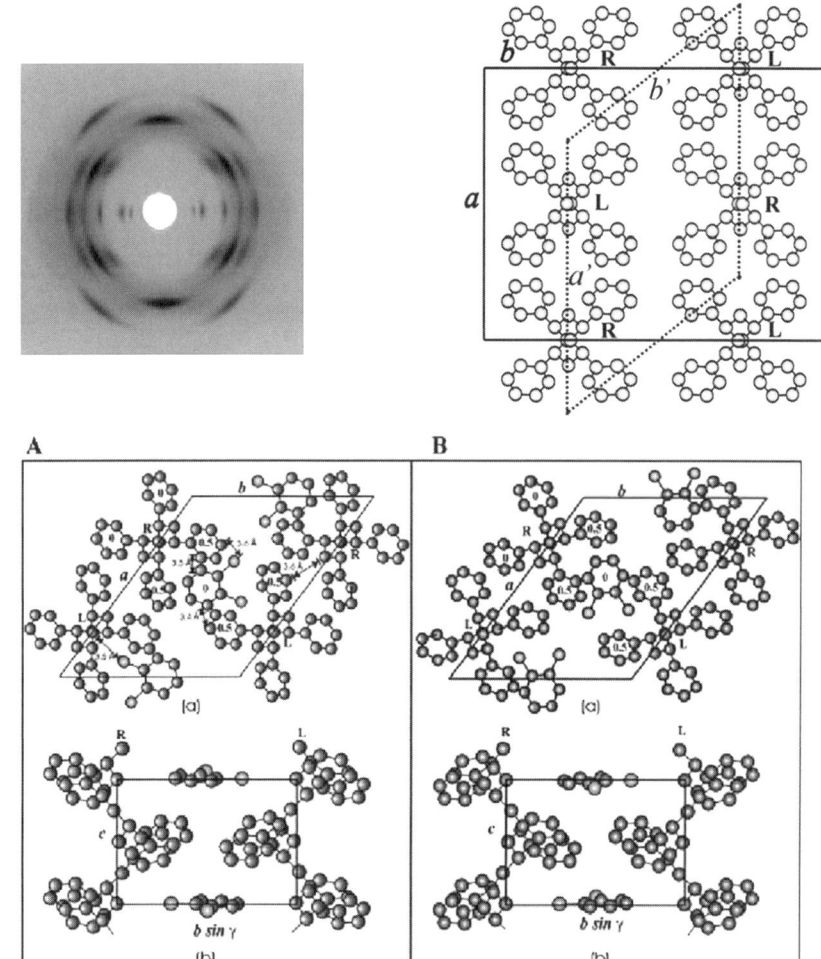

Fig. 16.12. Crystal lattice of sPS/*o*-dichlorobenzene. **Upper left:** diffraction pattern of oriented samples. **Upper right:** the orthorhombic unit cell. **Lower:** The two possible monoclinic unit cells of minimum energy: (a) *ab* projection; (b) *cb* sin γ projection. (From Tarallo and Petraccone 2005. Reprinted with permission from Wiley Interscience.)

then exposed to CS_2 vapours. Similarly, the oriented samples of sPS/I_2 compounds were obtained from oriented solid sPS crystallized under the α form and then exposed to I_2 vapours. The diffraction pattern together with the corresponding crystalline lattice sPS/CS_2 compounds are shown in Figure 16.13. The crystalline lattice sPS/CS_2 compounds is of space group $P2_1/a$ with $a = 1.73$ nm, $b = 1.27$ nm, $c = 0.79$ nm, and $\gamma = 120°$. Tarallo and Petraccone further report that sPS/I_2 compounds possess the same crystalline lattice as that found in sPS/CS_2 compounds.

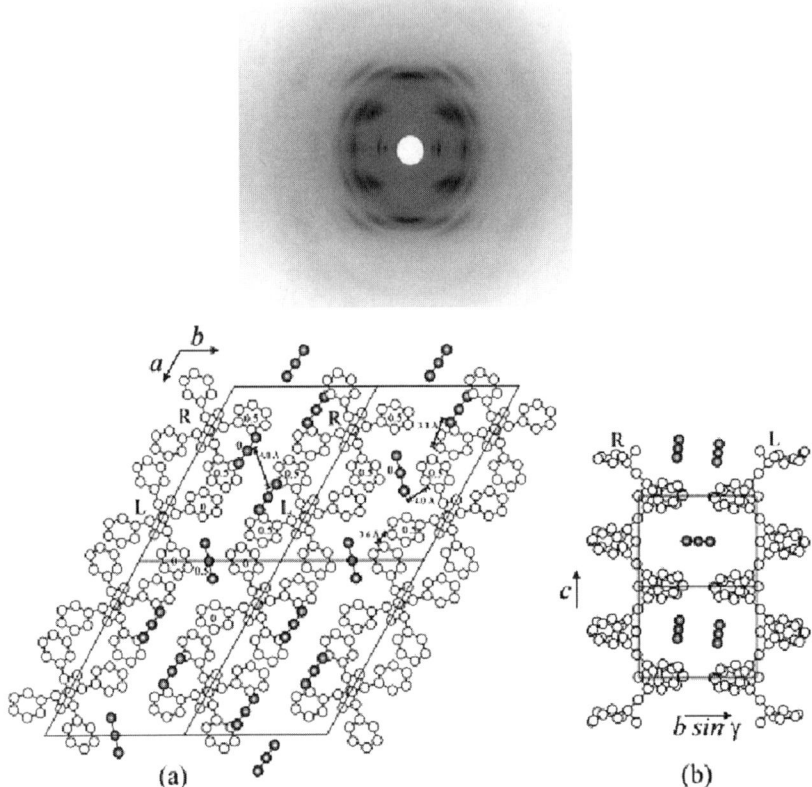

Fig. 16.13. Crystalline structure of sPS/CS_2 compounds. **Upper:** the diffraction pattern of oriented samples. **Lower left**: the lattice as seen parallel to the helical axis. **Lower right:** the lattice as seen perpendicular to the helical axis and to $b \sin \gamma$; the occupancy of the cavity between polymer stems can be of one or two CS_2 molecules. (From Tarallo and Petraccone 2004. Reprinted with permission from Wiley Interscience.)

It is worth noting that, in all the cases studied, the crystal structure is basically the same, so the use of δ_c *form* as proposed by Malik et al. (2006) for designating these systems is still relevant.

Recently, Tarallo et al. obtained the crystalline structures for other intercalates (of δ_i form) using the fluorescent molecules 1,3,5,-trimethyl benzene and 1,4-dimethyl naphthalene (Tarallo et al. 2006). The molecular compounds are prepared by immersing for 48 hours in the liquid solvent, solid amorphous sPS samples that have been stretched beforehand. The crystal packings are shown in Figure 16.14. For both systems Tarallo et al. propose a monoclinic unit cell in which the s(2/1)2 helices and guest molecules are packed according to the space group $P2_1/a$ with $a = 1.73$ nm, $b = 1.54$ nm, $c = 0.78$ nm, and $\gamma = 95.7°$ for sPS/1,3,5,-trimethyl benzene compound, and with $a = 1.74$ nm, $b = 1.72$ nm, $c = 0.78$ nm, and $\gamma = 116.4°$ for sPS/1,4-dimethyl naphthalene compound.

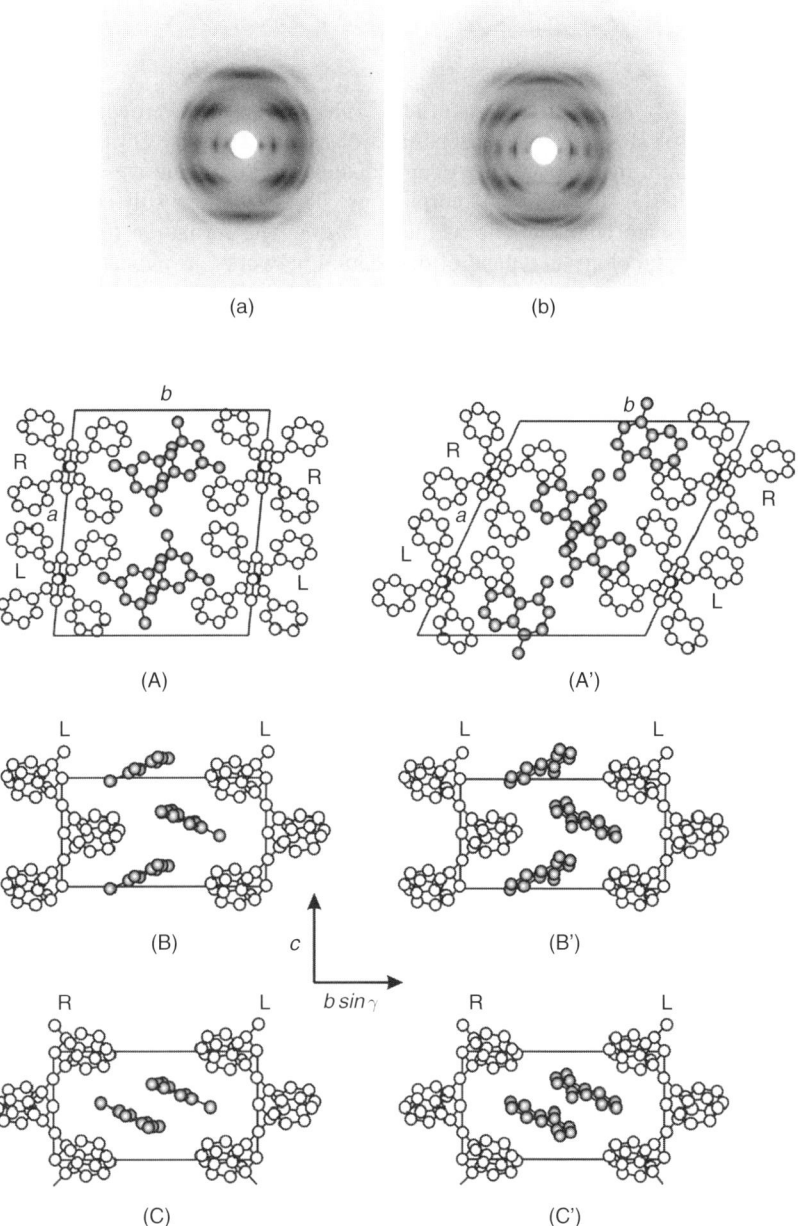

Fig. 16.14. Top: diffraction patterns for oriented samples, (a) = sPS/1,3,5-trimethyl-benzene; (b) = sPS/1,4-dimethyl-naphthalene. **Bottom:** packing mode proposed for the systems sPS/1,3,5-trimethyl benzene (A,B,C) and sPS/1,4-dimethyl naphthalene (A', B', C'). The case of random left-handed and right-handed helices is shown. (From Tarallo et al. 2006. Reprinted with permission from Elsevier.)

Solution-cast compounds

Single crystals

Wang et al. (1992) and later Moyses et al. (1998) succeeded in growing crystals from dilute solutions of sPS in ethyl benzene, and observing them by electron microscopy. Indeed, the vacuum in electron microscopy chamber tends to remove solvent from the crystals, thus altering the original morphology. These authors still consider that the observed features are characteristic of the δ_c *phase*. The formation of lamellar stacks or ribbon-like morphology is typical of dislocation networks, as indicated by the Moiré fringes. Both groups were unable to observed individual single crystals. Moyses and colleagues have noticed a change towards a more fibrous morphology at a crystallization temperature of 50°C.

On annealing above 100°C to achieve the δ_c to γ transformation, the above-mentioned authors noticed that the original morphology becomes more diffusive. Wang et al. observed a significant shrinkage of the b spacing – later confirmed by Moyses et al. (from 1.142 nm down to 1.077 nm) – that is clearly linked to the removal of solvent molecules from the crystalline lattice.

Conversely, Moyses et al. observed an important increase of the X-ray long spacing (from 6 nm up to 9 nm), which suggests crystal thickening. On further heating, the thickening process is reactivated at the transition from the γ *phase* to the β *phase*.

The folding habit of sPS in crystals grown from dilute solutions of ethyl benzene (the δ_c *phase*) has been extensively studied by Moyses and Spells (1998, 1999) by means of neutron scattering. Their neutron study by sample tilting allows determination of the in-plane and the out-of-plane radius of gyration. They observe two regimes depending upon the molecular weight. Comparison between the out-of-plane radius of gyration and the long spacing determined by SAXS shows close agreement for the 'low' molecular weight samples. Conversely, the out-of-plane radius of gyration is about twice as large for the 'high' molecular weight samples. As with other polymers such as polyethylene (Sadler and Spells 1984) and isotactic polystyrene (Guenet et al. 1990), this suggests that superfolding occurs for higher molecular weight chains, i.e. sheets containing the stems of one chain fold back on themselves but also occupy two adjacent lamellae. Studies at larger q-vectors led Moyses and Spells to consider a sheet-like model arising from adjacent re-entry in the a direction. The best agreement with the scattering curves is achieved by taking a probability of adjacent re-entry $P_A = 0.75$ (Figure 16.15). This implies that regular alternation in stem helicity takes place.

The neutron results obtained after completion of the transformation of the δ_c *phase* into the γ *phase* are consistent with the occurrence of significant lateral movement of stems during the transition.

Moyses et al. (1998) have studied the thermal stability from mats of single crystals grown from dilute solutions in ethyl benzene. The method of preparation is clearly different as the mats are allowed to dry in air instead of vacuum, but yields essentially similar results, albeit that certain differences can be observed. These authors have studied simultaneously

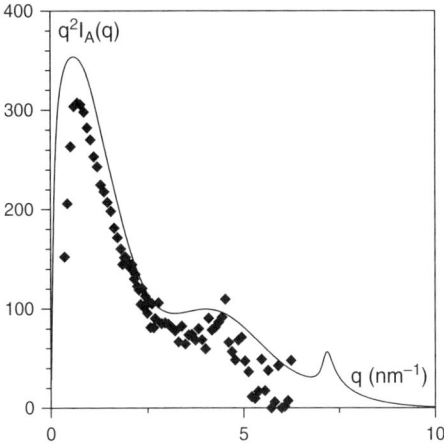

Fig. 16.15. Small-angle neutron scattering data for deuterated sample ($M_w = 144\,500$) in a hydrogenous matrix for crystals under the δ_c-phase. The solid line is calculated for 22 stems in 2 sheets with a probability of adjacent re-entry $P_A = 0.75$. (Data from Moyses and Spells 1998.)

the DSC thermogram, weight loss, lamellar thickness and infrared intensity of the doublet at 940 and 943 cm^{-1} normalized by the 1585 cm^{-1} (benzene ring stretching vibration). The peak at 940 cm^{-1} corresponds to the ethyl benzene-solvated helix while that at 943 cm^{-1} to the desolvated helix. Their results are given in Figure 16.16. In the DSC curve the first event occurs at about 70°C which, as suggested by these authors,

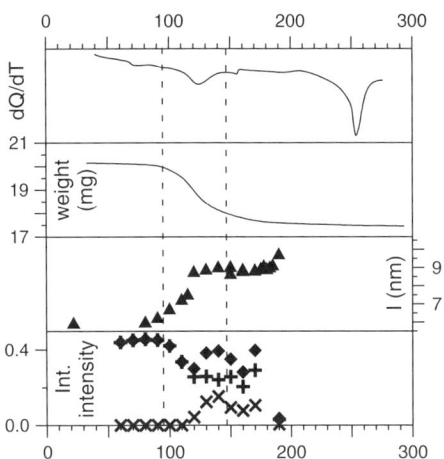

Fig. 16.16. Simultaneous study of the DSC response (**upper**), weight loss (**upper middle**), lamellar thickness (**lower middle**) and infrared integrated intensity for the band at 943 cm^{-1} of the ethyl benzene-solvated helix (+), and at 940 cm^{-1} of the desolvated helix (✖); (♦) corresponds to the sum of both bands. (Data from Moyses et al. 1998.)

may well correspond to the glass transition. The transformation of the δ_c *phase* into the γ *phase* evidently takes place by means of an endothermic event peaking at about 120°C and extending over some 40°C. At 158°C an exothermic event corresponds to the crystallization under the α *phase*. The endothermic peak at 120°C and its extension is clearly associated with the solvent loss and with the change of lamellar thickness (from 6 nm up to 9 nm).

Infrared measurements indicate that the helical conformation undergoes a slight, partial disorder just before the onset of the transition as seen by DSC but retrieves its initial value at the end of the transformation process. This again indicates that the transformation of the δ_c *phase* into the γ *phase* is achieved while keeping the 2_1 helix virtually unaltered.

Spherulitic textures. Phase diagrams

For some sPS/solvent systems, it is observed that quenching more concentrated solutions produces spherulitic structures, as shown in Figure 16.17 for sPS/*trans*-decalin systems.

Deberdt and Berghmans (1993), Daniel (1996), and later Malik et al. (2005b) determined the phase diagrams for sPS/*trans*-decalin systems, Moyses et al. (1997) for sPS/ethylbenzene systems. Roels et al. (1994) obtained these diagrams using a series of 'dichlorobenzenes'.

The phase diagrams obtained by Deberdt and Berghmans, and by Daniel for sPS/*trans*-decalin systems are basically similar and so are the interpretations. Daniel was able to measure with sufficient accuracy the enthalpies associated with the different thermal events; this phase diagram together with typical DSC traces is shown in Figures 16.18 and 16.19.

An incongruently melting compound (C_1) occurs, as revealed by the non-variant event at $T = 125 \pm 5°C$, which corresponds to the δ_c *form* (clathrate). The variation of the enthalpy associated with the first endotherm varies linearly on either side of a concentration which

Fig. 16.17. Spherulitic morphology of sPS/*trans*-decalin systems. Normarsky phase contrast. (From Malik et al. 2005a. Reprinted with permission from ACS.)

Syndiotactic polystyrene 229

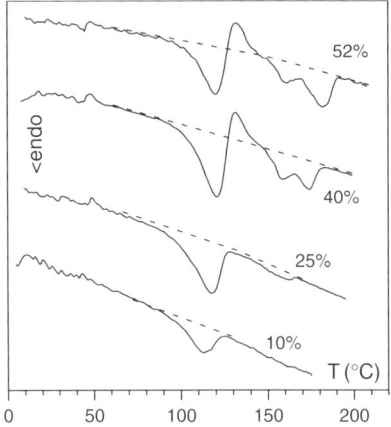

Fig. 16.18. DSC traces for sPS/*trans*-decalin systems. (After Daniel et al. 1996.)

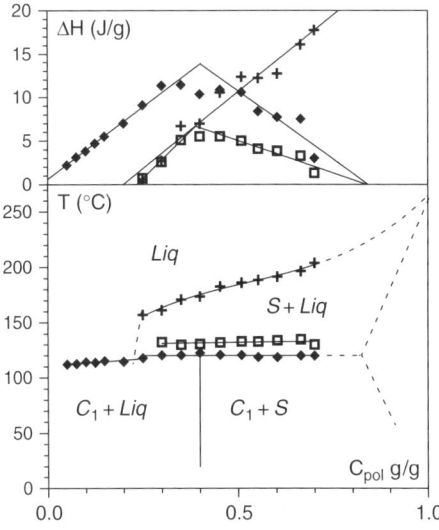

Fig. 16.19. Temperature–concentration phase diagram and Tamman's diagram for sPS/*trans*-decalin systems. (□) = exothermic event. (After Daniel et al. 1996.)

gives the thermodynamic stoichiometry, namely *1 trans-decalin molecule/monomer unit*. The incongruent melting at $T = 125 \pm 5°C$ of compound C_1 is immediately followed by an exothermic event, which suggests therefore the existence of a metastable phase.

According to Guenet (2003), the exotherm is not a recrystallization effect but corresponds to the transformation of the 2_1 helix, which is a metastable helical form in the absence of solvent, into the planar zig-zag conformation. The main argument for supporting this view relies on the variation with polymer concentration of the enthalpy associated with this exotherm. Its variation follows exactly the same behaviour as the enthalpy associated with the incongruent melting of compound C_1 (see Figure 16.17). If it were a recrystallization process this enthalpy should increase continuously with polymer concentration. Conversely, if this thermal event is related to the transformation $2_1 \Rightarrow$ *planar zig-zag*, then it is directly proportional to the content of 2_1 helix, which in turn is proportional to the amount of compound C_1. This is indeed what is experimentally observed.

To summarize, compound C_1 transforms into a non-solvated form, hence the endotherm, and then the 2_1 helix transforms into the planar zig-zag, hence the exotherm. This eventually produces the *β form*.

Time-resolved X-ray diffraction and neutron diffraction experiments give further support to the outcomes of the phase diagram (Malik et al. 2005b).

In Figure 16.20 are shown the diffraction patterns obtained for a polymer concentration $X_{pol} = 0.3$ w/w, wherein incongruent melting takes place. The transformation of the δ_c *form* into the *β form* is clearly seen. Yet, Malik et al. have not observed any intermediate form such as the *γ form*, which is consistent with the interpretation given by Guenet for the occurrence of an exotherm.

Neutron diffraction investigations carried out on this system are of special interest as they offer a convincing illustration of the potential of this technique in the case of polymer/solvent molecular compounds, particularly for differentiating them from non-solvated structures (see Chapter 2). As has been advocated by Guenet, this technique

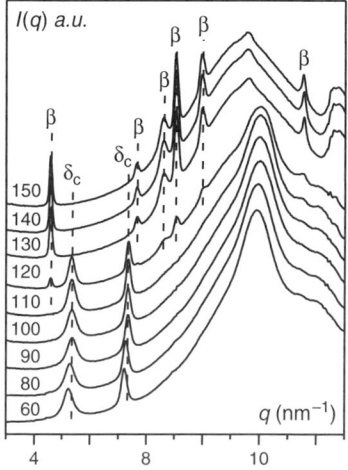

Fig. 16.20. Time-resolved X-ray diffraction experiments performed at a heating rate of 2°C/min for sPS/*trans*-decalin systems. $X_{pol} = 0.3$ w/w, temperatures in °C. The reflections of the different forms are indicated. (Data from Malik et al. 2005b.)

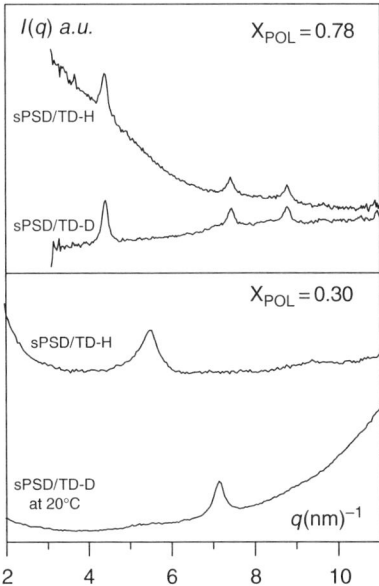

Fig. 16.21. Neutron diffraction patterns obtained. **Lower:** $X_{pol} = 0.3$ w/w; the system is under the δ_c *form*; the use of two differently labelled *trans*-decalins, hydrogenous *trans*-decalin (TD-H) and deuterated *trans*-decalin (TD-D), affects the diffraction pattern, and particularly the ratio between those reflections at $q = 5.5\,\text{nm}^{-1}$ and $q = 7.2\,\text{nm}^{-1}$. **Upper:** the system is under the β *form*, labelling of the solvent does not affect the ratio between the different reflections. (From Malik et al. 2005b.)

allows one to confirm, or to disregard, the occurrence of such compounds without previous knowledge of the crystalline lattice and also without applying destructive processes to the sample, such as stretching (Guenet 1986, 1992). It is particularly useful when the system exhibits a very low number of reflections.

As can be seen in Figure 16.21, the use of deuterated or hydrogenous *trans*-decalin significantly alters the diffraction pattern when sPS/*trans*-decalin forms a molecular compound, while the diffraction pattern is not sensitive to the solvent labelling in the case of the β *form* (Malik et al. 2005b).

In the case of *sPS/ethyl benzene* systems, Moyses (1997) has observed the occurrence of no fewer than three compounds that are all incongruently melting (Figure 16.22). Compound C_1 of thermodynamic stoichiometry 1.6 *ethyl benzene molecule/monomer* melts incongruently at $120° \pm 3°C$, compound C_2 of thermodynamic stoichiometry 0.5 *ethyl benzene molecule/monomer* melts incongruently at $150° \pm 3°C$ while compound C_3 of thermodynamic stoichiometry 0.25 *ethyl benzene molecule/monomer* melts incongruently at $185° \pm 3°C$. All these events are summarized as follows (see Figure 16.22):

$$C_1 + Liq \xrightarrow{T=120°C} C_2 + Liq \xrightarrow{T=150°C} C_3 + Liq \xrightarrow{T=183°C} \beta\ phase$$

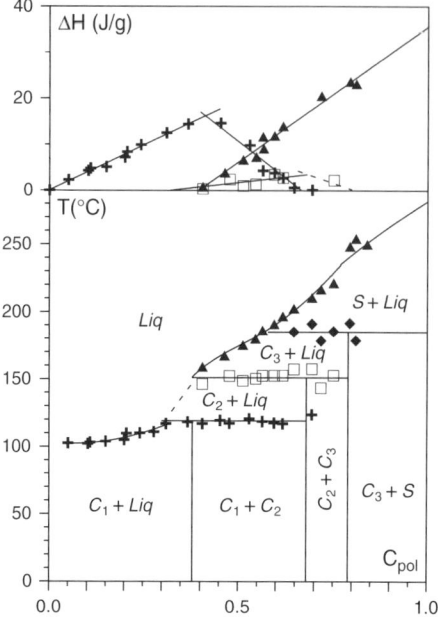

Fig. 16.22. Temperature–concentration phase diagram for sPS/ethyl benzene. (Data from Moyses 1997.)

Interestingly, compound C_3 possesses the same stoichiometry as that reported by Chatani and co-workers for sPS/toluene systems obtained by solvent vapour exposure (Chatani et al. 1993). Here it is worth emphasizing that those single crystals grown by Moyses et al. (1998) from dilute solutions in ethyl benzene certainly correspond originally to C_1, so any drying process in air or vacuum is liable to modify the original stoichiometry. As will be discovered in what follows, this may account for why the morphology of these single crystals is not so well defined (Wang et al. 1992; Moyses et al. 1998). The loss of solvent that takes place while drying takes place is liable to promote the creation of defects within the crystal mats.

As expected, the shape of the temperature–concentration phase diagrams is strongly dependent upon the solvent type. Additional diagrams shown in the section Thermoreversible gels, below, will again emphasize this point. The degree of solvation (stoichiometry) as well as the number of compounds varies considerably from one solvent to another.

Compounds from high-melting-point solvents

Until recently, sPS compounds were always prepared from solvents that are liquids at room temperature. Malik et al. (2005a, 2005b, 2006) have carried out a detailed study of systems in which the solvent, such as naphthalene, biphenyl, benzophenone and diphenyl methane, is a solid at room temperature. Following usual practice, they first mapped out the temperature–concentration phase diagram by DSC, and then examined the structure by time-resolved X-ray diffraction carried out at a heating rate equal to or close to that used in their DSC experiments.

The temperature–concentration phase diagram for sPS/naphthalene systems is shown in Figure 16.23.

The variations of the enthalpies associated with the different events together with the characteristic shape of the phase diagram suggest the existence of the two compounds C_1 and C_2 of different thermodynamic stoichiometries. C_1 is probably of the singularly melting type. Interestingly, C_1 transforms into another compound C_2 either by cooling at low temperature (when naphthalene crystallizes) in a polymer fraction range $0 < X_{pol} < 0.5$, or by heating at high temperature ($T \approx 160°C$) in a polymer fraction range $0.5 < X_{pol} < 0.7$. The thermodynamic stoichiometry of compound C_1 as derived from the Tamman diagram corresponds to the maximum of the enthalpy associated with the melting and transformation of the compound, namely at $X_{\gamma 1} = 0.5$, which yields about 4 *naphthalene molecules per* 5 *monomer units*.

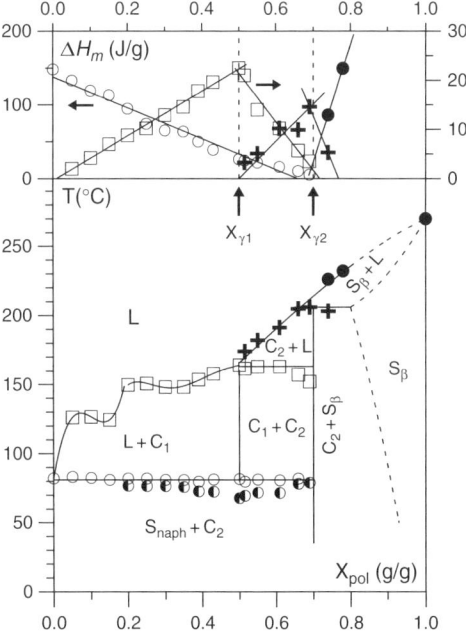

Fig. 16.23. Temperature–concentration phase diagram and Tamman's diagram for sPS-naphthalene systems. C_1 and C_2 are two compounds of different thermodynamic stoichiometries (see text). S_β is a solid polymer solution corresponding to the β form. $X_{\gamma 1}$ and $X_{\gamma 2}$ are the stoichiometric compositions of compounds C_1 and C_2, respectively. (From Malik et al. 2005a.)

On the other hand, compound C_2 is of the incongruently melting type. The thermodynamic stoichiometry of this compound is derived both from the polymer fraction at which free naphthalene molecules no longer remain in the system, and from the polymer fraction at which the melting enthalpy of compound C_2 passes through a maximum. This yields $X_{\gamma 2} = 0.7 \pm 0.02$, which corresponds to about 1 *naphthalene molecule per 3 monomer units*.

As will be discussed later in the section on sPS thermoreversible gels, the transformation from one compound to another while crystallizing the solvent has already been observed for sPS/benzene systems. This suggests that part of the solvent in compound C_1 is loosely bound.

Beyond $X_{pol} = 0.7$ g/g, the system consists of compound C_2 and a solid phase S_β (namely no solvent molecules participating in the crystalline lattice). This solid phase S_β consists of the *β form* of sPS.

Malik et al. have further examined the crystalline structure of the different phases as a function of polymer fraction by means of X-ray time-resolved experiments (Figure 16.24). The heating rate was nearly the same as that used for the DSC investigations that provided the T–C phase diagram, namely 2°C/min. The intense reflections seen up to $X_{pol} = 0.61$ at $q = 5.5$ nm^{-1} ($d = 1.13$ nm) and $q = 7.4$ nm^{-1} ($d = 0.85$ nm) are characteristic peaks for the δ_c *form*. At a polymer fraction $X_{pol} = 0.78$, the reflections at $q = 4.5$ nm^{-1} ($d = 1.42$ nm), 7.6 nm^{-1} ($d = 0.83$ nm), 8.6 nm^{-1} ($d = 0.73$ nm), 9.0 nm^{-1} ($d = 0.7$ nm) and 10.0 nm^{-1} ($d = 0.63$ nm) clearly indicate the overwhelming presence of the *β form* of sPS, although the reflections of the δ_c *form*, admittedly rather weak, are still observed.

These results show that there is no significant difference, as regards the position of the peaks in the diffraction pattern, in the concentration range from $X_{pol} = 0.05$ to

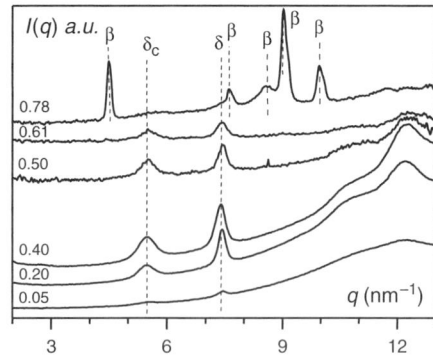

Fig. 16.24. X-ray diffraction patterns of sPS/naphthalene systems obtained at 90°C to eliminate the strong naphthalene diffraction for samples containing free solvent. Polymer concentration as indicated in w/w. The main reflections of the δ_c form and the β form are shown. (From Malik et al. 2005a.)

$X_{pol} = 0.61$ although the two different compounds C_1 and C_2 are involved. Only the ratio between the reflection at $q = 5.5\,\mathrm{nm}^{-1}$ and at $q = 7.4\,\mathrm{nm}^{-1}$ seems to be affected. This may indeed suggest that the only difference lies, therefore, in the value of their thermodynamic stoichiometries. It will be certainly worth shedding some light on this point in order to bridge the gap between the conclusions drawn from DSC and those drawn from structural investigations.

Neutron diffraction experiments shown in Figure 16.25 confirm the occurrence of compound C_2 at $X_{pol} = 0.74$. First, the position of the reflections correspond to the usual δ_c form, but, also, the relative intensities of these reflections are dependent upon the solvent labelling.

The morphology obtained with this solvent is of interest since a fibrillar network is obtained (Figure 16.26). Above the melting point of naphthalene, the sPS/naphthalene system displays all the characteristics of fibrillar thermoreversible gels, and particularly a relative transparency as compared with spherulitic assemblies. sPS/naphthalene systems will be further discussed below under Multiporous networks.

The compounds prepared from *sPS/biphenyl* solution have revealed another solvated form which was actually first observed in sPS/benzene gels by Daniel and colleagues as early as 1996 (Daniel et al. 1996). The temperature–concentration phase diagram shown in Figure 16.27 has been mapped out by Malik et al. (2006a).

Three non-variant events are observed: at $T = 70°C$ (solvent melting, endothermic), $T = 90°C$ (endothermic), and $T = 100°C$ (exothermic). The shape and the variation of the enthalpies in the Tamman diagram are consistent with the existence of two sPS/solvent compounds and one solid phase. The maximum of the melting enthalpy ΔH_{90} of the non-variant transformation at 90°C provides data on the thermodynamic stoichiometry of compound C_1, of about 1 *biphenyl/monomer*. The concentration at

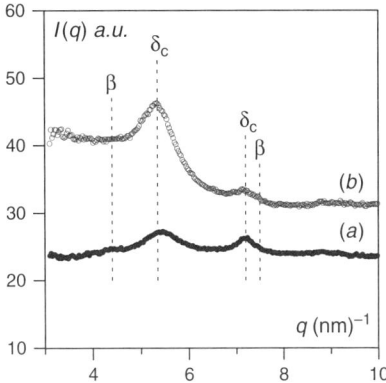

Fig. 16.25. Neutron diffraction pattern for sPS/naphthalene system for $X_{pol} = 0.73$ g/g for different isotopic labelling: (a) sPSD/NaphD and (b) sPSD/NaphH. The characteristic reflections of the δ_c *form* and of the β *form* are shown. (From Malik et al. 2006a.)

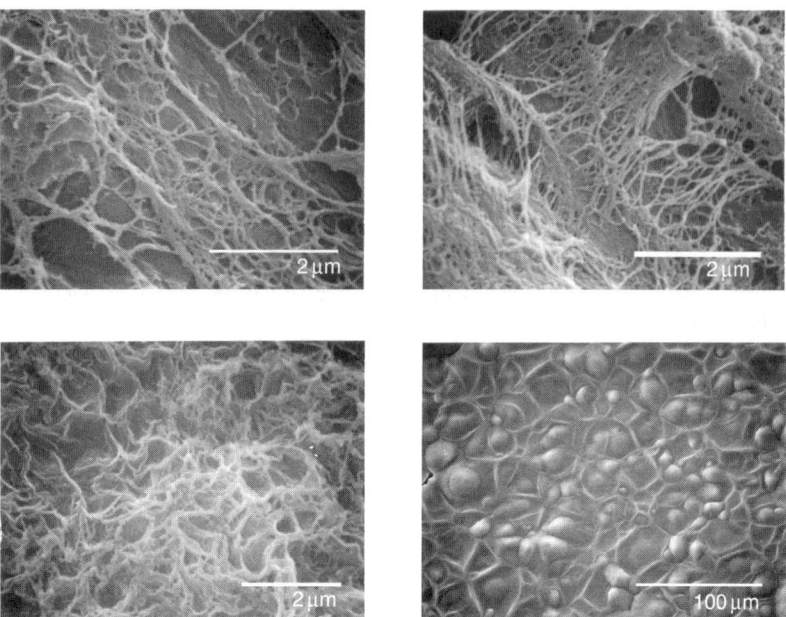

Fig. 16.26. Scanning electron micrographs of: **upper left**, sPS/naphthalene-dried gels for an initial concentration $X_{pol} = 0.30$ w/w; **upper right**, sPS/biphenyl system ($X_{pol} = 0.2$ w/w) after sublimation of biphenyl (from Malik et al. 2006a); **lower left**, sPS/benzophenone system $X_{pol} = 0.40$ w/w after sublimation of benzophenone; **lower right**, sPS/diphenyl methane system $X_{pol} = 0.30$ w/w. (From Malik et al. 2005a, 2006a. Reprinted with permission from ACS.)

which ΔH_{90} together with the melting enthalpy of the solvent becomes zero gives a thermodynamic stoichiometry for compound C_2 of about 0.5 *biphenyl/monomer*.

As with benzene and naphthalene, the transformation of compound C_1 into compound C_2 also occurs at the crystallization temperature of biphenyl. Finally, a narrow domain, containing $C_2 + liquid$ has to be considered so as to fulfil Gibbs phase rules. As with *trans*-decalin systems, the exothermic non-variant event at $T = 100°C$ is probably related to the transformation of the 2_1 helical form into the planar zig-zag conformation which occurs just after the desolvation of compound C_2. In fact, this transformation is exothermic since, in the absence of solvent, the 2_1 helix is *metastable*. At this stage of investigation the detail of the structural and conformational events taking place at this temperature is not known. In particular, it cannot be said whether this exotherm is due to the following sequence $\delta_c \rightarrow \delta_e \rightarrow \gamma \rightarrow \beta$.

Note that extrapolation to $X_{pol} = 1$ of the enthalpy associated with the melting of the solid phase S_β gives $\Delta H = 35$ J/g. Considering that the melting enthalpy for 100% crystallized sPS should be of about $\Delta H_{sPS} \approx 75$ J/g, a value valid for all non-solvated crystals of polyolefins, a degree of crystallinity of about 50% is found.

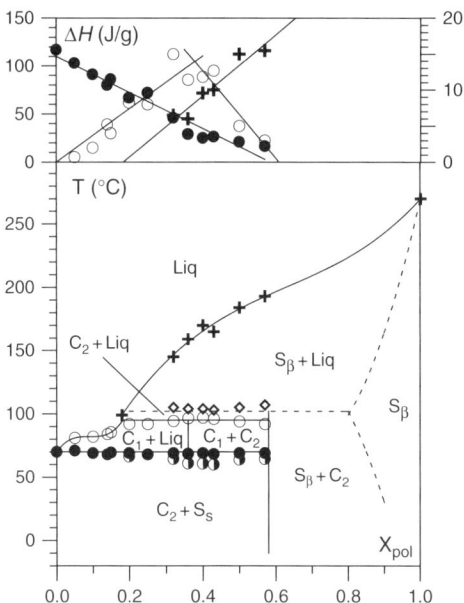

Fig. 16.27. Temperature–concentration phase diagram and Tamman's diagram for sPS/biphenyl systems. Diamond symbols (◇) stand for an exothermic event taking place at $T = 100°C$; all other symbols represent endothermic events. C_1 and C_2 are incongruently melting compounds, and S_s is a solid solution which is chiefly composed of biphenyl crystals. S_β is a solid phase composed of the sPS β *form*. In Tamman's diagram the symbols (+) and (○) are related to the right ordinate, while the symbol (●) is related to the left ordinate axis. (From Malik et al. 2006a.)

The structural investigations by means of time-resolved X-ray diffraction have confirmed the outcomes from the temperature–concentration phase diagram (Figure 16.28). Moreover, the δ_i (intercalate) form, first reported for sPS/benzene gels, is also observed in this solvent. In particular, the first reflection is seen at $q = 4 \pm 0.1\,\mathrm{nm}^{-1}$, while the δ_c form exhibits a first reflection at $q = 5.5 \pm 0.1\,\mathrm{nm}^{-1}$. Interestingly, this form occurs with a solvent of molecular size significantly larger than that of benzene or of norbornadiene ($V_{biphenyl} = 156\,\mathrm{cm}^3/\mathrm{mol}$; $V_{benzene} = 89\,\mathrm{cm}^3/\mathrm{mol}$; $V_{norbornadiene} = 101\,\mathrm{cm}^3/\mathrm{mol}$). The structural stoichiometry proposed by Petraccone et al. (2005b), namely 1 *solvent molecule for 2 monomer units*, corresponds to the thermodynamic stoichiometry derived from the phase diagram.

Yet, the appearance of what is probably a mesophase, which Malik and colleagues have named the δ_N *form*, remains puzzling. They suggest that this phase possesses a rather nematic-like structure as the first peak at $q = 6.86\,\mathrm{nm}^{-1}$ ($d = 0.91\,\mathrm{nm}$) is quite narrow while the second peak at $q = 11.9\,\mathrm{nm}^{-1}$ ($d = 0.53\,\mathrm{nm}$) is much broader. This phase is clearly observed only at high concentrations (see Figure 16.28). Malik et al. have reasoned that this phase possibly corresponds to C_2 since, for lower polymer concentrations above the melting point of biphenyl, the reflections related to this phase are conspicuously absent. This assumption remains, however, to be confirmed.

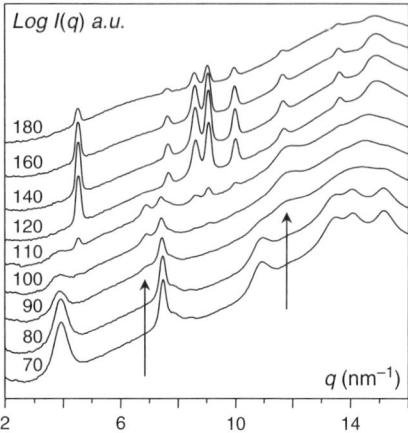

Fig. 16.28. Time-resolved X-ray diffraction patterns for sPS/biphenyl systems at $X_{pol} = 0.5$ w/w. The diffraction pattern of the δ_i *form* (intercalate) is seen at low temperature while that of the β *form* is seen at high temperature. Temperatures are indicated in °C. Arrows indicate two reflections that correspond to the δ_N *form* (see text for details). (From Malik et al. 2006a.)

The morphology of the sPS/biphenyl systems is fibrillar, as shown in Figure 16.26. The fibrils have a cross-section of about 50 nm diameter while the mesh size lies in the micrometre range for concentrations around $X_{pol} = 0.2$ w/w.

In the case of *sPS/benzophenone* systems, although the temperature–concentration phase is quite similar to that of sPS/biphenyl systems, Malik et al. (2006a) suspect the occurrence of a third compound. They designated it as C_o as its existence relies only on the absence of linear variation of the enthalpy associated with a non-variant thermal event (Figure 16.29). In addition, the behaviour is somewhat more complex at the solvent melting temperature, which might well result from the existence of this hypothetical third compound. Two non-variant events can be observed within 10°C near this temperature which can be summarized as:

$$C_2 + S \rightarrow C_1 + L \quad \text{and} \quad C_1 + L \rightarrow C_o + L$$

This is certainly an interesting case, which deserves deeper study.

Investigations by time-resolved X-ray diffraction are consistent with the outcomes of the temperature–concentration phase diagram, and have also evidenced the occurrence of the δ_i (intercalate) form (Figure 16.30). Note that benzophenone has about the same molecular size as biphenyl ($V_{benzophenone} \approx 168$ cm^3/mol).

As with sPS/biphenyl systems, the δ_N *form* is observed above the benzophenone melting point. Here too, it may correspond to C_2. One may, however, wonder whether this is not a transformation of the original lattice of C_2 into a nematic-like structure while preserving the stoichiometry.

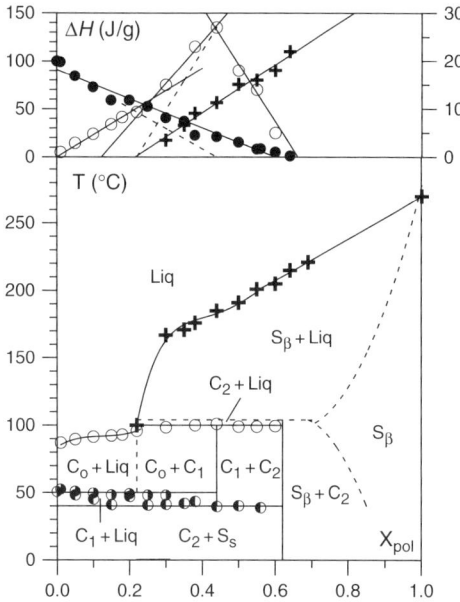

Fig. 16.29. Temperature–concentration phase diagram and Tamman's diagram for sPS/benzophenone systems. C_1 and C_2 are incongruently melting compounds, and S_s is a solid solution which is chiefly composed of benzophenone crystals. S_β is a solid phase composed of the sPS *β form*. C_o is a hypothetical compound whose existence is inferred from the non-linear variation of the enthalpy associated with the event at 90–100°C (○ symbols), but also from the existence of two-non-variant events near the solvent melting temperature. The dotted line in Tamman's diagram highlights the possible variation of the enthalpies if the endotherm could be resolved in two endotherms. In Tamman's diagram the symbols (+) and (○) are related to the right ordinate, while the symbol (●) is related to the left ordinate axis and corresponding to the total melting enthalpy related to the benzophenone melting. (Data from Malik et al. 2006a.)

The morphology of sPS/benzophenone systems is not straightforwardly definable. The sample surface after benzophenone elimination resembles a crumpled cloth (see Figure 16.26). This peculiar morphology may arise from the sublimation process, which might not be as efficient as it is with naphthalene and biphenyl. It seems that fibrils are present in the sample and are therefore responsible for this surface structure. Electron microscopy investigations on freeze-fractured samples should cast some light on this point.

Malik and colleagues have investigated a third system in this series of bulky solvents, namely *sPS/diphenyl methane*. Diphenyl methane also has a high molar volume, $V_{diphenylmethane} \approx 166$ cm^3/mol.

The temperature–concentration phase diagram, together with the Tamman diagram, is shown in Figure 16.31. These diagrams are quite similar to those of *sPS/biphenyl* and *sPS/benzophenone* systems. Here too, the occurrence of two compounds is suggested.

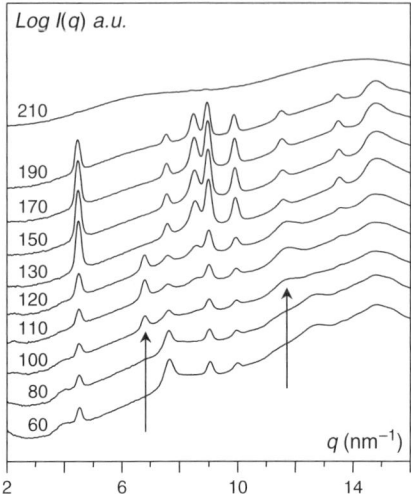

Fig. 16.30. Time-resolved X-ray diffraction patterns for sPS/benzophenone systems at $X_{pol} = 0.59$ w/w. Temperatures are indicated in °C. The diffraction pattern of the δ_i *form* (intercalate) is seen at low temperature together with that of the β *form*. Here the system is probably already slightly beyond the thermodynamic stoichiometry, hence the coexistence of the β *form* and the δ_i *form*. Arrows indicate two reflections that correspond to the δ_N *form* (see text for details). (From Malik et al. 2006a.)

Compound C_1 has a thermodynamic stoichiometry of about 1 *diphenyl methane/monomer unit* against 1 *diphenyl methane/3 monomer units* for compound C_2. Also, compound C_1 transforms into compound C_2 while the solvent crystallizes.

Time-resolved X-ray experiments have also been carried out on these systems. Typical diffraction patterns are displayed for a polymer concentration of $X_{pol} = 0.43$, which corresponds approximately to the stoichiometric composition of compound C_1 (Figure 16.32). As with sPS/biphenyl and sPS/benzophenone systems, reflections that differ from the usual δ_c *form* are observed, and show that the crystal structure is, rather, of the δ_s *form*. Typical reflections are seen at $q = 4 \pm 0.1$ nm^{-1}, 7.6 ± 0.1 nm^{-1}, 9.6 ± 0.1 nm^{-1}, 11.4 ± 0.1 nm^{-1}, and 13.1 ± 0.1 nm^{-1}. They slightly differ at the highest q values from the reflections observed for sPS/benzophenone and sPS/biphenyl, which indicates that there are variants of this δ_s-form. Also, the δ_N *form* appears above 90°C and vanishes at about 120°C. At this temperature the transformation into the β *form* is completed.

Unlike sPS/biphenyl and sPS/benzophenone systems, the morphology of sPS/diphenylmethane systems is not fibrillar but looks spherulitic instead (see Figure 16.26). The dissimilarity in morphology between sPS/biphenyl and sPS/diphenyl methane systems is therefore rather surprising as they exhibit very similar phase diagrams.

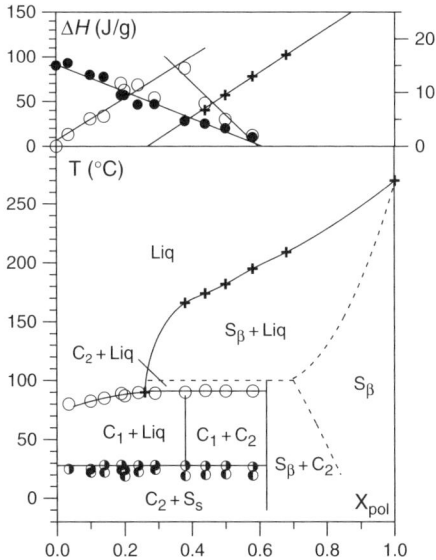

Fig. 16.31. Temperature–concentration phase diagram and Tamman's diagram for sPS/diphenyl methane systems. C_1 and C_2 are incongruently melting compounds, and S_s is a solid solution which is chiefly composed of diphenyl methane crystals. S_β is a solid phase composed of the sPS β form. In Tamman's diagram the symbols (+) and (○) are related to the right ordinate, while the symbol (●), corresponding to the total biphenyl melting enthalpy, is related to the left ordinate axis. (From Malik et al. 2006a.)

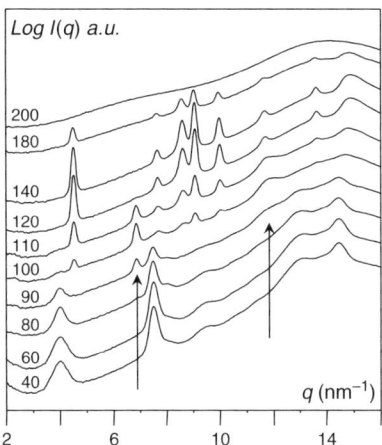

Fig. 16.32. Time-resolved X-ray diffraction patterns for sPS/diphenyl methane systems at $X_{pol} = 0.43$ w/w. Temperatures are indicated in °C. The diffraction pattern of the δ_i form (intercalate) is seen at low temperature. Here, the system is at a polymer concentration slightly above the thermodynamic stoichiometry of compound C_1. Arrows indicate the two reflections that belong to the δ_N form (see text for details). (Data from Malik et al. 2006a.)

Thermoreversible gels

As does its isotactic counterpart, syndiotactic polystyrene forms gels – and only gels – in a few solvents including benzene, chloroform, and tetraline (Kobayashi et al. 1990; Daniel et al. 1994, 1996; Malik et al. 2005b). In others, such as toluene, it may produce a gel provided the cooling rate is high enough, otherwise spherulitic structures are obtained (Daniel et al. 1997). Extensive studies of these gels by several methods have been carried out by Kobayashi et al. (1990) and Daniel et al. (1994, 1996, 1997), and more recently by Malik et al. (2005b). In particular, these authors have established the temperature–concentration phase diagrams in benzene, toluene, chloroform and tetraline.

Kobayashi and co-workers, as well as Daniel et al. (Kobayashi et al. 1990; Daniel et al. 1996, 1997), have shown by infrared spectroscopy that, in all these gels, the chains take on a 2_1 helical conformation.

In most cases the solvents promoting gelation are quite volatile, so rheological investigations turn out to be exceedingly difficult to accomplish using classic rheometers or mechanical set-ups. To the best of our knowledge, no systematic studies have been reported to date.

sPS/benzene

A system can be considered a gel when the solvent is the main constituent, a condition not fulfilled for high polymer concentrations. A better understanding of the gel requires, however, the establishment of the temperature–concentration phase diagram in all the accessible concentration domains. In this section the phase diagrams determined in three solvents, namely benzene, toluene and chloroform, will be presented.

As has been mentioned above, in benzene only a gel is produced independent of the cooling rate. In fact, this system shows perfect reversibility of the thermal events (Daniel et al. 1996). Two exotherms are seen on cooling and the corresponding endotherms do appear on heating. There is only a shift of temperature, as is expected for first-order transitions. In addition, the enthalpies associated with each event are the same whether measured on the basis of the exotherms or the endotherms. The existence of such a reversibility accounts for why the cooling rate plays no role in the gelation of sPS in this solvent. Here we are dealing with what may be termed a system at 'equilibrium', which is seldom the case when dealing with solutions of crystalline polymers.

The temperature–concentration phase diagram shown in Figure 16.33 allows one to unravel the mechanisms involved, and particularly the significance of the occurrence of two endotherms (and correspondingly the exotherms observed on cooling). As can be seen, the low-melting endotherm is temperature-invariant and occurs at $T = 75 \pm 3°C$ while the high-melting endotherm increases continuously. At this point, it is worth emphasizing that Daniel et al. performed their study in a tightly sealed sample pan, which allowed them to heat well beyond the boiling point of benzene.

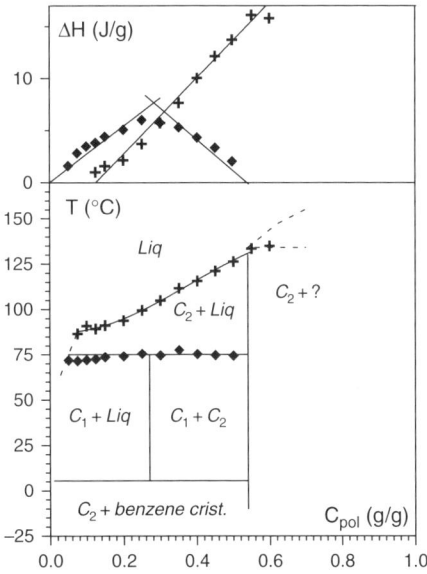

Fig. 16.33. Temperature–concentration phase diagram for sPS/benzene gels. (Data from Daniel et al. 1996.)

The shape of the diagram is consistent with, and can be interpreted through, the existence of two compounds. This statement is confirmed by the Tamman plot of the associated enthalpies. The enthalpy of the low-melting endotherm increases linearly up to $C_p = 0.26$ (w/w), and then decreases to reach zero for $C_p = 0.57$. Such a variation allows one to derive the thermodynamic stoichiometry of each compound: the first compound (C_1), which is undoubtedly an *incongruently melting compound*, possesses a thermodynamic stoichiometry of about 4 ± 1 *benzene molecules/monomer unit* while the second compound (C_2) has a thermodynamic stoichiometry of 1 *benzene molecule/monomer unit*. At $T = 75°C$, compound C_1 melts incongruently and transforms into compound C_2:

$$C_1 \xrightarrow{T=75°C} C_2 + \text{Liquid}$$

Interestingly, this temperature is close to the boiling point of benzene, i.e. 80°C. This has led Daniel et al. to suggest that three benzene molecules of the four involved in compound C_1 are loosely bound within the crystalline lattice, as they are expelled readily in the vicinity of the temperature at which benzene boils. The same occurs at the crystallization point of benzene: the concentration at which there are no longer free solvent molecules corresponds to the stoichiometry of compound C_2. Here we observe the inverse situation of an incongruent melting: compound C_1 transforms into compound C_2 on cooling.

Experiments at higher concentrations were not possible, and therefore it cannot be said whether another hypothetical, less-solvated compound exists.

The conclusions drawn from the temperature–concentration phase diagram receive further confirmation by neutron diffraction experiments. The very existence of compounds is demonstrated unambiguously by altering the type of labelling in the solvent. As can be seen in Figure 16.34, the relative ratio of intensities of the first and second reflections (at $q = 3.9\,\text{nm}^{-1}$ and $q = 7.7\,\text{nm}^{-1}$) differs, whether using hydrogenous or deuterated benzene (see Part I). Also, the position of the first reflection (at $q = 3.9\,\text{nm}^{-1}$ for $C_p = 0.26$ and at $q = 5.7\,\text{nm}^{-1}$) clearly points to two different crystalline lattices. The crystalline lattice of compound C_1 appears to resemble that of compound C_2 except for the b parameter, which is larger: in fact, compound C_1 is close to the δ_i *form* (intercalate), the crystalline lattice of which was recently described by Petraccone et al. (2005b) for sPS/norbornadiene systems.

Finally, the incongruent melting (non-variant event at $T = 75°C$) is clearly evidenced as the reflection originally at $q = 4\,\text{nm}^{-1}$ shifts towards a higher value close to that given by compound C_2 when a sample prepared at the stoichiometric concentration of compound C_1 is heated above 75°C (see Figure 16.33).

Very few morphological studies of these gels have been published. Daniel et al. (1994) have observed the morphology of sPS aggregates prepared from benzene solutions and discovered the presence of fibrils of about 20 nm diameter.

Daniel et al. (1996, 1997) have further studied the chain trajectory in the *sol state* and in the *gel state* by means of small-angle neutron scattering in systems produced from benzene and toluene. Typical scattering curves for the sol state are shown in Figure 16.35. According to Daniel et al., these curves can be fitted by considering a worm-like chain of a persistence length of about $l_p = 4.8\,\text{nm}$. At larger q-values, they

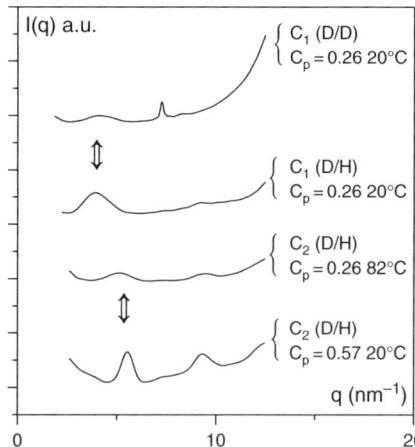

Fig. 16.34. Neutron diffraction experiments on various gel samples. Concentration, temperature and type of labelling are indicated for each curve. The double arrows highlight the change of position for the first reflection either between C_1 ($C = 0.26$) and C_2 ($C = 0.57$) or after transformation from C_1 into C_2. (Data from Daniel et al. 1996.)

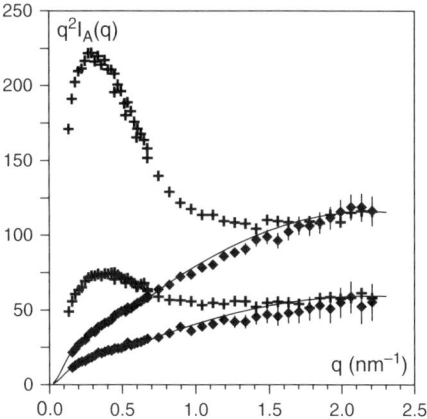

Fig. 16.35. Small-angle neutron scattering curve ($q^2 I_A(q)$ vs q) in absolute units (see Part I) for sPS/benzene sol at 110°C (♦) and gel (✦). Upper curves, $C_D = 6\%$; lower curves, $C_D = 3\%$. Total polymer concentration $C_{pol} = 12\%$ (w/w). The solid line is a best fit obtained by using a worm-like statistics (see text). (Data from Daniel et al. 1996.)

account for the departure from the $1/q$ regime by introducing a cross-section radius $r_H = 0.6$ nm. Both the mass per unit length, $\mu_L = 450 \pm 45$ g/mol per nm, and the cross-section radius, $r_H = 0.6$ nm, correspond, to within experimental uncertainties, to the molecular characteristics of the 2_1 helix. These authors therefore conclude that in the sol state the chains are rather rigid and possess a conformation close to the 2_1 helix. As with isotactic polystyrene, the gel state is therefore obtained in place of assemblies of spherulites because the chains are rigid, and therefore cannot fold on themselves, which eventually impedes the formation of lamellae, the elementary structure of spherulites. Conversely, chain rigidity promotes the growth of fibres.

Determination of the chain trajectory in the gel state has unfortunately been unsuccessful on account of the strong interferences present at small q-values (see Figure 16.35). These interferences vanish on lowering the content of deuterated material yet not to the extent required to be able to observe the 'single chain' behaviour. These observations have led Daniel and co-workers to draw two main conclusions:

1. The strong interferences are not due to isotopic segregation, as was observed for instance with polyethylene (Sadler and Keller 1977), but most probably arise from a very high parallelism between chains within the fibres, which gives a strong additional signal (see Chapter 2). Although it seems very likely that no folding occurs at the sol–gel transition, the experiments are not conclusive on this point.

2. The intensities scattered by the chains in either the sol state or the gel state are identical to within experimental uncertainties at the largest q-values; this confirms the conclusions drawn from the theoretical fit, namely that the chains possess a conformation close to the 2_1 helix in the *sol state*.

This outcome is in line with Guenet's suggestion that the 2_1 helix forms in solvent medium in order to expose only phenyl rings to the solvent molecules (Guenet 2003).

sPS/toluene

The phase diagram established by Daniel et al. (1997) for gels prepared from toluene differs significantly from that in benzene (Figure 16.36). After recent X-ray time-resolved experiments, it turns out that the shape of the phase diagram together with variation of the enthalpies associated with the different thermal events can be accounted for by considering one compound whose thermodynamic stoichiometry is significantly lower than that in benzene, namely 0.8 *toluene molecules/monomer unit*. This compound melts incongruently at 142° ± 3°C:

$$C_1 \xrightarrow{T=142\pm3°C} C_2 + \text{Liq}$$

Here too, the existence of compound C_1 is confirmed by neutron diffraction experiments.

The same type of small-angle neutron scattering experiments have been performed by Daniel et al. (1997) on gels containing a small number of deuterated chains. The behaviour is exactly the same as that observed with benzene: at large q-values the

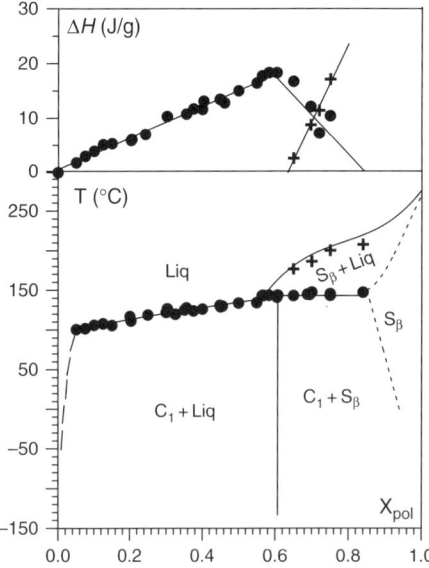

Fig. 16.36. Temperature–concentration phase diagram and Tamman's diagram for sPS/toluene gels. (Results after Daniel et al. 1997; the phase diagram has been corrected in the light of recent time-resolved X-ray experiments. Malik and Guenet, unpublished.)

scattering curves from the gel and from the solution merge, which implies that sPS chains take on a conformation close to the 2_1 helix in the sol state.

sPS/chloroform

Daniel et al. have also established the temperature–concentration phase diagram in chloroform (Figure 16.37). A compound, designated as C_1, occurs for a stoichiometric composition of about 1 *chloroform molecule/monomer unit*. This compound is an *incongruently melting* compound, although in a wide concentration range it melts directly without transforming into another phase. This differs from benzene, in which, in all the concentration domains investigated, compound C_1 never melts but transforms into compound C_2.

Beyond the stoichiometric concentration, an exothermic event shows the existence of a metastable structure which can be the 2_1 helical form as discussed in the previous section, namely a *desolvation* process (incongruent melting) which is endothermic, and a *conformational change* of the 2_1 helix into the planar zig-zag, which is exothermic.

It is worth noting that sPS/chloroform systems can reach their vitreous state at high concentrations as ascertained by the observation of the glass transition. Crystallization occurs on reheating.

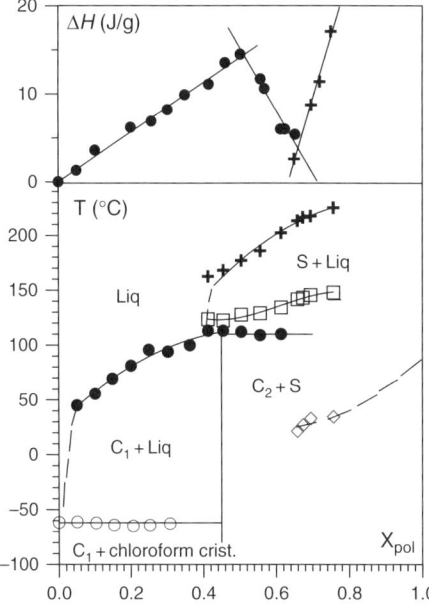

Fig. 16.37. Temperature–concentration phase diagram for sPS/chloroform gels. (□) symbols stand for an exothermic event, (◇) symbols correspond to the glass transition. (After Daniel et al. 1997.)

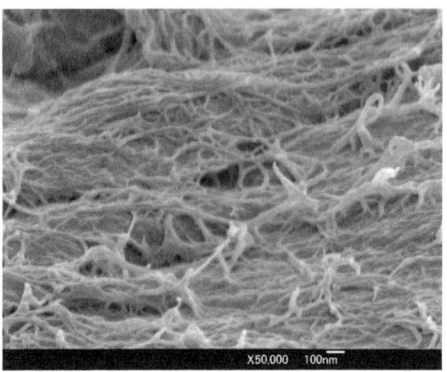

Fig. 16.38. Electron micrograph of an sPS/chloroform gel ($C_{sPS} = 0.03\,\text{g/cm}^3$). (After Itagaki et al. 2005. Reprinted with permission from ACS.)

Itagaki et al. (2005) have obtained picture of the morphology of sPS/chloroform gels by scanning electron microscopy. Their micrographs reveal the fibrillar structure of these gels with fibrils of cross-sections of about 20 nm diameter, a figure in the usual range for these gels (Figure 16.38).

sPS/tetrahydronaphthalene (tetraline)

The temperature–concentration phase diagram of *sPS/tetraline* systems, which is shown in Figure 16.39, points to the occurrence of a congruently melting compound C_1 whose thermodynamic stoichiometry is about *one tetraline molecule per monomer* as derived from the Tamman diagram (Malik et al. 2005b). As will be shown below, this compound corresponds to the δ_c *form* (clathrate). Increasing the polymer fraction further eventually gives a solid phase S_β which consists of the sPS β *form*. It is worth emphasizing that the *sPS/tetraline* system is the only one studied so far which gives a congruently melting compound with syndiotatic polystyrene.

In Figure 16.40 are displayed time-resolved X-ray diffractograms of sPS/tetraline systems at $X_{pol} = 0.40\,\text{g/g}$ and $X_{pol} = 0.48$ w/w. These polymer fractions stand on either side of the stoichiometric composition of the complex ($X_{pol} = 0.42$ w/w). At $X_{pol} = 0.40$ w/w, the reflections at $q = 5.5\,\text{nm}^{-1}$ and $q = 7.4\,\text{nm}^{-1}$ of the δ_c *form* (clathrate) of sPS are seen in the entire temperature range as expected from the phase diagram. At $X_{pol} = 0.48$ w/w, the reflections at $q = 5.5\,\text{nm}^{-1}$ and $7.4\,\text{nm}^{-1}$ of the δ_c *form* are present up to $T = 140°\text{C}$ while, above $T = 140°\text{C}$, the reflections at $q = 4.5\,\text{nm}^{-1}$, $7.6\,\text{nm}^{-1}$, $8.6\,\text{nm}^{-1}$, $9.0\,\text{nm}^{-1}$ and $10.0\,\text{nm}^{-1}$ characteristic of the β *form* replace those due to the δ *form*, thus confirming the outcomes from the T–C phase diagram.

The morphology has been observed by electron microscopy on freeze-fractured samples. The morphology is definitely fibrillar in a large range of polymer concentrations (Figure 16.41). Fibril cross-sectional diameters range between 8 and 14 nm as derived by image processing.

Syndiotactic polystyrene

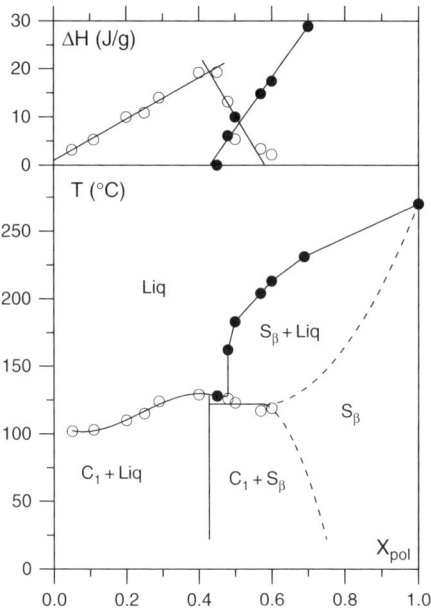

Fig. 16.39. Temperature–concentration phase diagram and Tamman's diagram for sPS/tetrahydronaphthalene system. (Data from Malik et al. 2005b.)

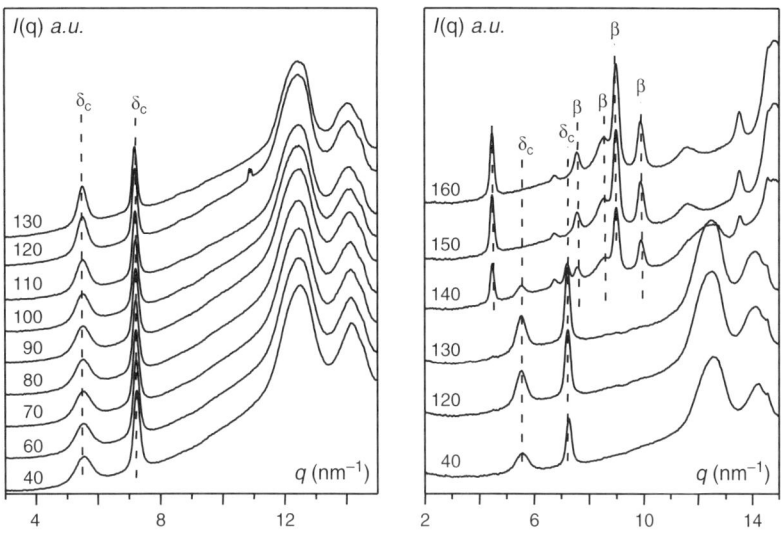

Fig. 16.40. Time-resolved X-ray diffractograms obtained at a heating rate of 2°C/min for sPS/tetraline systems. **Left**, $X_{pol} = 0.40$ w/w; **right**, $X_{pol} = 0.48$ w/w. (Data from Malik et al. 2005b.)

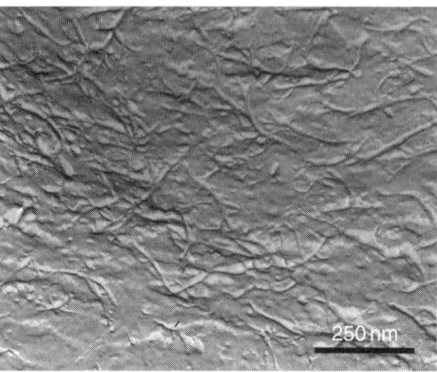

Fig. 16.41. Transmission electron micrograph of sPS/tetralin system. The sample was prepared using the freeze-fracture method. Fibrils show up with an average diameter of about 8–14 nm. (From Malik et al. 2005b. Reprinted with permission from ACS.)

Effect of formation path on melting behaviour

The capability of sPS of forming compounds has led to many thermodynamic studies that have included the determination of the temperature–concentration phase diagram. A major question concerns the relevancy of these diagrams since, strictly speaking, one is not dealing with what is usually termed *systems at thermodynamic equilibrium*. One characteristic of such a system is that its melting behaviour should be independent of the path followed to reach a given T_o, C_o coordinate of the phase diagram. For instance, cooling a hot solution of concentration C_o down to temperature T_o (solution-cast) should be equivalent to decreasing the polymer concentration from $C = 1$ to $C = C_o$ at constant temperature T_o (solvent-exposed).

Interestingly enough, this turns out to be the case with sPS/toluene and sPS/benzene systems as reported by Ray et al. (2002). As shown in Figure 16.42, the melting behaviour in sPS/toluene and sPS/benzene systems is independent of the path followed for the sample preparation: quenching hot, homogeneous solutions down to a given temperature T_o at constant concentration C_o, or exposing solid amorphous polymer samples to liquid solvent at temperature T_o in order to reach a concentration C_o gives basically the same melting temperatures and also the same melting enthalpies within experimental uncertainties.

These results indicate that Gibbs phase rules originally derived for systems at thermodynamic equilibrium can actually be also employed in polymer/solvent systems. In fact, it seems that these rules apply so long as one is dealing with stable phases. In other words, if the proportions of the different phases are equal to those that would be achieved by having equality of the chemical potential of the different constituents in the different phases, then the system is equivalent to a system at thermodynamic equilibrium (Guenet 2003). No further evolution should be observed with time. If this is not the case, then exothermic events should be observed.

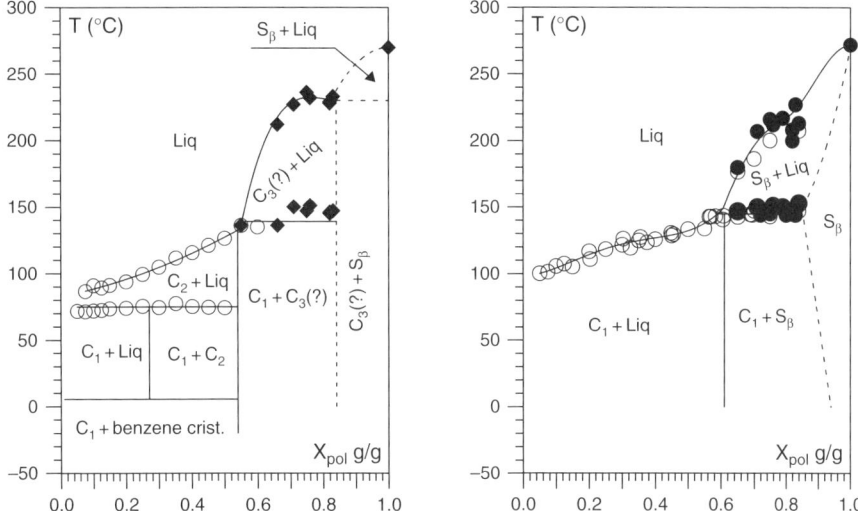

Fig. 16.42. Temperature–concentration phase diagrams: (∘) = melting temperature for solution-cast samples; (•), (♦) = melting temperature for solvent-exposed samples. **Right:** sPS/toluene. **Left:** sPS/benzene. (Results from Ray et al. 2002; the labelling of the different domains of the phase diagram of sPS/toluene has been corrected on the basis of time-resolved X-ray investigations. Malik and Guenet, unpublished.)

Multiporous networks

As has been discovered by Guerra and co-workers, the solvent molecules in the δ_c form (clathrate) can be removed in some systems without affecting the parameters of the crystal unit cell in the case of Chatani's lattice (Guerra et al. 1994, 1998; Manfredi et al. 1995). This results in the existence of cavities between the rows of polymer stems, thus giving birth to the occurrence of a *nanoporosity*. The resulting structure was first designated as an *emptied chlathrate*, and is now referred to as the δ *nanoporous phase* by the Italian school, and as the δ_e *form* by Malik and colleagues.

Guerra's group has further shown that *emptied chlathrates* possess the propensity to absorb various organic solvents dispersed in water in very low quantities. For instance, experiments carried out with water containing 100 ppm of dichloroethane have revealed that most of this solvent can be removed from water within 1 hour (see Figure 16.43).

Guerra's group has also determined the structure of emptied chlathrates obtained after absorption of 1,2 dichloroethane (De Rosa et al. 1999). They have found $a = 1.711$ nm, $b = 1,217$ nm, and $c = 0.77$ nm, with $\gamma = 120°$. The density is $\rho = 1.23$ g/cm^3 and the stoichiometry that found by Chatani, namely 1/4.

Here it is worth mentioning that Guerra's experiments were carried out with powder samples dispersed in water. The material therefore acts as a sponge but must be recovered from water through filtration.

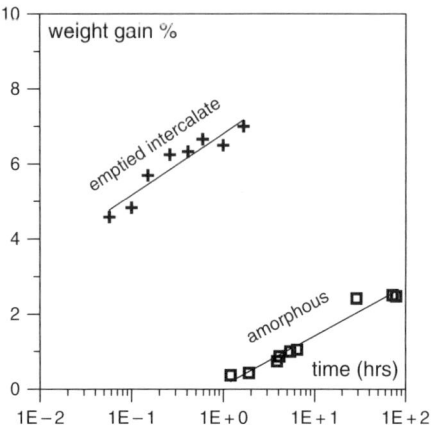

Fig. 16.43. Absorption properties shown as the weight gain vs time for emptied chlathrates obtained from sPS compound (**upper**) and amorphous sPS (**lower**) when left in water containing 100 ppm of dichloroethane (0.5% w/w). (Results from Guerra et al. 1998.)

Guenet et al. (2003) later suggested that filtering membranes could be produced if one were using thermoreversible gels prepared from sPS solutions, as these gels possess a network structure of mesh size in the micrometre range with fibrils of a few nanometres in cross-section. These fibrils are of the δ_c *form*, and so emptied chlathrates could be produced with a fibrillar morphology instead of the spherulitic morphology. These authors reasoned that these membranes would be capable of retaining large particles, and at the same time capturing small pollutants due to their very high specific surface.

Guerra's and Guenet's groups have succeeded in creating such fibrillar networks independently by means of two different processes.

Daniel et al. (2005) have used supercritical CO_2 for replacing the solvent molecules, i.e. chloroform, and then obtained the δ_e *form* by elimination of CO_2 (nanoporous δ), as shown in Figure 16.44.

Daniel et al. have further tested the absorption properties of these materials by exposing samples to chloroform vapours. Their results are shown in Figure 16.45. As can be seen, the aerogels are extremely efficient for capturing chloroform vapours, and far more efficient than films.

Guenet's group has used another route, which entails sublimating naphthalene from sPS/naphthalene fibrillar networks (Malik et al. 2006b). An AFM picture of such a system is shown in Figure 16.46.

These systems have been further characterized by X-rays, and FTIR has allowed the estimation of the extent of naphthalene removal. By the use of X-rays, it has been ascertained that sublimation does not alter the crystalline lattice, while FTIR has allowed these authors to estimate how much naphthalene remains in the δ_e *form*. They have found

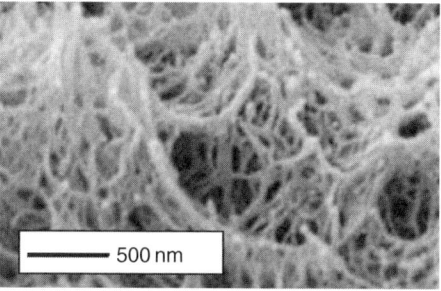

Fig. 16.44. Scanning electron micrograph of an sPS gel prepared in chloroform after replacement and extraction with supercritical CO_2. (After Daniel et al. 2005. Reprinted with permission from Wiley Interscience.)

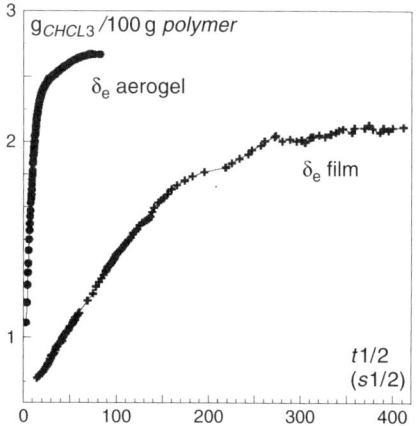

Fig. 16.45. Chloroform-vapour sorption kinetics carried out at a pressure of $P_{CHCl3} = 667$ Pa and $T = 56°C$. (•) = aerogels of 10 mm length and 5 mm diameter and (+) = films of 8.5 μm obtained under the usual conditions. (From Daniel et al. 2005.)

that there is about 1 naphthalene molecule for every 9 to 10 monomer units. While not all the solvent is removed, over 50% of the cavities are emptied. Increasing this figure can certainly be achieved by preparing thinner films.

Malik et al. have further investigated the microporosity of these systems by means of mercury intrusion porosimetry (Figure 16.47). The specific surface depends upon polymer concentration, in relation to the smallest pores. Of particular interest in the concentration $C_{pol} = 0.3$ (w/w), for which the specific area is the largest. Figure 16.47 also presents a histogram of a system prepared at a concentration $C_{pol} = 0.3$ g/g. There are typically two ranges of mesoporosity, centred around diameters of about 0.05 μm and 8–10 μm.

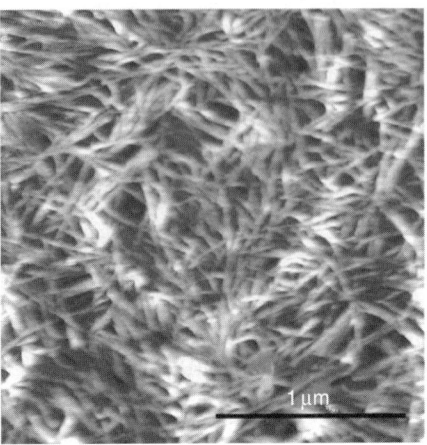

Fig. 16.46. AFM picture taken from an sPS/naphthalene system ($C_{pol} = 0.30$ g/g) after sublimation of naphthalene through vacuum extraction for 10 days. (From Malik et al. 2005a. Reprinted with permission from ACS.)

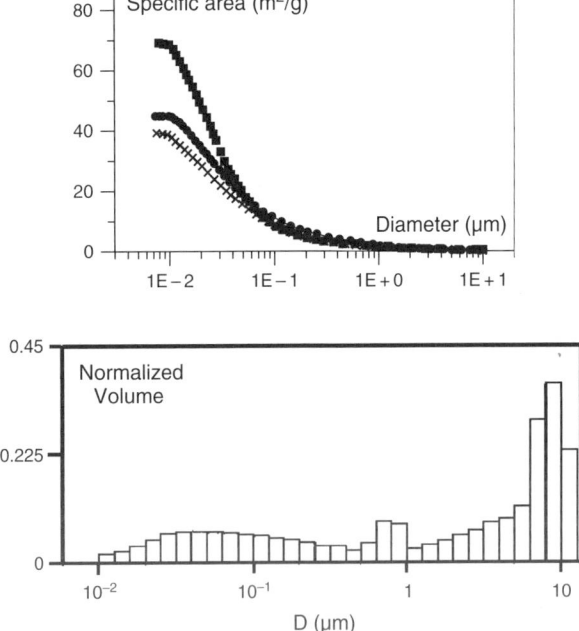

Fig. 16.47. Upper: specific surface vs diameter of pores of sPS/naphthalene systems: (■) for $C_{pol} = 0.30$, (●) for $C_{pol} = 0.20$ and (✱) for $C_{pol} = 0.05$ g/g. **Lower:** Normalized volume vs pore diameters of sPS/naphthalene systems for $C_{pol} = 0.30$. Results were obtained after sublimation of naphthalene through vacuum extraction for 10 days. (Data from Malik et al. 2006.)

CHAPTER 17

Atactic polystyrene

Atactic polystyrene does not crystallize from the bulk. Yet, in some solvents it can form thermoreversible gels but in no case spherulitic assemblies. Atactic polystyrene is a 'copolymer' with regard to stereoregular sequences, which is consistent with gelation but definitely hinders spherulitic formation as in the case of atactic PVC. As aPS belongs to the polystyrene family (aPS, sPS, iPS), it appeared more judicious to place the discussion about its gelation propensity after that of its stereoregular counterparts.

As is the case for the PMMAs (see Chapter 19), stereoregular sequences (iso or syndio) must possess a minimum length for gelation to take place (Spěváček and Schneider 1987). Although, to the best of the author's knowledge, no quantitative data are available about this minimum length, the experiments by François et al. (1988) on epimerized isotactic polystyrene are instructive on this point. By means of a chemical modification, it is possible to obtain the usual atactic polystyrene while starting from the isotactic species. Obviously, the degree of modification can be controlled. François et al. have shown that for a degree of conversion of about 50% the resulting epimerized polystyrene does not form a gel.

Thermodynamics

The synthesis of aPS is rather straightforward, particularly by anionic polymerization whereby samples with relatively low polydispersity can be prepared. Accordingly, this polymer has frequently served as a model polymer for testing former theories (Flory 1953) as well as new concepts (de Gennes 1979). It has been extensively used for demonstrating the Gaussian conformation of polymer chains in the bulk state (Cotton et al. 1974) or for assessing the validity of scaling theories in semi-dilute solutions (Daoud et al. 1975).

The report by Baer and co-workers that aPS could form thermoreversible gels in a variety of organic solvents came as a surprise, not to mention disbelief (Wellinghof et al.

1979; Tan et al. 1983). Many scientists questioned the reality of this phenomenon because common sense dictated that thermoreversible gelation was a crystallization phenomenon, and was therefore the privilege of crystallizable polymers. Actually, many scientists had already witnessed the phenomenon of aggregation/gelation of atactic polystyrene but it went unnoticed. In the 1960s, reports were made on light scattering investigations into concentrated solutions of atactic polystyrene for which an unexpected phenomenon was observed (Hyde and Taylor 1963; Benoit and Picot 1966; Dautzenberg 1970). In principle, a homogeneous solution of Gaussian chains should scatter light in the concentrated regime as (Daoud et al. 1975):

$$I(q) \propto \frac{1}{q^2 + \xi^{-2}} \tag{17.1}$$

where ξ is the so-called screening length as defined by Edwards (1966). This relation holds for $q\xi < 1$, which is the case in light-scattering experiments. Plotting $I^{-1}(q)$ vs q^2 should therefore yield a straight line.

Actually, in many systems, a pronounced downturn can be observed (designated as *enhanced low-angle scattering, ELAS*), an effect which hints at the existence of aggregates (Figure 17.1). This effect was first believed to arise from an undissolved fraction of the polymer sample. However, heating the solution led to no change in any parameter. Clearly, this phenomenon remained puzzling and unexplained for almost 20 years.

Note 17.1 Scattering from polymer solutions

The scattering intensity can be written (Debye 1915):

$$I(q) \propto \int 4\pi r^2 \gamma(r) \frac{\sin qr}{qr} dr$$

where $\gamma(r)$ is the correlation function. In homogeneous polymer solutions below a given screening length ξ the monomer–monomer correlations are described by an excluded-volume statistics while above they obey a Gaussian statistics. For $q\xi < 1$, the correlation function is written:

$$\gamma(r) \propto r^{-1} e^{-r/\xi}$$

which gives relation 17.1. If inhomogeneities of average size a are present in the solution, such as crystallites, then $\gamma(r)$ reads (Debye and Bueche 1949):

$$\gamma(r) \propto e^{-r/a}$$

which yields:

$$I(q) \propto \left(q^2 + a^{-2}\right)^{-2}$$

which is responsible for the downturn when $I^{-1}(q)$ is plotted against q^2.

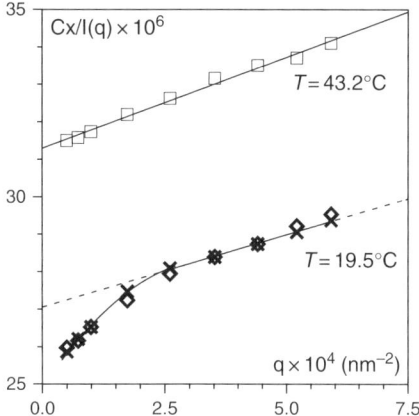

Fig. 17.1. Light scattering experiments (C/I(θ) as a function of $\sin^2(\theta/2)$) showing the enhanced low-angle scattering at 19.5°C and its vanishing at 43.2°C. (✗) = experiment after heating to 43.2°C and cooling down again to room temperature (aPS/tetrahydrofuran solutions; $C = 5.6 \times 10^{-3}$ g/cm^3, $M_w = 1.7 \times 10^6$). (Data from Gan et al. 1986.)

As mentioned above, the decisive step came from Baer's group in Cleveland. They reported a study on a series of solvents which showed that gelation did occur in some of those known for giving the ELAS effect. As a rule the gelation temperatures are very low and only one solvent, so far, stands out: carbon disulphide (CS$_2$). As we shall decribe in more detail below, gelation in this solvent occurs through a first-order transition, which undoubtedly hints at the creation of molecular order. Although some solvents said to be 'gelling' solvents proved to be in the wrong category, as shown by Plazek and Altares (1986), this work did provide an interesting list of solvents for which gelation could not take place. Willmott, Ellsmore and Guenet suggested that the enhanced light scattering at room temperature might well arise from remnants of the former gel structure, i.e. aggregates of finite size (Guenet et al. 1983). In fact, as has been shown in the previous chapter for PVC systems, the macroscopic gel melting does not correspond to the disappearance of all organized structures but simply to the loss of a three-dimensional, infinite network. These authors showed that ELAS is absent whenever they used a 'non-gelling' solvent. A conclusive experiment was eventually reported by Gan et al. (1986), who reasoned that if a first-order transition is involved, then there exists a temperature above which these aggregates must disappear, and the scattering by the solution must be described by equation (17.1). As is shown in Figure 17.1, the ELAS effect does vanish above $T \approx 40°C$. It is worth stressing that in this experiment the moderately concentrated solution was investigated at 45°C, which was at variance with previous studies (Dautzenberg 1970; Koberstein et al. 1985). At that time nobody would have thought of the existence of some local order in solution of atactic polystyrene, so solutions were effectively heated to get rid of what was thought to be non-dissolved material, but systematically cooled and subsequently studied at room temperature (Koberstein et al. 1985). This procedure was doomed to failure, as was clearly shown by Gan and colleagues: *the aggregates reform reversibly on cooling back to room temperature.*

Baer and co-workers had claimed that a formation exotherm could be detected by means of differential scanning calorimetry (DSC), which therefore hinted at the occurrence of a first-order transition. François et al. (1986), using more elaborated procedures for sample preparation, confirmed these results: gel formation and gel melting could be clearly differentiated from the glass transition (T_G). These authors evidenced a hysteresis between gel formation (T_{gel}) and gel melting once proper extrapolation to a zero-heating rate had been achieved. In comparison, after such extrapolation, the glass transition temperature was found to be virtually identical whether measured on cooling or on heating. The differentiation between the gel temperature and the glass transition was later confirmed by Izumi et al. (1992a) by quasielastic neutron scattering. From this DSC study the temperature–concentration phase diagram was established for various polymer samples of different molecular weights (Figure 17.2).

Two salient features are apparent from this phase diagram: the melting temperatures do not depend upon the molecular weight although the formation/melting enthalpies do; the melting temperatures and the formation/melting enthalpies display a maximum near $C = 0.5$–0.55. François et al. checked that this maximum did not arise from kinetics effects that are liable to take place near the glass transition, and which may considerably slow down the gelation process. Both maxima therefore hint at the existence

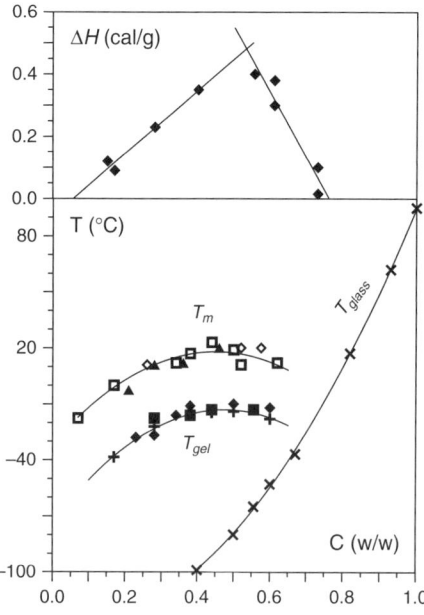

Fig. 17.2. Temperature–concentration phase diagram of aPS/CS_2 gels for samples of different molecular weights. (Data from François et al. 1986.)

of a *polymer–solvent compound* between polystyrene and carbon disulphide. This is not really surprising as the gelation phenomenon of this polymer is not linked to solvent quality, something very similar to what occurs in isotactic polystyrene. For some reasons that are yet not clear but most certainly related to the solvent molecular size and to the 'stereoregular' sequence involved in the formation of these ordered physical junctions, carbon disulphide is the most appropriate solvent. Does there exist yet another such solvent? So far, to the best of our knowledge, no report has been made on a similar solvent which would produce gels possessing melting points as high as room temperature.

The absence of a molecular weight effect on the melting temperature of aPS/CS_2 gels contrasts with the behaviour of PVC gels (Chapter 10). Yet, here too, the fraction and length of long stereoregular sequences involved in the gelation process most probably increase with increasing molecular weight, as actually hinted at by the melting enthalpies. Admittedly, the terminal melting of PVC gels is not related to the disappearance of solvated crystals but, instead, of non-solvated crystals. Whether this dissimilarity has any bearing upon the melting properties remains to be demonstrated.

Here, it is worth emphasizing that gelation in carbon disulphide does not proceed through a liquid–liquid phase separation via spinodal decomposition. This mechanism was once put forward on the basis of turbidity experiments. On this line, some authors advocated a mechanism whereby the gel structure was obtained after the polymer-rich phase created at the early stage of spinodal decomposition was pinned down by the occurrence of the glass transition (Frank and Keller 1988). While this mechanism can be effectively obtained in some systems, it does not pertain to the present case. Indeed, Gan et al. showed that properly dried CS_2 was a good solvent to polystyrene down to very low temperature and that no turbidity could then be observed. Izumi et al. (1992b) have also observed by small-angle neutron scattering that aggregation takes place well above the θ-temperature reported by Tan et al. (1983). This means that the solution does not need to be quenched within a hypothetical miscibility gap to exhibit aggregation properties. Also, the existence of aggregation in various solvents (usually good) in solutions of relatively low concentration at room temperature is totally inconsistent with the view developed by Frank and Keller: no spinodal decomposition or glass transition takes place under these conditions. Conversely, aggregation through compound formation is not impaired at all and a good solvent can even promote it.

Attempts to measure the quantity of chains participating in the physical junctions were made by using forced Rayleigh scattering (FRS) (Lee et al. 1988). This technique consists in studying the relaxation behaviour of a solution containing dispersed dye-labelled aPS chains that are submitted to photochromic excitation through a grating of spacing $\Delta\delta$. If two types of chains with different relaxation times, τ, are present in the medium, then the output signal reads:

$$V(t) = \left[A_s \exp(-t/\tau_s) + A_f \exp(-t/\tau_f)\right]^2 + C^2 \qquad (17.2)$$

where A is an amplitude and C a constant, and where the subscripts s and f stand for the slow and the fast mode, respectively. The quantity of chains belonging to the junctions, $X_{junctions}$, is then written:

$$X_{junctions} = \frac{A_s}{A_s + A_f} \quad (17.3)$$

Lee et al. reported from the polymer study, concentrations ranging from 18 to 25% at $T = 250$ K. They obtained $X_{junctions} \approx 20\%$. Here it is necessary to stress that this figure does not imply that only 20% of the material is capable of forming junctions (or of being incorporated into junctions, as can happen for amorphous parts). As is evident from the phase diagram, the gel is a two-phase system: a polymer-rich phase (i.e. the gel framework) and a polymer-poor phase. At this temperature and starting concentration the phase diagram indicates that the concentration of the polymer-poor phase is still 10%. As a result, by either decreasing the temperature or increasing the polymer concentration, the proportion of polymer-rich phase will automatically increase by incorporation of chains from the polymer-poor phase. In the end, it is quite possible that the fraction of chains engaged in the gel is greater than 20%. Only experiments at higher concentrations and lower temperature would definitely settle this point.

Molecular structure

The term 'atactic' is usually misleading as one inevitably tends to view aPS as being totally disordered at the level of the monomer arrangement. In fact, NMR characterization has shown that 65% of the triads are syndiotactic, which makes it possible to find long syndiotactic sequences. Guenet and co-workers suggested, therefore, that these sequences were responsible for the gelation of atactic polystyrene (Guenet et al. 1983).

In the early 1980s, syndiotactic polystyrene was soon to be synthesized for the first time but was still not available, so these authors could only think of one conformation for these sequences, i.e. the planar zig-zag. Nakaoki and Kobayashi (1991) later solved this question by means of infrared spectroscopy studies. They confirmed that short sequences were involved, as no significant change was observed in the mid-infrared region on cooling, unlike what is seen with iPS/CS_2 gels. To these authors this suggested that the construction of long, regular sequences of *tg* or *ttgg* was restricted. As a reminder, *tg* is the arrangement giving the iPS 3_1 helix, *tt* the planar zig-zag of sPS and *ttgg* the arrangement producing the 2_1 helix of sPS (see Chapters 15 and 16 for further details). However, in the $500-600$ cm^{-1} range a detectable change in the absorption profile is observed (Figure 17.3). The most conspicuous increase in intensity occurs for 572 cm^{-1}, which is the band associated with the *ttgg* form, while the 556 and 562 cm^{-1} bands, associated with the *tg* form, display a slight and continuous increase during gelation. Conversely, the 540 cm^{-1} band, which is a superposition of *ttgg* and *tt* components, drops quite markedly. According to Nakaoki and Kobayashi this depression may result from the decrease of the *tt*/*ttgg* molar ratio of the syndiotactic part. They eventually come to the conclusion that the formation of short *ttgg* (i.e. 2_1 helix) promotes gelation of this polymer. Clearly, while these sequences are too short to allow crystallization from the

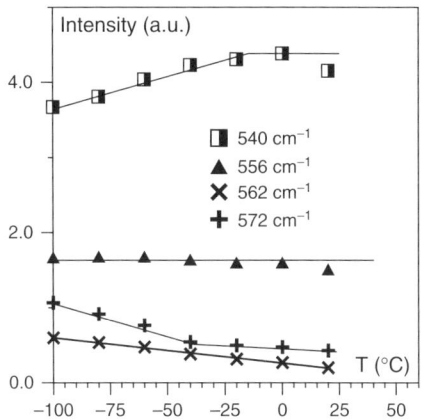

Fig. 17.3. Evolution of the different infrared integrated intensities for IR bands in the range 500–600 cm^{-1} for aPS/CS$_2$ systems. (Data from Nakaoki and Kobayashi 1991.)

bulk their length is certainly long enough to promote the formation of polymer–solvent compound, and, correspondingly, gelation.

That order is created in the gels of this atactic polymer has been demonstrated by neutron diffraction experiments, thus confirming the involvement of a first-order transition in the gelation process (Guenet et al. 1989). Neutron diffraction was used rather than X-rays for two major reasons: (1) the temperature–concentration phase diagram points to the presence of a solvated structure, and it is therefore mandatory to study it in the wet state; (2) under these conditions X-rays would be mostly diffracted by CS$_2$, so the diffraction maxima would be buried in the noise. Conversely, as the neutron scattering length of sulphur ($a_s = 0.214$ barns) is significantly lower than that of deuterium ($a_D = 0.662$ barns), diffraction by a deuterated polymer should not be smeared by the solvent diffraction halo.

As is apparent in Figure 17.4, a diffraction peak is clearly seen at $q = 5.7 \pm 0.05$ nm^{-1}, which corresponds to a Bragg distance of $d = 1.095$ nm. If this peak is related to the distance between neighbouring polymer stems that would adopt a nematic-like order, then it would correspond to an actual distance of $d \approx 1.095 \times 1.115 = 1.22$ nm, a value consistent with 2$_1$ helices in close contact. An unexpected second peak at $q = 19.5$ nm^{-1} can be observed (i.e. Bragg distance of $d_B = 0.32$ nm). All these results were later confirmed by Izumi and co-workers (Izumi et al. 1992c).

It is worth emphasizing that these two reflections are also observed in the δ_c form (see Chapter 16). Does this imply the presence of a 'crystalline lattice' close to the δ_c form, which would further confirm the involvement of the syndiotactic sequences? At any rate, the observation of several diffraction peaks of such intensity points toward the existence of a large collection of chains participating in the junctions. Paradoxically and most

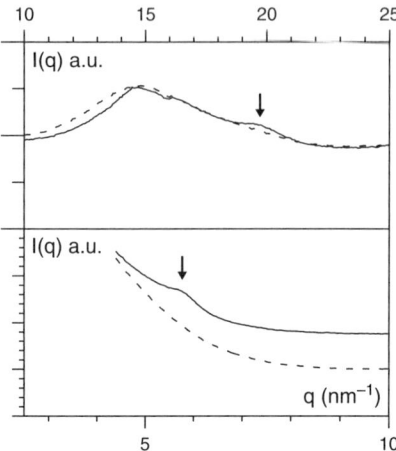

Fig. 17.4. Neutron diffraction data for aPS/CS$_2$ solutions (broken line) and gel at $-60°C$ (solid line). Arrows highlight the appearance of diffraction maxima. (Data from Guenet et al. 1989.)

amusingly, aPS gels eventually appear to possess a higher molecular organization than do iPS gels, for which only one peak is seen.

The long-range structure of the gel has been studied by Izumi et al. (1992b) using small-angle neutron scattering (all the chains are labelled). They observe a clear departure from the single-chain behaviour (which possesses an excluded-volume statistics above T_{gel} as already reported by Daoud et al. 1975). These authors interpret their results by means of star-branched chains models. However, the scattering curves could just as well be interpreted on the basis of fibre-like structures such as those found for PVC gels. Also, if aggregates were star-branched, how would they then be able to connect with one another to form the infinite, three-dimensional network? It is our feeling that this point remains to be elucidated.

The results obtained by the same authors on systems where some labelled chains are embedded in a gel matrix (deuterated chains + hydrogenous chains + solvent) may cast some light on this point (Izumi et al. 1992c). When the fraction of labelled chains is low ($C_D = 0.01\,\text{g/cm}^3$ for a total concentration $C_T = 0.23\,\text{g/cm}^3$), the chains scatter as approximately $1/q^2$ (Figure 17.5a). Conversely, when the fraction of deuterated material is relatively high ($C_D = 0.014\,\text{g/cm}^3$ for a total concentration $C_T = 0.068\,\text{g/cm}^3$), there is a strong forward scattering indicative of segregation (Figure 17.5b). As aPSD and aPSH are known to be compatible, the origin of this segregation must be sought out elsewhere. We suspect that it might be due to the formation of fibre-like structures whereby chains are concentrated in a restricted volume. This results in enhancing intermolecular scattering between labelled chains. Such an effect has been observed for syndiotactic polystyrene, as can be seen in Figure 16.35 (Daniel et al. 1996). Random aggregation, as would occur in a star-branched system, would not produce such a forward scattering, as has been shown with syndiotactic PMMA (see Chapter 19) (Saiani and Guenet 1997). But again, additional experiments are vital in order to settle this issue.

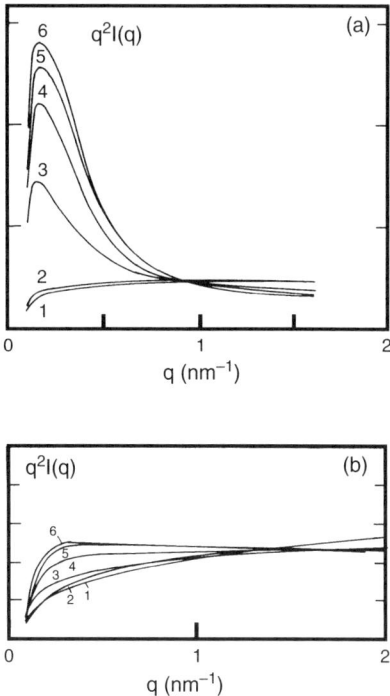

Fig. 17.5. Temperature variations of the scattered intensity (Kratky plot) of an aPSD sample ($M = 1.11 \times 10^5$) in CS_2/aPS systems; (a) $C_D = 0.014\,\text{g/cm}^3$ and $C_T = 0.068\,\text{g/cm}^3$; (b) $C_D = 0.01\,\text{g/cm}^3$ and $C_T = 0.23\,\text{g/cm}^3$; 1 = 294 K, 2 = 273 K, 3 = 254 K, 4 = 231 K, 5 = 208 K, 6 = 171 K. (From Izumi et al. 1992c. Reprinted with permission from Elsevier.)

Mechanical properties

The gel status of aPS/solvent systems has been a subject of controversy because some solutions said to exhibit gelling properties have proven not to be gels (Plazek and Altares 1986). However, as shown by Clark et al. (1983) and later by Gan et al. (1985), Koltisko et al. (1986) and Dammer (1991), aPS/CS_2 solutions do form gels on cooling. Below T_{gel} the storage modulus G' remains constant in a large range of frequencies ($\omega = 0.1$ to 100 rad/sec), and is larger than the loss modulus G'' by at least one order of magnitude.

Kotilsko et al. have further studied this system as a function of temperature and polymer concentration by means of a coaxial shear apparatus. They report that the shear E modulus varies as:

$$E \propto C^2 \tag{17.4}$$

Also, by using a reduced temperature $T_r = T/T_{gel}$ they were able to draw a master curve.

From these results Kotilsko et al. have concluded that their data is analysable in the framework of the rubber elasticity theory. However, at the time of carrying out their research, these authors were not aware of the existence of a strong order and the implications thereof. They essentially viewed the gel as a medium resembling an amorphous solution except for some strong, point-like interactions between chains. That a master curve could be drawn is not a proof of such a structure nor is the C^2 behaviour of the shear modulus another indication. For instance, rigid gels where the connecting objects are cylinder-like must behave in the same way (see for instance κ-carrageenan gels; Rochas and Landry 1988).

Again, determination of the mechanical properties is certainly not the best method to use to cast light on gel structure as the outcomes are model-dependent, and ambiguity can arise since two different models may exhibit a similar behaviour. As is detailed in Part II in relation to PVC and agarose gels, a C^2 behaviour may also arise from an array of rigid fibres exhibiting enthalpic elasticity (see Part I; Jones and Marquès 1990).

CHAPTER 18

Syndiotactic poly[p-methyl styrene]

The catalytic systems giving rise to high syndiotactic polystyrene have also been applied to styrene derivatives such as *para*-methyl styrene in order to synthesize poly[p-methyl styrene] (sPpMS) (Ishihara et al. 1988; Grassi et al. 1987, 1989). In spite of its close resemblance to syndiotactic polystrene, poly[p-methyl styrene] (sPpMS) differs significantly with regard to crystallization behaviour. Unlike sPS, sPpMS does not crystallize from the melt or after the annealing of amorphous samples. Crystallization always requires the presence of solvent either by precipitation from solution or by solvent-induction (Iuliano et al. 1992). Several crystalline forms have been reported: forms I and II, which contain 2_1 helices but no solvent molecules; forms III, IV and V, which are made up with chains taking on a trans-planar conformation (planar zig-zag); and several clathrate forms. These clathrate forms are subdivided into three classes: α, β, and γ. The α and β classes contain 2_1 helices while the γ class contains chains with another conformation, $t_6g_2t_2g_2$, which has not to date been observed for sPS. Again, these solvated systems could just as well be designated as intercalates; however, in this chapter the term clathrates will be used.

Non-solvated forms

Here a short description of non-solvated forms is given, as all examples thereof derive from solvated (clathrate) forms. The crystalline structure of only two forms has been elucidated: that of form I and of form III.

Form I can be obtained by crystallization from solution, by thermal treatment of the α class of clathrates in the temperature range 120–160°C or by acetone extraction of the solvent from the α class. According to Esposito et al. (2006) acetone treatment produces only form I, unlike the case with the other treatments. The investigations of Esposito and colleagues have shown that form I possesses a monoclinic unit cell with parameters $a = 2.45$ nm, $b = 1.24$ nm, $c = 0.81$ nm and $\gamma = 143.5°$, and belongs to space group $P2_1/a$ (Figure 18.1). The chains accordingly take on an $s(2/1)2$ helical conformation. Form I is therefore nothing but a desolvated form of clathrate, as is the γ form of

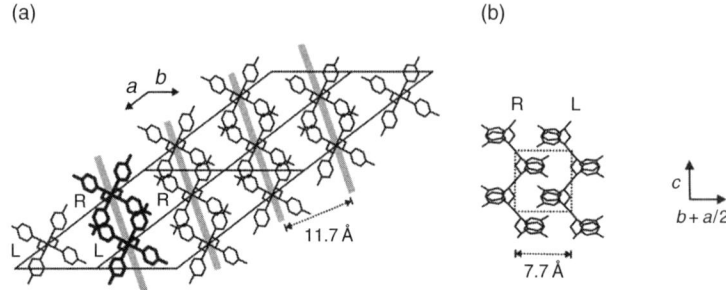

Fig. 18.1. Packing of the 2_1 chains in form I of sP*p*MS as seen parallel to the helix axis (a) and perpendicular to the *bc* plane. The grey stripes highlight the direction along which polymer stems are supposed to move when the α class clathrates transform into form I (R = right-handed helices; L = left-handed helices). (From Esposito et al. 2006.)

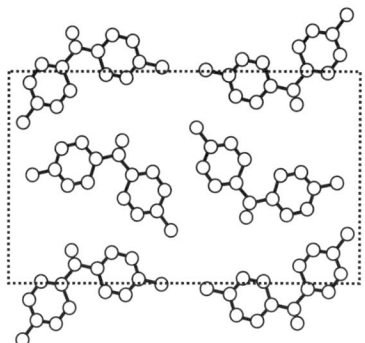

Fig. 18.2. Packing of the trans-planar chains in form III of sP*p*MS as seen parallel to the chain axis with $a = 1.336$ nm, $b = 2.321$ nm, $c = 0.512$ nm. (After De Rosa et al. 1999.)

sPS. Form II also contains chains with an $s(2/1)2$ helical conformation, yet to date the structure has not been described.

Form III is prepared through the recrystallization process of form I at high temperature (210°C). De Rosa and co-workers have established that form III possesses an orthorhombic unit cell with parameters $a = 1.336$ nm, $b = 2.321$ nm, and $c = 0.512$ nm, and therefore the chains take on a trans-planar conformation (Figure 18.2). Form IV is a mesophase that contains trans-planar chains with a packing close to that of form III (Ruiz de Ballesteros et al. 1998). It can be obtained through the stretching of form I at 135°C, but other techniques can be used. Again, form V is made up of trans-planar chains but no description is presently available.

Clathrates

As already mentioned, three classes of clathrates have so far been identified. The α class tends to be produced when using benzene derivatives with groups in the *ortho* position, although N-methyl-2-pyrrolidone also produces this form. With *ortho*-dichlorobenzene (*o*-DCB), solution-cast samples give birth to the unoriented structure, while the oriented samples are obtained by exposing to *o*-DCB fibres prepared under the mesomorphic form IV beforehand (Petraccone et al. 2000).

The crystal structure of the clathrates from the α class prepared from sP*p*MS/*ortho*-dichlorobenzene has been established by Petraccone et al. (2000). These authors propose a monoclinic unit cell with parameters $a = 2.34$ nm, $b = 1.18$ nm, $c = 0.77$ nm, and $\gamma = 115°$ which belongs to space group $P2_1/a$. The unit cell contains two polymer chains (eight monomer units) and two *ortho*-dichlorobenzene molecules, which means that the stoichiometry of the compound is *1 solvent molecules/4 monomer units*. This gives a crystal density of 1.07 g/cm^3 (Figure 18.3).

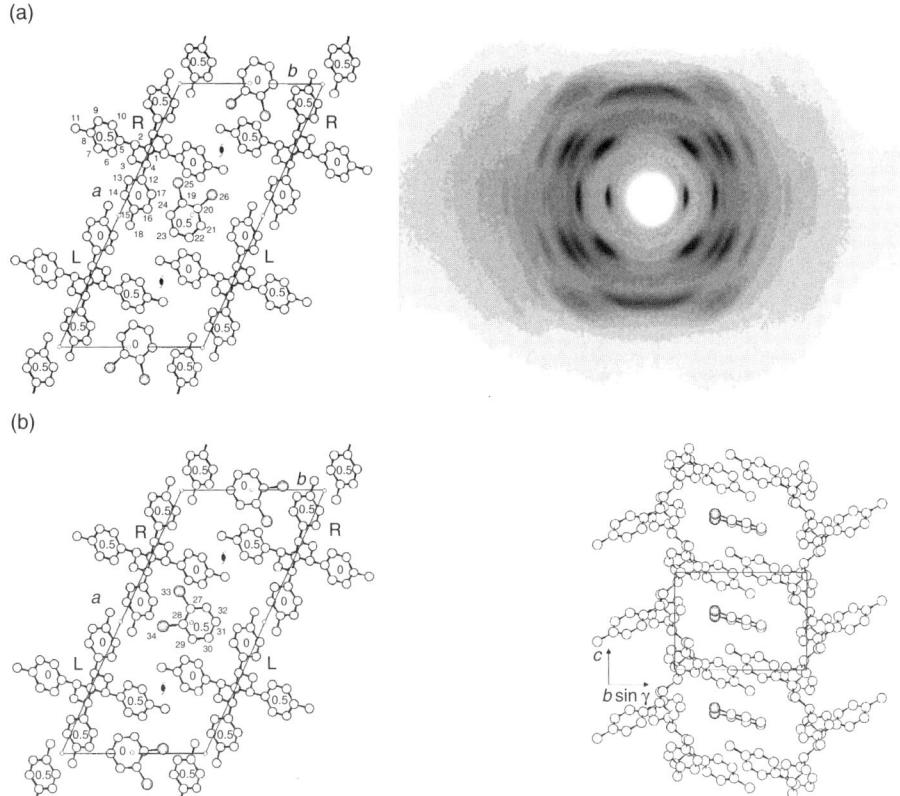

Fig. 18.3. **Top right:** diffraction pattern for the oriented form of sP*p*MS/*ortho*-dichlorobenzene (α class). **Left:** two possible models for the packing of the crystal structure (a and b) as seen parallel to the helical axis. Both positionings of the *ortho*-dichlorobenzene molecules are probable and a statistical arrangement certainly occurs. **Bottom right:** $cb \sin \gamma$ projection of the chain + solvent packing. (From Petraccone et al. 2000. Reprinted with permission from ACS.)

Note that single crystals grown from solvent diffusion within thin films have a rounded lozenge shape (Rizzo et al. 2000).

The β class tends to grow in the presence of non-substituted molecules such as tetrahydrofuran (THF) and benzene. The main difference from the α class lies in the stoichiometry, which is *1 solvent molecule/2 monomer units* in the case of the β class. This type of molecular compound can also be obtained by solution-cast procedures from moderately concentrated solutions (about 10%). Oriented fibres can be prepared from oriented fibres of form IV that are subsequently exposed to THF vapours for a few minutes. Petraccone et al. (1998) determined the crystal unit cell of the compound sP*p*MS/THF. They found a monoclinic unit cell with parameters $a = 1.88$ nm, $b = 1.27$ nm, $c = 0.77$ nm and $\gamma = 100°$ that belongs to space group $P2_1/a$ (Figure 18.4). The crystal density calculated is 1.13 g/cm^3, taking into account 2 polymer chains (8 monomer units) and 4 solvent molecules. Accordingly, the stoichiometry is *1 THF molecule/2 monomer units*.

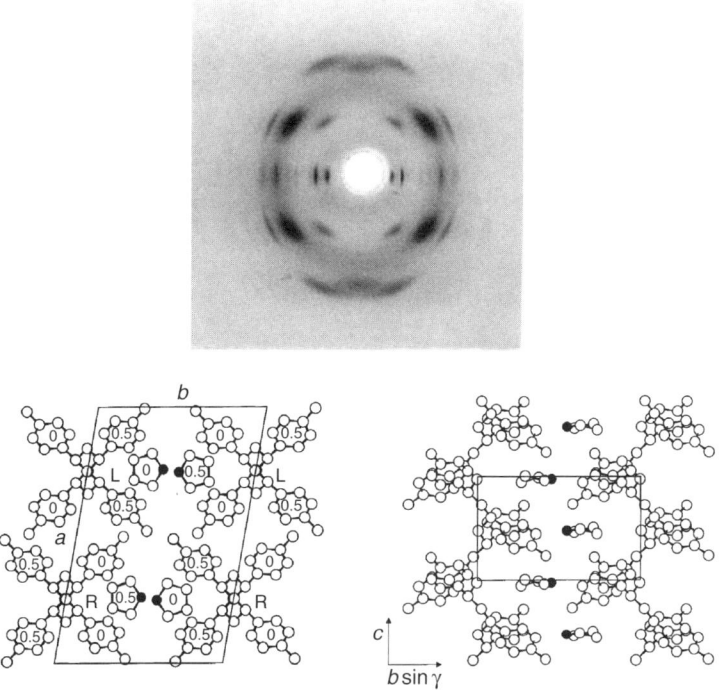

Fig. 18.4. **Top:** diffraction pattern for the oriented form of sP*p*MS/THF (β class). **Bottom left:** packing of the crystal structure as seen parallel to the helical axis. Figures indicate the positioning with respect to the c axis. L and R stand for left-handed and right-handed helices respectively. **Bottom right:** $cb \sin \gamma$ projection of the chain + solvent packing. The oxygen atom of the THF molecule is shown as a filled circle. (From Petraccone et al. 1998. Reprinted with permission from ACS.)

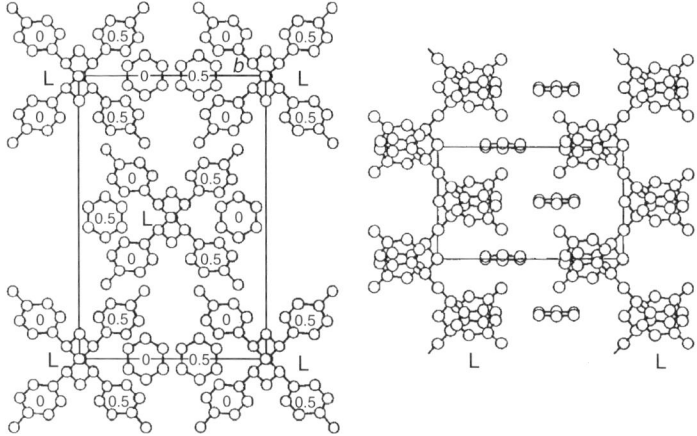

Fig. 18.5. sP*p*MS/benzene (β class). **Left:** packing of the crystal structure as seen parallel to the helix axis. Figures indicate the positioning with respect to the c axis. L stands for left-handed helices. **Right:** cb projection of the chain + solvent packing. (From La Camera et al. 2001. Reprinted with permission from ACS.)

Another molecular compound of the β class has been obtained with benzene. The crystal structure was determined by La Camera et al. (2001). According to these authors the unit cell is orthorhombic with parameters $a = 1.95$ nm, $b = 1.33$ nm, $c = 0.77$ nm and $\gamma = 115°$, and belongs to space group $C222_1$ (Figure 18.5). Here again, chains take on an $s(2/1)2$ helical conformation.

Quite recently, another type of clathrate was discovered by Petraccone et al. (2005a) that they designated as the γ class. Unoriented specimens were obtained by solution-cast procedure from solutions in cyclohexanone or in cyclohexane, while the oriented samples were prepared from oriented mesophase form IV exposed to cyclohexanone or cyclohexane vapours. To date these are the only solvents that produce this new form. This form melts at 150°C and transforms into form II after subsequent annealing.

The crystalline unit cell has still not been determined yet the chain conformation differs significantly from the 2_1 helix. The repeat distance turns out to be 1.18 nm, a value clearly at variance with the repeat distance of the trans-planar conformation (0.51 nm) or the 2_1 helix (0.77 nm) (Figure 18.6). Note that the pitch and the repeat distance are the same for the last two conformations.

By analogy with other crystalline polymers, such as syndiotactic polypropylene (e.g. see Chatani et al. 1991), and by performing energy calculations of the conformation, Petraccone et al. have proposed a conformational sequence of the type $t_6g_2t_2g_2$ (see Figure 18.6), which differs significantly both from the trans-planar conformation (tt) and from the 2_1 helix ($ttgg$).

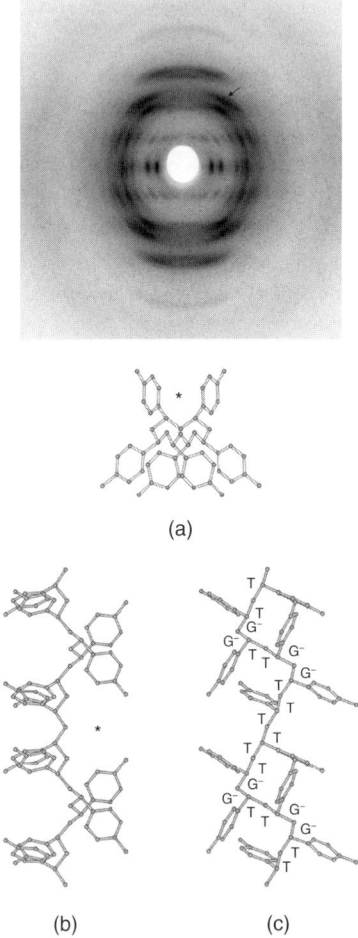

Fig. 18.6. Top: diffraction pattern for the oriented form of sP*p*MS/cyclohexanone (γ class); the arrow indicates the presence of form IV. **Bottom:** the helical conformation with repeat unit 1.18 nm (a) as seen parallel to the helix axis; (b) and (c) as seen perpendicular to the helix axis with different views. The star highlights the possible cavity for housing solvent molecules. (From Petraccone et al. 2005a. Reprinted with permission from ACS.)

CHAPTER 19

Stereoregular poly[methyl methacrylate]s

Stereoregular poly[methyl methacrylate]s (PMMAs) come in two varieties: isotactic (iPMMAs) and syndiotactic (sPMMAs). Stereoregular PMMAs stand out in the realm of synthetic polymers as they have the propensity to form double helices (Kusunagi et al. 1976; Dybal et al. 1986; Saiani and Guenet 1997), and in so doing share the property of some polysaccharides such as carrageenans and amylose. In addition to this unusual behaviour, Watanabe and colleagues further discovered in the early 1960s that mixtures of sPMMA and iPMMA in solution produce the so-called stereocomplex (Watanabe et al. 1961).

Further work has enabled the determination of the stoichiometry of the stereocomplex, i.e. sPMMA/iPMMA = 2/1 (Liquori et al. 1965; Vorenkamp et al. 1979). The value of the stoichiometry agrees quite well with the helical structure put forward by Schomaker and Challa (1989) where one sPMMA chain wraps around one iPMMA chain as does ivy around a tree branch (Figure 19.1).

The complex double helix is asymmetric: the helical structure of the iPMMA chain is a 9_1 *helix* around which the 18_1 *helix* of the sPMMA chain wraps. The resulting double-stranded helical form is an 18_1 *helix* with a repeat unit of 1.84 nm.

X-ray diffraction patterns obtained on iPMMA systems have been interpreted by Kusunagi et al. (1976) by means of intertwined 10_1 *single helices* that eventually produce a symmetric double-helical structure with a periodicity of 2.08 nm (see Figure 19.1).

As regards the helical structure of sPMMA, conflicting views arose. According to Kusuyama and colleagues, organized sPMMA systems consist of single helices, wherein 74 monomers occur in 4 turns (see Figure 19.1). This model is derived from X-ray diffraction patterns on solid sPMMA samples crystallized by exposure to solvent vapour (Kusuyama et al. 1983). The same authors further propose a crystalline lattice which appears, however, very dubious. An alternative view was later put forward by Dybal and colleagues, who suggested the occurrence of a double helix on the basis of the value of nearly 2 of the order of reaction of the aggregation process in solution as measured by

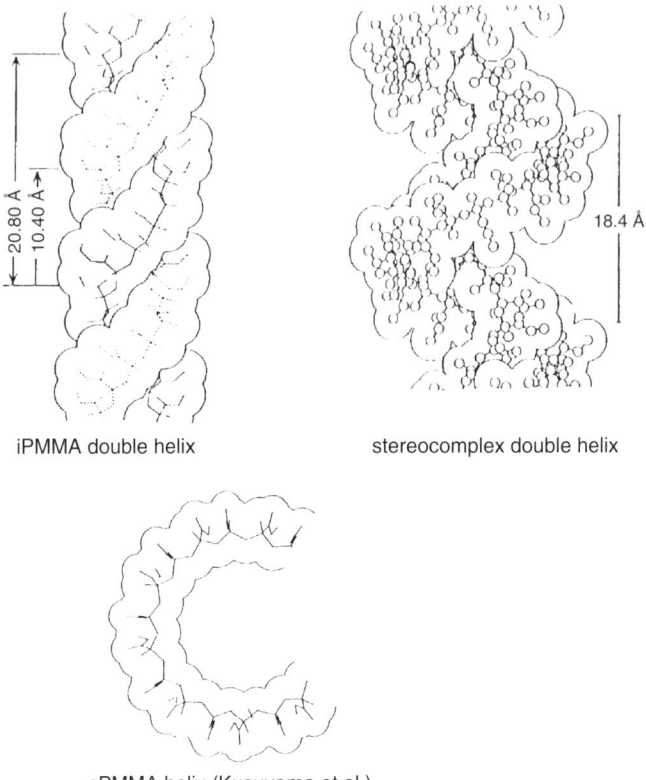

Fig. 19.1. Different types of helix for stereoregular PMMAs. The iPMMA and stereocomplex double helices are seen perpendicular to their axis while the sPMMA helix of Kusuyama et al. is seen parallel to its axis. Reprinted with permission from ACS and from Elsevier.

NMR (Dybal et al. 1986). No detailed model could be, however, given except that the chains should take on locally a *near-tt* conformation. More recently, Saiani and Guenet have proposed an asymmetric double helix on the basis of small-angle neutron scattering experiments (Saiani and Guenet 1997). This helix is, to some extent, reminiscent of that of the stereocomplex, yet, both strands are sPMMA chains and the outer helix is said to be irregular. From their analysis, Saiani and Guenet come to the conclusion that the inner helix certainly corresponds to Kusuyama et al.'s helical form. The merit of Saiani and Guenet's model is to reconcile two sets of apparently conflicting data: *the helix of Kusuyama et al. would eventually be involved in a double-helical structure*. Further details will be given below in the section headed Gel molecular structure.

The occurrence of polymer–solvent compounds in organized PMMA systems has been suspected for years as molecular organization only takes place when solvents are involved. Do as one might, PMMAs do not crystallize spontaneously from the solid state as is the case, for instance, for stereoregular polystyrenes when placed under appropriate conditions. In a thorough review of the formation of aggregates from solution Spěváček and Schneider (1987) highlight the seemingly erratic role of the solvent in that the behaviour is not at all controlled by solvent quality. For instance, aggregation occurs strongly in toluene for sPMMA while it is quite low for iPMMA. Similarly, molecular aggregates are seen to form in iPMMA/acetonitrile solutions but are absent when sPMMA is used instead. Interestingly, acetonitrile is known to be a θ-*solvent* for atactic PMMA at $T = 30°C$, which, according to views still prevailing until recently, should promote aggregation when solutions are quenched to room temperature. Again, solvent quality and gel formation are not necessarily linked. Gel formation does not unavoidably require liquid–liquid phase separation to take place, and similarly *liquid–liquid phase separation*, even through *spinodal decomposition*, does not systematically give rise to a gel.

Evidence for compound formation

Until recently, evidence for the existence of PMMA/solvent compounds was chiefly circumstantial. Additional studies, such as determination of the T–C phase diagrams or solvent mobility by NMR, have been carried out in an attempt to put this issue on safer ground. Most of these investigations deal either with thermoreversible gels, which are commonly obtained while cooling moderately concentrated solutions, or with highly concentrated solutions. In actuality, to date no report has been forthcoming on the observation of chain-folded crystals, either from the bulk or from dilute solutions (single crystals).

Temperature–concentration phase diagrams

Saiani et al. (1998) have established the T–C phase diagrams for sPMMA systems in toluene, chlorobenzene and bromobenzene. The phase diagrams are quite distinct in spite of the similarity in molecular size and shape, and solvent quality, of these three solvents.

In *toluene* the T–C phase diagram and the Tamman diagram both hint clearly at the existence of a polymer–solvent compound of stoichiometry *2 toluene/3 monomers* (Figure 19.2). Beyond the stoichiometric composition appears the outline of a eutectic melting, which is eventually impaired by the occurrence of the glass transition. This compound is designated here as C_2 for reasons dealing with the stoichiometry, which will become apparent in what follows.

The T–C diagram differs conspicuously when using *chlorobenzene* instead (Figure 19.3). This time two compounds can be identified with the following stoichiometries: compound C_1 = *10 chlorobenzene/6 monomer units* and compound C_2 = *2 chlorobenzene/3 monomers*. In addition, Saiani and colleagues have observed the existence of a mesophase in a concentration domain bracketed by these two stoichiometric values. Indeed, in the domain of concentration and temperature where this mesophase occurs, the system is no longer a gel but is a fluid (further evidence for this mesophase is provided by small-angle neutron scattering data, which is discussed below).

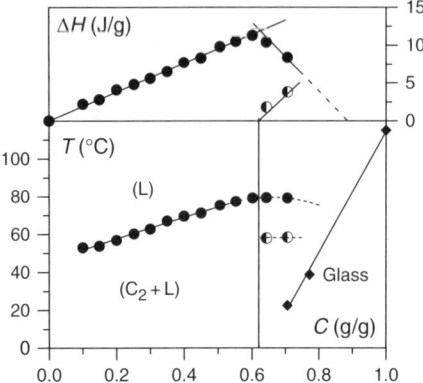

Fig. 19.2. Temperature–concentration phase diagram (**lower**) and Tamman's diagram (**upper**) for sPMMA/toluene systems. (After Saiani et al. 1998.)

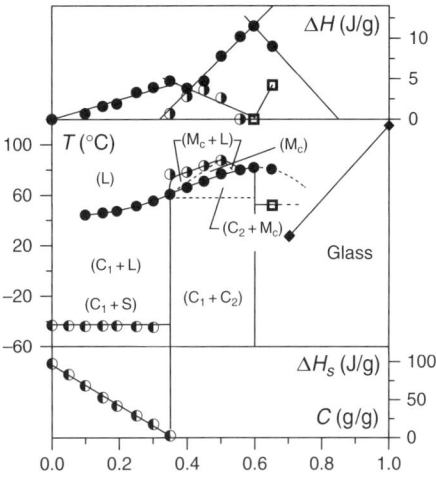

Fig. 19.3. Temperature–concentration phase diagram (**middle**) and Tamman's diagram (**above**, for polymer system melting; and **below**, for free solvent melting) for sPMMA/chlorobenzene systems. (After Saiani et al. 1998.)

In bromobenzene, two compounds have also been identified by Saiani et al. The existence of the first compound, C_1, is supported by making use of the *free solvent crystallization method* (see lower Tamman's diagram in Figure 19.4). The only hint would be, otherwise, the non-linear variation of the melting enthalpy as a function of the polymer concentration. These authors have indeed come to the conclusion that the melting endotherm is actually the convolution of two endotherms, hence the non-linear variation. They emphasize that slower heating rates do not allow resolution of the two supposed endotherms. Again, the crystallization of the free solvent is essential in revealing a

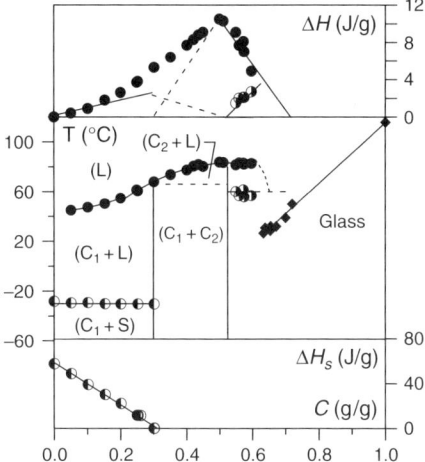

Fig. 19.4. Temperature–concentration phase diagram (**middle**) and Tamman's diagram (**above**, for polymer system melting; and **below**, for free solvent melting) for sPMMA/bromobenzene systems. (After Saiani et al. 1998.)

concentration at which no free solvent remains, which differs from the concentration at which the melting enthalpy goes through a maximum. Compounds C_1 and C_2 possess stoichiometries close to those determined for chlorobenzene (*3 bromobenzene/2 monomers* and about *2 bromobenzene/3 monomers*). Unlike the case with chlorobenzene, no mesophase is observed.

In the three solvents the same compound is observed (compound C_2 about *2 solvent molecules/3 monomers*) while a second compound of a higher degree of solvation is identified in bromobenzene and chlorobenzene. According to Saiani et al. this may arise from the difference in polarization of the solvent molecules. A CH_3 group, as occurs in toluene, repels electrons while halogens attract them.

Saiani et al. also further emphasize that for polymer concentrations lower than the stoichiometric concentration of compound C_1, the terminal melting, i.e. the liquidus line, lies in toluene some 10°C above that in chlorobenzene and bromobenzene, whereas it is nearly the same for higher concentrations (this can be highlighted by plotting the terminal melting as a function of the polymer molar fraction). In the concentration range below the stoichiometric concentration of compound C_1, the terminal melting does not correspond to that for compounds of the same degree of solvation: compound C_2 in toluene against compound C_1 in bromobenzene and in chlorobenzene. Since compound C_1 is thought to possess a higher degree of solvation than compound C_2 it is unsurprisingly expected to melt at a lower temperature. Conversely, above the stoichiometric concentration of compound C_1 the liquidus line corresponds this time to the melting of compounds of the same degree of solvation, namely C_2. The behaviour of the terminal melting provides additional support for the existence of a more solvated compound in bromobenzene and chlorobenzene.

The case of several iPMMA gels may seem rather puzzling if the notion of polymer–solvent compound is not considered. In many instances iPMMA gels possess melting points significantly below room temperature, as has been observed by Klein and Guenet for iPMMA/acetonitrile gels (Klein and Guenet 1989) and by Berghmans for iPMMA/2-butanone (Van den Broecke and Berghmans 1990). Recently, Spěváček and colleagues have investigated the temperature–concentration phase diagram in butyl acetate, a solvent wherein gels melt at about $T \approx -5°C$. From the variation of the gel melting enthalpy as a function of concentration, and from the crystallization behaviour of the free solvent, these authors have come to the conclusion that a compound is most probably formed whose stoichiometry is about *3 butyl acetate molecules/2 monomers* (Spěváček et al. 1996). Such a high degree of solvation would account for the low melting temperature of these gels. As stressed by Spěváček et al., the compound is only formed when the system is cooled down rapidly to very low temperature (typically $-60°C$). When the solution is allowed to stand at room temperature for several hours macroscopic phase separation occurs and the melting endotherm corresponding to this state spreads between 50°C and 70°C. In this case there is no evidence for compound formation.

Spectroscopic methods

NMR and neutron diffraction have given direct eveidence of the occurrence of compounds in PMMA systems. As has been discussed in Part I, NMR can determine solvent mobility. It was first believed that the mobility of solvent molecules involved in the compound should drop dramatically (Pérez et al. 1988). While this is so in the case of *PEO/dihalogenobenzene* compounds (see Part IV), the effect is barely visible in the case of thermoreversible gels when using the only determination of T_1. In this instance, Pérez et al. concluded hastily that there was no compound formed in the case of an iPS/decalin system in spite of the thermodynamic evidence (see Part IV). In fact, mobility is considerably reduced in the case of *PEO/dihalogenobenzene* compounds because the system is investigated at a temperature significantly lower than the solvent melting point. This is usually not so in the case of thermoreversible gels. The free solvent in the polymer-poor phase is highly mobile and can therefore exchange with the solvent 'bound' in the compound structure (the *polymer-rich phase*).

As shown by Spěváček and Suchoparek, the difference in mobility between the free solvent and the 'bound' solvent can be highlighted by comparing non-selective $R_1(NS)$ and selective $R_1(SE)$ proton spin–lattice relaxation rates (Spěváček and Suchoparek 1997). The principle of this type of study is described in more detail in Part I. The results obtained by Saiani et al. for gels prepared in bromobenzene are summarized in Table 19.1. While in pure bromobenzene and in aPMMA/bromobenzene solutions $R_1(NS)$ and $R_1(SE)$ do not significantly differ, $R_1(NS)$ being somewhat lower with respect to $R_1(SE)$ as expected for $\omega_o \tau_c < 1$, the reverse situation occurs in gel systems where $R_1(SE) > R_1(NS)$. In addition, this discrepancy increases on decreasing temperature. These results definitely show that bromobenzene molecules form a compound with sPMMA. Note that the corresponding relaxation times for the 'bound' solvent molecules are rather short, which suggests fast exchange with the free solvent. The increase of τ_c on decreasing temperature in the gel systems as opposed to what takes place in pure bromobenzene and aPMMA/bromobenzene solutions also points to compound formation.

Table 19.1. Non-selective and selective proton spin-lattice relaxation rates and correlation times of bromobenzene (2,6-protons) in PMMA solutions and gels. aPMMA stands for atactic PMMA, a variety that does not produce any aggregates in bromobenzene. [a]here, τ_c corresponds to the bound solvent.

sample	$T(K)$	$R_1 \, (NS)s^{-1}$	$R_1 \, (SE)s^{-1}$	τ_c ns
bromobenzene	250	0.235	0.196	0.37
	300	0.125	0.102	0.34
aPMMA C = 12.3 wt%	250	0.392	0.370	0.52
	300	0.139	0.123	0.45
sPMMA C = 13.7 wt%	250	0.526	1.724	3.1[a]
	280	0.286	0.435	1.6[a]
	300	0.168	0.189	1.1[a]
sPMMA C = 27.4 wt%	300	0.333	0.465	1.05[a]

This means that decreasing the Brownian motion allows 'bound' solvent molecules to spend more time within the compound.

Another way of showing the existence of compound consists in studying the neutron diffraction patterns by making use of the isotopic labelling. Four labelling possibilities are usually available for a binary system, which gives access to four structure factors.

In non-solvated systems, cross-terms are negligible, so intensities of the various diffraction maxima are always in the same ratio independent of isotopic labelling. Alternatively, in the case of a polymer–solvent compound, these cross-terms cannot be ignored, so the relative intensities are sensitive to the isotopic nature of the constituents.

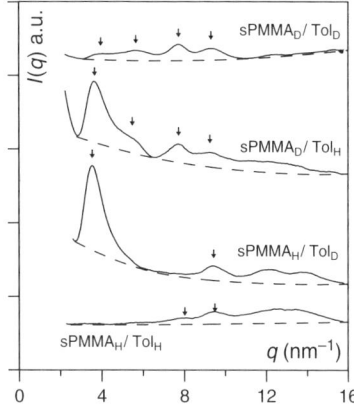

Fig. 19.5. Neutron diffraction patterns obtained by using different isotopic labelling. Polymer concentration is 0.25 g/cm³. (After Saiani et al. 1998.)

As can be seen in Figure 19.5, the intensities of the different maxima do not vary in the same way when using different isotopic labelling (Saiani et al. 1998). For instance, the ratio of the intensities of the first two maxima differs whether using hydrogenous or deuterated toluene with deuterated sPMMA. These diffraction patterns clearly demonstrate the existence of a PMMA/toluene compound. The same behaviour is seen with bromobenzene and chlorobenzene.

Gel molecular structure

Berghmans et al. (1987) have shown by infrared spectroscopy that sPMMA helices grow before gel formation. This is therefore a two-step process, unlike what usually takes place during crystallization, when the two processes are concurrent. The gel is an array of these helices. Fazel et al. (1994) have investigated the gel nanostructure of a *partially syndiotactic sPMMA* (syndio = 0.66) by small-angle neutron scattering.

In Figure 19.6 are shown the scattering curves obtained at two different temperatures for PMMA/bromobenzene mixtures whose concentration $C_p = 0.03$ g/cm^3 is slightly above the gelation threshold. At $T = 45°C$ in bromobenzene a sol is obtained. The corresponding scattering curve can be fitted by considering one type of cylinder (i.e. one type of helix; see Chapter 1, equation (1.11)) of cross-section $r_H \approx 1.5$ nm (see Figure 19.6). This agrees with the findings of Berghmans et al. that helices are already formed just above the gelation threshold. In view of the PMMA tacticity and the experimental cross-sectional radius, Fazel et al. suggest that one is, rather, dealing with the stereocomplex helical form (theoretical radius ≈ 1.2 nm). At $T = 20°C$ a gel is formed which has a significant effect on the scattering curve (see Figure 19.6). Here, the fit can only be achieved by considering a binary population of cylinders with cross-section $r_{H1} \approx 1.5$ nm and $r_{H2} \approx 3.0$ nm. The theoretical relation is:

$$q^2 I(q) \propto \sum_{i=1}^{i=2} \pi q w_i \mu_{Li} \frac{4 J_1^2 (q r_{Hi})}{(q r_{Hi})^2} + Cte \qquad (19.1)$$

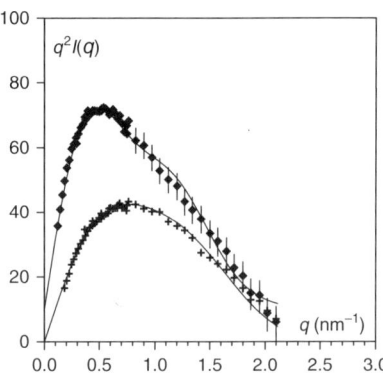

Fig. 19.6. Neutron scattering curve for sPMMA (syndio = 0.66) systems in deuterated bromobenzene: (×) for $T = 45°C$ (sol state); (♦) for $T = 20°C$ (gel state); $C_p = 0.03$ g/cm^3. (After Fazel et al. 1994.)

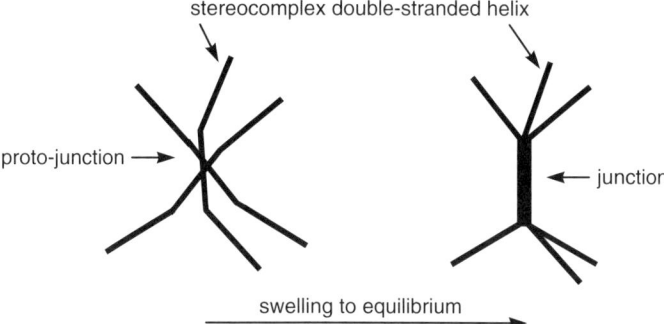

Fig. 19.7. Schematic model for a proto-junction as may occur in nascent gel (**left**), and a junction in gels once swelling equilibrium has been reached (**right**). (After Fazel et al. 1994.)

where w_i is the weight fraction of cylinders of cross-sectional radius r_{Hi} and of mass per unit length μ_{Li}. The last term is approximated to a constant, and stands for the intermolecular scattering.

It is worth stressing that the fit with two radii gives a unique answer as the maximum in the Kratky representation of the scattered intensity imposes the largest radius, while the absence of oscillations fixes both the value of the smallest radius and the weight fraction of the corresponding cylinder. Fazel et al. have accordingly assumed that the network is composed essentially of two entities: *fibrils consisting of the double-stranded helix* of the stereocomplex and *junctions*. Taking into account the value of the largest radius, it can be inferred that junctions are made up through the association of about three double helices (Figure 19.7, right).

Unexpectedly enough, the scattering curve of nascent gels prepared from higher polymer concentration (namely $0.17 \,\text{g/cm}^3$) essentially can be fitted with only one radius. According to Fazel et al., this may suggest that, unlike in gels prepared close to the critical gelation concentration, junctions are not perfectly formed (they name these junctions *proto-junctions*). *Once the gel has been aged in an excess of preparation solvent the resulting scattering curve is again fitted with two radii, of essentially the same values as those considered for the 3% gel.* This led these authors to put forward the mechanism portrayed in Figure 19.7 where a proto-junction is converted into an effective gel junction after swelling to equilibrium, hence the presence of two cross-sectional radii. As will be discovered in the next section, the rheological behaviour is also sensitive to this effect.

Note that the scattering transfer momenta explored do not give access to the mesh size. They are certainly larger than 20 nm but so far no reliable information is available on this aspect of the network structure.

For *highly syndiotactic sPMMA* (syndio $= 0.89$), Saiani and Guenet (1999) propose essentially the same nanostructure as that found for the 3% gel: three double helices meet at the gel junction. This is derived from a neutron scattering study where only a small fraction of the chains are deuterium-labelled. As a result, the scattering curve reflects only the 'single' chain structure (Figure 19.8).

In order to fit the scattering curve for the deuterated sPMMA chains in the gel state, Saiani and Guenet have contemplated the existence of a double helix as two radii are needed for accounting for the data. The situation is portrayed in Figure 19.9 in the case of a double helix with only a fraction of the chains being labelled.

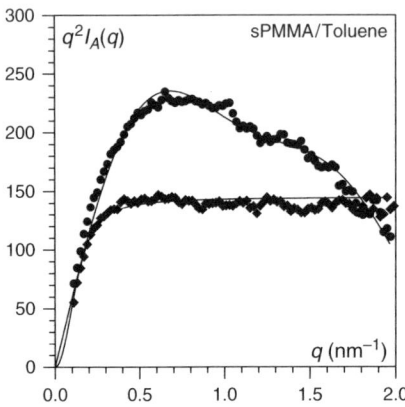

Fig. 19.8. Neutron scattering curves for sPMMA/toluene systems wherein the total polymer concentration is $C_{pol} = 0.35$ with a deuterated chain concentration $C_{sPMMAD} = 0.06$. (◆) = gel state, the solid lines being theoretical fits (see text); (●) = solution state at $T = 106°C$ fitted with Debye's relation for Gaussian chains with $R_G \approx 10$ nm $I(q) \propto 2/q^2 R_G^2 \times [exp - q^2 R_G^2 + q^2 R_G^2 - 1]$. (From Saiani and Guenet 1997.)

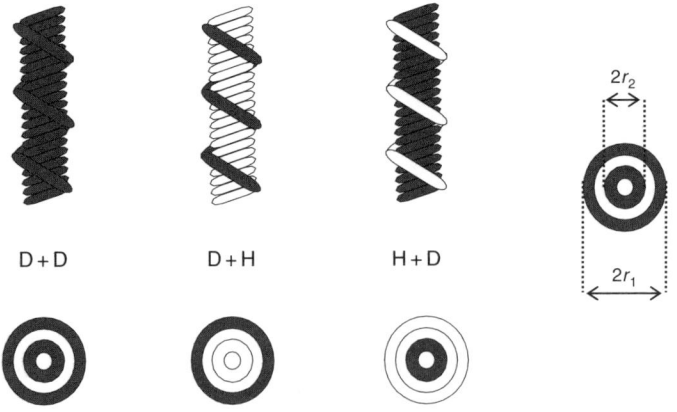

Fig. 19.9. Representation of the three different possibilities of labelling when dealing with a double helix and a mixture of hydrogenous chains (white) and deuterated chains (grey). (After Saiani and Guenet 1997.)

As a result, the scattering function is written:

$$I(q) = W_1 I_1(q) + W_2 I_2(q) + W_{12} I_{12}(q) \tag{19.2}$$

in which W_1, W_2 and W_{12} are the fraction of deuterated chains (see Figure 19.9) and $I_1(q)$, $I_2(q)$ and $I_{12}(q)$ are the corresponding intensities. In this range of transfer momenta q these intensities are essentially due to the helix cross-section and are therefore written: for D+H

$$q^2 I_1(q) = 4\pi q C_p \mu_{L1} \left[\frac{J_1(qr_1) - \gamma_1 J_1(q\gamma_1 r_1)}{qr_1(1-\gamma_1^2)} \right]^2 \tag{19.3}$$

for H+D

$$q^2 I_2(q) = 4\pi q C_p \mu_{L2} \left[\frac{J_1(qr_2) - \gamma_2 J_1(q\gamma_2 r_2)}{qr_2(1-\gamma_2^2)} \right]^2 \tag{19.4}$$

and for D+D

$$q^2 I_{12}(q) = 4\pi q C_p \mu_{L12} \left[\frac{r_1 J_1(qr_1) - \gamma_1 r_1 J_1(q\gamma_1 r_1) + r_2 J_1(qr_2) - \gamma_2 r_2 J_1(q\gamma_2 r_2)}{qr_1^2(1-\gamma_1^2) + qr_2^2(1-\gamma_2^2)} \right]^2$$

$$\tag{19.5}$$

Where r_1 anr r_2 are the outer radii of each strand (see Figure 19.9), and γ_i are related to the strand 'thickness' e trough:

$$e = r_i - \gamma_i r_i \tag{19.6}$$

The different weight fractions W_1, W_2 and W_{12} are calculated by means of a combinatorial analysis. This shows that about 10% of the double helices are totally deuterium-labelled, a figure that is far from being negligible in the analysis of the neutron scattering curves. The following cross-sectional radii are thus obtained for the double helix.

inner radius $r_1 = 1.1 \pm 0.2$ nm with $\mu_{L_1} = 2410 \pm 300$ g/mol × nm

outer radius $r_2 = 2.3 \pm 0.3$ nm with $\mu_{L_2} = 1580 \pm 200$ g/mol × nm

The parameters of the inner helix are those expected from the 74/4 helical form proposed by Kusuyama et al. (1983). Saiani and Guenet conclude that the inner helix probably corresponds to that helix described by Kusuyama and colleagues, while the second helix wraps around the inner one. Clearly, the two strands do not possess the same pitch, so one is dealing here with an asymmetric double helix, something highly unusual.

Allegedly, one sPMMA chain is capable of forming two different helices: the 74/4 (the helix of Kusuyama et al.) and the 18_1 helix in the stereocomplex. In all cases the monomer arrangement is close to a *tt* configuration. Dybal et al. also conclude that, in the double helix, configurations are essentially *tt* (Dybal et al. 1986). There is therefore no reason why sPMMA could not produce an irregular helix of *tt* or *nearly tt* configuration. Clearly, further investigations are needed in order to test the notion of an asymmetric double-stranded helix, and to gain information on the location of the solvent molecules within the polymer structure.

The radius of the outer helix implies the existence of a gap with respect to the inner helix. Saiani and Guenet have considered the possibility that this gap is filled with solvent molecules, an assumption which would agree with the occurrence of a compound. The double helix would be consistent with the two-step process of gel formation: Kusuyama's helix would form first and the irregular helix would then grow around it.

Wide-angle neutron diffraction on totally labelled samples (Sainai and Guenet 1997, 1999) has provided further support for this unusual type of helical form. As can be seen from Figure 19.10, Saiani and Guenet come to this conclusion because they only have to consider the *higher-order layer lines* of Kusuyama et al.'s helix plus the *zeroth order layer line* of their asymmetric helix for reproducing satisfactorily the diffraction pattern of the gel samples (Saiani and Guenet 1999). Owing to the irregularity of the outer helix, *higher-order layer lines* are obviously absent. Their approach allows one to understand the diffraction pattern of these systems that had otherwise puzzled Kusuyama et al. In fact, these authors attempted to derive a crystalline lattice from their results, but failed. Note a simple packing of double helices such as those proposed by Saiani and Guenet can account for the crystal unit cell put forward by Kusuyama et al.

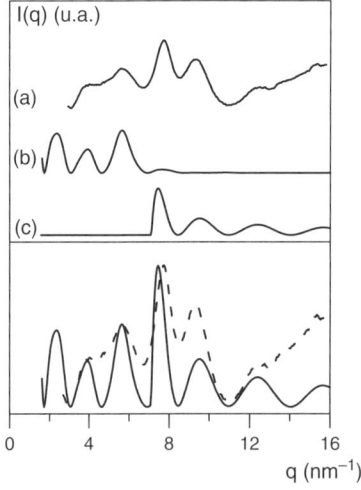

Fig. 19.10. Upper panel: (a) intensity diffracted by a 25% (v/v) sPMMA$_D$/toluene$_D$ gel; (b) zeroth layer line calculated from diffraction by the network; (c) first layer line calculated by considering Kusuyama's helix. **Lower panel:** (broken line) intensity diffracted by a 25% (v/v) sPMMA$_D$/toluene$_D$ gel and (solid line) sum of the zeroth and first layer lines as calculated above (see text for details). (From Saiani and Guenet 1999.)

Gel rheology

This section is focused only upon the relation modulus–concentration, which has a direct bearing upon the gel structure. PMMA gels display a significant relaxation rate when submitted to a compressive deformation (Fazel et al. 1992). The rate $m = dLog\sigma/dLog\,t$ is independent of polymer concentration, and is close to $m = 0.15$, a value quite similar to that observed in isotactic polystyrene gels. The use of an isochronal elastic modulus is therefore needed.

For partially syndiotactic sPMMA, Fazel et al. (1992) obtained two experimental relations depending upon whether the nascent gel state or the swollen gel state was investigated:

nascent gel $\qquad E = 2.08 \times 10^2\,C^{1.86\pm0.1}$ kPa

swollen gel $\qquad E = 1.94 \times 10^3\,C^{1.99\pm0.1}$ kPa

These relations indicate that, upon swelling, both the exponent and the front factor are altered. Moduli increase about 10-fold after swelling, which may seem paradoxical. Usually, the modulus is proportional to the number of junctions per unit volume. Swelling entails, in principle, dilution of the junctions, and correspondingly a decrease of gel modulus is expected. The mechanism described in Figure 19.7, which was derived from the study of the nanostructure, can account for this behaviour. This mechanism implies some kind of transition from an *entropic gel* (proto-junctions are certainly flexible and act more or less as 'entanglements') to an *enthalpic gel* (see Part I). Under these conditions the elastic modulus can increase owing to the global stiffening of the system although the number of junctions per unit volume decreases. Note that the exponent 1.99 is somewhat consistent with *enthalpic elasticity* when the elements of the network have a fractal dimension $D_f = 1$ (see Part I). This also requires the structure to be self-similar from the microscopic scale to the mesoscopic scale.

The case of *highly syndiotactic* sPMMA appears to be at variance, which does suggest that the helical structure differs from that considered in *partially syndiotactic* sPMMA. First, swelling tends to disaggregate the gels: the infinite three-dimensional network gives birth to aggregates of finite size. As a result, swelling to equilibrium is meaningless. Second, the exponents are far larger than those given above. For non-swollen gels aged 1 week prior to observation, Saiani has obtained the following relations for the 40s-isochronal modulus (Saiani 1997):

sPMMA/toluene gels $\qquad E = 1.33 \times 10^7\,C^{3.7\pm0.3}$ kPa

sPMMA/bromobenzene gels $\qquad E = 1.32 \times 10^8\,C^{6.1\pm05}$ kPa

In view of the rod-like nanostructure, exponents between 1.5 (*entropic*) and 2 (*enthalpic*) would have been expected. The discrepancy is large enough to imply that the microscopic and the mesoscopic structures are not self-similar. In other words these gels are likely to be highly inhomogeneous. Note that the strongest departure from the theoretical exponents occurs for bromobenzene. This may be linked to the existence of two compounds in this solvent instead of one in toluene.

Le savant doit ordonner; on fait de la science avec des faits comme une maison avec des pierres; mais une accumulation de faits n'est pas plus une science qu'un tas de pierres n'est une maison.

Henri Poincare, *La Science et l'Hypothèse*

Conclusion

One of the aims of this book was to highlight the wealth of polymer–solvent molecular compounds that exist, from biopolymers to synthetic polymers. Almost every crystallizable polymer possesses the propensity to organize intimately with solvent molecules. It should, however, be clear after reading this book that there are still a great many questions that remain to be solved. While the goal of the characterization of these compounds at the molecular level has been achieved these past 20 years, the origin of their formation is still not understood. The main questions that arise are probably: what is the physics behind the formation of these compounds? Can we develop a theory, of the type of Flory's theory for amorphous polymers, which could be able to give some hints as to the shape of the temperature–concentration phase diagram? Or will we be restricted to make some predictions through the use of computer simulations only? The latter approach may become more efficient in the coming years for clathrates and intercalates in view of the important role played by the shape of the solvent and of the chain microstructure. Conversely, for systems wherein only hydrogen bonds or electrostatic interactions are the main driving process of compound formation, an analytical theory with a minimum of adjustable parameters might be achievable.

Other issues deal with compound morphology. The occurrence of a compound needed to rigidify chains otherwise flexible so as to produce a fibrillar morphology seems required, yet the question remains as to why the formation of a compound does not always enhanced chain rigidity. A striking case concerns iPS and sPS in *trans*-decalin (see Part IV): fibrils are formed in the former while a spherulitic morphology is observed in the latter. Another pending question deals with the formation and the puzzling 'stability' of the emptied chlathrates from sPS (δ_e or δ nanoporous form): which physical process is involved in stabilizing this form – the amorphous domains that stand below their glass transition T_g and, as such, pin down the entire structure, or a few solvent molecules still trapped in the crystal structure and acting as struts, or something else? This undoubtedly remains an important issue for further development of these multiporous materials, especially if other polymers are to be used for extending the field of application.

This book also highlights the fact that molecular compounds are often produced with organic solvents, particularly when synthetic polymers are involved. Indeed, current concern for the earth environment is acute, and the use of organic solvents is often considered as outrageous. One may then wonder whether these systems may have any future in industrial applications. The answer should tend to be positive provided that the organic solvents are properly recovered and reused. Producing molecular compounds could then be a step in the manufacture of materials, such as those multiporous systems obtained from syndiotactic polystyrene (see Multiporous networks in Chapter 16). A known example, which supports this view, is the making of PVC fibres. These fibres are produced through the spinning of concentrated PVC physical gels obtained in a mixture of carbon disulphide/acetone. The solvent is totally recovered after the drying of the fibres, and, in spite of the strong and nauseating smell of the carbon disulphide, visiting the factory is not at all an ordeal because the solvent is adequately recycled!

The reader may therefore wonder about current applications and possible applications in the near future. While the knowledge of the molecular structure and the observation of some properties of these molecular compounds have allowed one to improve some existing processes, particularly in the case of cellulose or of starch, at the present time new applications are not numerous. Nanoporous gels are certainly promising materials for filtration purposes but also for storage systems (storage of hydrogen for instance). In view of the unexpected and unpredicted fate of other systems that had no apparent future, molecular compounds may soon be of industrial interest. Who would have bet on the extensive use of laser in hi-fi systems when it was invented in the early 1960s? And yet, they are now part of everyday life. Who would have guessed that NMR would become a non-invasive tool in medical investigations? Clearly, molecular compounds may have applications in unexpected domains. In other words, the question is: should the scientist involved in the study of these systems think simultaneously about potential applications, or will applications be discovered by persons foreign to this field? The spirit of this book is, rather, to act as a kind of 'old-fashioned hardware store', at least of the type that once existed in France (a few still do), where one might be able to find strange, unexpected objects that would eventually trigger new ideas for use. One would pay a visit to these stores with no preconceived ideas but simply out of sheer curiosity, and in many cases one would stumble onto the item that would change everything. It is simply hoped that this book will prove to be this kind of old-fashioned hardware store.

References

Abied, H., Brûlet, A. and Guenet, J.M. *Colloid Polym. Sci.* **1990** *268* 403.

Aiken, W., Alfrey, T., Jr, Janssen, A., Mark, H. *J. Polym. Sci.* **1947** *2* 178.

Ajayaghosh, A. and George, S.S. *J. Am. Chem. Soc.* **2001** *123* 5148.

Ajayaghosh, A. and George, S.S. *Angew. Chem.* **2003** *115* 346.

Albert, R., Jeong, H. and Barabási, A.L. *Nature* **1999** *401* 130.

Alfrey, T. Jr, Wiederhorn, N., Stein, R.S., Tobolsky, A. *Ind. Eng. Chem.* **1949** *41* 701.

Anderson, J.O. *Calphad* **1987** *11* 271.

Anderson, N.S., Campbell, J.W., Harding, M.M., Rees, D.A., Samuel, J.W.B. *J. Mol. Biol.* **1969** *45* 85.

Andreatta, A., Cao, Y., Chiang, J.-C., Smith, P., Heeger, A.J. *Synth. Met.* **1988** *26* 383.

Andress, K. *Z. Physik. Chem. (Leipzig)* **1928** *136* 279.

Andress, K. *Z. Physik. Chem.* **1929** *B4* 190.

Andress, K. and Rheinhardt, L. *Z. Physik. Chem. (Leipzig)* **1930** *A151* 425.

Angelopoulos, M., Asturias, G.E., Ermer, S.P., Ray, A., Scherr, E.M., MacDiarmid, A.G. *Mol. Cryst. Liq. Cryst.* **1988** *160* 151.

Arai, S., Chatake, T., Minezaki, Y., Niimura, N. *Acta Cryst.* **2002** *D58* 151.

Arai, S., Chatake, T., Ohhara, T., Kurihara, K., Tanaka, I., Suzuki, N., Fujimoto, Z., Mizuno, H. and Niimura, N. *Nucleic Acid. Res.* **2005** *33* 3017.

Arnauts, J. and Berghmans, H. *Polym. Commun.* **1987** *28* 66.

Arnott, S., Scott, W.E., Rees, D.A., McNab, C.G.A. *J. Mol. Biol.* **1974a** *90* 253.

Arnott, S., Fulmer A., Scott, W.E., Dea, I.C.M., Moorhouse, R., Rees, D.A. *J. Mol. Biol.* **1974b** *90* 269.

Arnott, S., Chandrasekaran, R., Birdsall, D.L., Leslie, A.G.W., Ratliff, R.L. *Nature* **1980** *283* 743.

Atkins, E.D.T. *Int. J. Biol. Macromol.* **1986** *8* 323.

Atkins, E.D.T., Hill, M.J., Jarvis, D.A., Keller, A., Sarhene, E., Shapiro, J.S. *Colloid Polym. Sci.* **1984** *262* 22.

Atkins, P.W. *Physical Chemistry*, 3rd edn, **1986**. Oxford University Press.

Bachmann, M.A., Gordon, W.L., Weinhold, S., Lando, J.B. *J. Appl. Phys.* **1980** *51* 5095.

Bachmann, M.A. and Lando, J.B. *Macromolecules* **1981** *14* 40.

Bacon, G.E. *Neutron Diffraction*, **1975**. Clarendon Press, Oxford.

Bailey, F.E. Jr. and France, H.G. *J. Polym. Sci.* **1961** *49* 397.

Barabási, A.L. and Albert, R. *Science* **1999** *286* 509.

Barham, P.J. In: *Crystallization of Polymers*, **1993**, M. Dosiére (ed.). Kluwer Academic Publishers: Dordrecht; p. 153.

Barry, A.J., Peterson, F.C. and King, A.J. *J. Am. Chem. Soc.* **1936** *58* 333.

Belfiore, L.A. and Ueda, E. *Polymer* **1992** *33* 3833.

Benoit, H.C. and Picot, C. *Pure Appl. Chem.* **1966** *12* 545.

Berghmans, H., Govaerts, F. and Overbergh, N. *J. Polym. Sci. Polym. Phys. Ed.* **1979** *17* 1251.

Berghmans, H., Donkers, A., Frenay, L., De Schryver, F.E., Moldenaers, P., Mewis, J. *Polymer* **1987** *28* 97.

Blackwell, J. *Biopolymers* **1969** *7* 281.

Blackwell, J., Sarko, A. and Marchessault, R.H. *J. Mol. Biol.* **1969** *42* 379.

Bleaney, B. and Bowers, K. *Proc. R. Soc. Lond.* **1952** *A214*, 451.

Bonvin, A.M.J.J., Sunnerhagen, M., Otting, G., van Gusteren, W.F. *J. Mol. Biol.* **1998** *282* 859.

Bourne, E.J., Donnison, G.H., Haworth, N., Peat, S. *J. Chem. Soc.* **1948** 1687.

Brûlet, A., Boué, F. and Cotton, J.P. *J. Phys. II (France)* **1996** *6* 885.

Buléon, A., Delage, M.M., Brisson, J., Chanzy, H. *Int. J. Biol. Macromol.* **1990** *12* 25.

Buléon, A., Duprat, F., Booy, F.P., Chanzy, H. *Carbohyd. Polym.* **1984** *4* 161.

Buléon, A., Gérard, C., Riekel, C., Vuong, R., Chanzy, H. *Macromolecules* **1998** *31* 6605.

Cahn, J.W. *J. Am. Ceram. Soc.* **1961** *52* 118.

Cahn, J.W. and Hilliard, J.E. *J. Chem. Phys.* **1959** *31* 688.

Cahn, J.W. and Hilliard, J.E. *J. Chem. Phys.* **1965** *42* 93.

Candau, S.J., Dormoy, Y., Mutin, P.H., Debeauvais, F., Guenet, J.M. *Polymer* **1987** *28* 1334.

Cao, Y., Smith, P. and Heeger, A.J. *Synth. Met.* **1989** *32* 263.

Cao, Y., Smith, P. and Heeger, A.J. *Synth. Met.* **1992** *48* 91.

Cao, Y. and Smith, P. *Polymer* **1993** *34* 3139.

Cao, Y., Smith, P. and Yang, C.Y. *Synth. Met.* **1995** *69* 191.

Carbonnel, L., Guieu, R. and Rosso, J.C. *Bull. Soc. Chim.* **1970** *8–9* 2855.

Cardoso, M.B., Putaux, J.L., Nishiyama, Y., Helbert, W., Hÿtch, M., Silveira, N.P., Chanzy, H. *Biomacromolecules* **2007** *8* 1319

Cates, M.E. *Macromolecules* **1987** *20* 2289.

Cates, M.E. *J. Phys. France* **1988** *49* 1593.

Chanzy, H.D. Structure of Fibrous Biopolymers, Colston papers No. 26, **1975**, E.D.T. Atkins and A. Keller (eds). Butterworths: London; p. 417.

Chanzy, H.D. and Roche, E. *Appl. Polym. Symp.* **1976** *28* 701.

Chanzy, H.D., Henrissat, B., Vuong, R., Revol, J.F. *Holzforschung* **1986** *40 (Suppl)* 25.

Chatani, Y. In: *Crystallization of Polymers*, **1993**, M. Dosiére (ed.). Kluwer Academic Publishers: Dordrecht; p. 267.

Chatani, Y., Fuji, Y., Shimane, Y., Ijitsu, T. *Polym. Prep. Jpn (Engl. edn)* **1988** *37* 428.

Chatani, Y., Maruyam, H., Asanuma, T., Shiomura, T. *J. Polym. Sci., Part B* **1991** *29* 1649.

Chatani, Y. and Nakamura, M. *Polymer* **1993** *34* 1644.

Chatani, Y., Shimane, Y., Inagaki, T., Ijitsu, T., Yukinari, T., Shikuma, H. *Polymer* **1993** *34* 1620.

Chédin, J. *Kolloid-Z.* **1952** *125* 65.

Chédin, J. *Chim. Ind. (Milan)* **1955** *37* 560.

Chédin, J. and Marsaudon, A. *Chim. Ind. (Paris)* **1954** *71* 55.

Chédin, J. and Marsaudon, A. *Makromol. Chem.* **1955** *15* 115.

Chédin, J. and Marsaudon, A. *Makromol. Chem.* **1956** *20* 57.

Chenite, A. and Brisse, F. *Macromolecules* **1991** *24* 2221.

Chevalier, P.Y. *Thermochim. Acta* **1989** *166* 211.

Cho, K. and Park, S.H. *Macromolecular Symposia* **2001** *166* 93.

Clark, G.L. and Parker, E.A. *J. Phys. Chem.* **1937** *41* 777.

Clark, J., Wellinghoff, S.T. and Miller, W.G. *Polym. Prep. ACS* **1983** *24* 86.

Cleven, R., van den Berg, C. and van der Plas, L. *Starch-Stärke* **1978** *30* 223.

Cochran, W., Crick, F.H.C. and Vand, V. *Acta Crystallogr.* **1952** *5* 581.

Cohen, Y. *J. Polym. Sci., Polym. Phys. Ed.* **1996** *34* 57.

Cohen, Y. and Cohen, E. *Macromolecules* **1995** *28* 3631.

Cohen, Y. and Dagan, A. *Macromolecules* **1995** *28* 7638.

Cohen, Y., Saruyama, Y. and Thomas, E.L. *Macromolecules* **1991** *24* 1161.

Cohen, Y. and Wade Adams, W. *Polymer* **1996** *37* 2767.

Corradini, P., Guerra, G., Petraccone, V., Pirozzi, B. *Eur. Polym. J.* **1980** *16* 1089.

Cortili, G. and Zerbi, G. *Spectrochim. Acta* **1967** *23A* 2216.

Cotton, J.P., Decker, D., Benoit, H., Farnoux, B., Higgins, J., Jannink, G., Ober, R., Picot, C., des Cloiseaux, J. *Macromolecules* **1974** *7* 863.

Cowie, J.M.G. and Toporowski, P.M. **1961** *Can. J. Chem. 39* 2240.

Creely, J.J., Segal, L. and Loeb, L. *J. Polym. Sci.* **1959** *36* 205.

Creely, J.J. and Wade, R.H. *Textile Res. J.* **1978a** *48* 336.

Creely, J.J. and Wade, R.H. *J. Polym. Sci. Polym. Lett. Ed.* **1978b** *16* 291.

Dahmani, M., Fazel, N., Munch, J.P., Guenet, J.M. *Macromolecules* **1997** *30* 1463.

Dahmani, M., Ramzi, M., Rochas, C., Guenet, J.M. *Int. J. Biol. Mol.* **2003** *31* 147.

Dahmani, M., Skouri, M., Much, J.P., Guenet, J.M. *Europhysics letter* **1994** *26* 19.

Damman, P. and Point, J.J. *Macromolecules* **1994** *27* 3919.

Damman, P. and Point, J.J. *Macromolecules* **1995** *28* 2050.

Dammer, C. *Diplomarbeit* **1991** Universität Ulm (FRG).

Dammer, C., Terech, P., Maldivi, P., Guenet, J.M. *Langmuir* **1995** *11* 1500.

Daniel, C. *Thesis* **1996** Université Louis Pasteur Strasbourg (F).

Daniel, C., Alfano, D., Venditto, V., Cardea, S., Reverchon, E., Larobina, D., Mensitieri, G., Guerra, G. *Adv. Mater.* **2005** *17* 1515.

Daniel, C., Brûlet, A., Menelle, A., Guenet, J.M. *Polymer* **1997** *38* 4193.

Daniel, C., Dammer, C. and Guenet, J.M. *Polymer* **1994** *35* 4243.

Daniel, C., De Luca, M.D., Brûlet, A., Menelle, A., Guenet, J.M. *Polymer* **1996** *37* 1273.

Daoud, M., Cotton, J.P., Farnoux, B., Jannink, G., Sarma, G., Benoit, H., Duplessix, R., Picot, C., de Gennes, P.G. *Macromolecules* **1975** *8* 804.

Daoud, M., Family, F. and Jannink, G. *J. Phys. Lett. France* **1984** *45* 199.

Dasgupta, D., Malik, S., Thierry, A., Guenet, J.M., Nandi, A.K. *Macromolecules* **2006** *39* 6110.

Dasgupta, D., Manna, S., Malik, S., Rochas, C., Guenet, J.M., Nandi, A.K. *Macromolecules* **2005** *38* 5602.

Dautzenberg, H. *Faserforsch. Textiltech.* **1970** *21* 117.

Davis, G.T., McKinney, J.E., Broadhurst, M.G., Roth, S.C. *J. Appl. Phys.* **1978** *49* 4998.

Davis, W.E., Barry, A.J., Peterson, F.C., King, A.J. *J. Am. Chem. Soc.* **1943** *65* 1294.

Deberdt, F. and Berghmans, H. *Polymer* **1993** *34* 2192.

Debye, P. *Ann. Phys. Lpz.* **1915** *46* 809.

Debye, P. and Bueche, A.M. *J. Appl. Phys.* **1949** *20* 518.

de Candia, F., Carotenuto, M., Guadagno, L., Vittoria, V. *Macromolecules* **1992** *25* 6361.

de Candia, F., Guadagno, L. and Vittoria, V. *J. Macromol. Sci. Phys.* **1995** *B34* 95.

de Candia, F., Guadagno, L. and Vittoria, V. *J. Macromol. Sci. Phys.* **1996** *B35* 265.

de Gennes, P.G. *J. Physique Lett.* **1976** *37* 1.

de Gennes, P.G. In: *Scaling Concepts in Polymer Science*, **1979**. Cornell University Press: NY.

Delaite, E., Point, J.J., Damman, P., Dosière, M. *Macromolecules* **1992** *25* 4768.

De Rosa, C., Rizzo, P., Ruiz de Ballesteros, O., Petraccone, V., Guerra, G. *Polymer* **1999** *40* 2103.

Des Cloiseaux, J. *Macromolecules* **1973** *6* 403.

Des Cloiseaux, J. and Jannink, G. In: *Les Polymères en Solutions*, **1988**. Editions de Physique Les Ulis (France).

Dikshit, A.K. and Nandi, A.K. *Macromolecules* **2000** *33* 2616.

Dobbins, R.J. *TAPPI* **1970** *53* 2284.

Doll, W.W. and Lando, J.B. *J. Macromol. Sci.-Phys.* **1968** *B2* 219.

Doriomedoff, M., Hautier-Cirstofini, F., De Surville, R., Jozefowicz, M., Yu, L.T., Buvet, R. *J. Chim. Phys.* **1971** *68* 1055.

Dorrestijn, A., Keijzers, A.E.M. and te Nijenhuis, K. *Polymer* **1981** *22* 305.

Dorrestijn, A. and te Nijenhuis, K. *Colloid Polym. Sci.* **1990** *268* 895.

Dreher, M.L. and Berry, J.W. *Starch-Stärke* **1983** *35* 76.

Drew, H.R. and Dickerson, R.E. *J. Mol. Biol.* **1981** *151* 535.

Dweltz, N.E. *Biochim. Biophys. Acta* **1961** *51* 283.

Dybal, J., Spevacek, J. and Schneider, B. *J. Polym. Sci. Polym. Phys. Ed.* **1986** *24* 657.

Edwards, S.F. *Proc. Phys. Soc. Lond.* **1966** *88* 265.

ElHasri, S., Ray, B., Thierry, A., Guenet, J.M. *Macromolecules* **2004** *37* 4124.

Esposito, G., Tarallo, O. and Petraccone, V. *Macromolecules* **2006** *39* 5037.

Falshaw, R., Furneaux, R.H. and Stevenson, D.E. **1998** *Carbohyd. Res.* *308* 107.

Fazel, N., Brûlet, A. and Guenet, J.M. *Macromolecules* **1994** *27* 3836.

Fazel, Z., Fazel, N. and Guenet, J.M. *J. Phys. II (Les Ulys)* **1992** *2* 1745.

Feke, G.T. and Prins, W. *Macromolecules* **1974** *7* 527.

Flory, P.J. *J. Chem. Phys.* **1942** *10* 51.

Flory, P.J. In: *Principles of Polymer Chemistry*, **1953**. Cornell University Press: NY.

Flory, P.J. *Proc. R. Soc., Lond. Ser A* **1956** *234* 60.

Foord, S.A and Atkins, E.D.T. *Biopolymers* **1989** *28* 1345.

Fournet, G. *Bull. Soc. Franç. Minèr. Crist.* **1951** *74* 39.

François, J., Gan, Y.S. and Guenet, J.M. *Macromolecules* **1986** *19* 2755.

François, J., Gan, Y.S., Sarazin, D., Guenet, J.M. *Polymer* **1988** *29* 898.

Frank, F.C. and Keller, A. *Polym. Commun.* **1988** *20* 186.

Franklin, R.E. and Gosling, R.G. *Acta Cryst.* **1953a** *6* 673.

Franklin, R.E. and Gosling, R.G. *Nature* **1953b** *171* 740.

Frederickson, G.H. *Macromolecules* **1993** *26* 2825.

French, A.D. and Zobel, H.F. *Biopolymers* **1967** *5* 457.

Fuller, W., Forsyth, T. and Mahendrasingam, A. *Philos. T. Roy. Soc. B* **2004** *359* 1237.

Fuller, W., Forsyth, T., Mahendrasingam, A., Pigram, W.J., Greenall, W.J., Langan, R.J., Bellamy, K.A., Al Hayalee, Y., Mason, S.A. *Physica B* **1989** *156–157* 468.

Gallant, D.J., Bewa, M., Buy, Q.H., Bouchet, B., Szylit, O., Sealy, L. *Starch-Stärke* **1982** *34* 255.

Gal'perin, Ye. L., Kosmynin, B.P. and Bychkov, R.A. *Vysokomol. Soed.* **1970** *B12* 555.

Gal'perin, Ye. L., Strogalin, Yu.V. and Mlenik, M.P. *Vysokomol. Soed.* **1965** *7* 933.

Gamini, A., Toffanin, R., Murano, E., Rizzo, R. *Carbohyd. Res.* **1997** *304* 293.

Gan, Y.S., François, J. and Guenet, J.M. *Macromolecules* **1986** *19* 173.

Gan, Y.S., François, J., Guenet, J.M., Allain, C., Gauthier-Manuel, G. *Makromol. Chem. Rapid Commun.* **1985** *6* 225.

Garcia, A.I., Muñoz, M.E., Peña, J.J., Santamaria, A. *Macromolecules* **1990** *23* 5251.

Gardner, K. and Blackwell, J. *Biopolymers* **1974** *13* 1975.

Gardner, K. and Blackwell, J. *Biopolymers* **1975** *14* 1581.

George, S.S. and Ajayaghosh, A. *Chem. Eur. J.* **2005** *11* 3217.

Gidley, M.J. and Bulpin, P.V. *Macromolecules* **1989** *22* 341.

Girolamo, M., Keller, A., Miyasaka, K., Overbergh, N. *J. Polym. Sci., Polym. Phys. Ed.* **1976** *14* 39.

Gomez, M.A. and Tonelli, A.E. *Macromolecules* **1991** *24* 3533.

Görlitz, M., Minke, R., Trautvetter, W., Weisgerber, G. *Angew. Makromol. Chem.* **1973** *29/30* 137.

Gowd, E.B., Nair, S.S., Ramesh, C., Tashiro, K. *Macromolecules* **2003** *36* 7388.

Grassi, A., Longo, P., Proto, A., Zambelli, A. *Gazz. Chim. Ital.* **1987** *117* 249.

Grassi, A., Pellechia, P., Longo, P., Zambelli, A. *Macromolecules* **1989** *22* 104.

Green, A.G. and Woodhead, A.E. *J. Chem. Soc. Trans.* **1910** *97* 2388.

Greiss, O., Asano, T., Xu, Y., Peterman, J. *Z. Kristallogr.* **1988** *182* 58.

Greiss, O., Xu, Y., Asano, T., Peterman, J. *Polymer* **1989** *30* 590.

Griess, G.A., Guiseley, K.B. and Serwer, P. *Biophys. J.* **1993a** *65* 138.

Griess, G.A., Edwards, D.M., Dumais, M., Harris, R.A., Renn, D.W., Serwer, P. *J. Struct. Biol. J.* **1993b** *111* 39.

Guenet, J.M. *Macromolecules* **1980** *13* 387.

Guenet, J.M. *Macromolecules* **1986** *19* 1960.

Guenet, J.M. *Macromolecules* **1987** *20* 2874.

Guenet, J.M. In: *Thermoreversible Gelation of Polymers and Biopolymers*, **1992**. Academic Press: London.

Guenet, J.M. *J. Phys. II* **1994** *4* 1077.

Guenet, J.M. *Trends Polym. Sci.* **1996a** *4* 6.

Guenet, J.M. *Thermochim. Acta.* **1996b** *284* 67.

Guenet, J.M. *J. Rheol.* **2000** *44* 947.

Guenet, J.M. *Macromolecular Symposia* **2003** *203* 1.

Guenet, J.M., Benoit, H.C. and Picot, C. *Macromolecules* **1979** *12* 86.

Guenet, J.M., Brûlet, A. and Rochas, C. *Int. J. Biol. Macromol.* **1993** *15* 131.

Guenet, J.M., Klein, M. and Menelle, A. *Macromolecules* **1989** *22* 494.

Guenet, J.M., Lotz, B. and Wittmann, J.C. *Macromolecules* **1985** *18* 420.

Guenet, J.M. and McKenna, G.B. *J. Polym. Sci. Polym. Phys. Ed.* **1986** *24* 2499.

Guenet, J.M. and McKenna, G.B. *Macromolecules* **1988** *21* 1752.

Guenet, J.M., Menelle, A., Schaffhauser, V., Terech, P., Thierry, A. *Colloid and Polymer Science* **1994** *272* 36.

Guenet, J.M. and Nandi, A.K. (eds) Fibrillar Networks as Advanced Materials, J.M. Guenet, J.M. and A.K. Nandi, (eds). In: *Macromol. Symp.* **2006** *241* papers therein.

Guenet, J.M. and Picot, C. *Macromolecules* **1981** *14* 309.

Guenet, J.M. and Picot, C. *Macromolecules* **1983** *16* 205.

Guenet, J.M., Poux, S. and Thierry, A. *Macromol. Symp.* **2006** *235* 25.

Guenet, J.M., Ray, B., ElHasri, S., Marie, P., Thierry, A. In: *Role of Interfaces in Environmental Protection*, **2003**, S. Barany (ed.), NATO ASI series. Kluwer Academic Publishers; p. 191.

Guenet, J.M., Sadler, D.M. and Spells, S.J. *Polymer* **1990** *31* 195.

Guenet, J.M., Willmott, N.F.F. and Ellsmore, P.A. *Polym. Commun.* **1983** *24* 230.

Guerra, G., Manfredi, C., Musto, P., Tavone, S. *Macromolecules* **1998** *31* 1329.

Guerra, G., Manfredi, C., Rapacciulo, M., Corradini, P., Mensitieri, G., Del Nobile, M.A. *Italian Patent* **1994** (CNR).

Guerra, G., Vitagliano, V.M., De Rosa, C., Petraccone, V., Corradini, P. *Macromolecules* **1990** *23* 1539.

Guerrero, S.J. and Keller, A. *J. Macromol. Sci.* **1981** *B(20)2* 167.

Guerrero, S.J., Keller, A., Soni, P.L., Geil, P.H. *J. Polym. Sci. Polym. Phys. Ed.* **1980** *18* 1533.

Guinier, A. In: *Théorie et Technique de la Radiocristallographie*, **1956**. Dunod Eds: Paris.

Guiseley, K.B. *Carbohyd. Res.* **1970** *13* 247.

Guttmann, C.M., Hoffmann, J.D. and Di Marzio, E.A. *J. Farad. Disc.* **1979** *68*.

Halle, F. *Kolloid Z. Z. Polym.* **1934** *69* 324.

Han, C.C., Shacklette, L.W. and Eisenbaumer, R.L. Symposium: *Electric, Optical and Magnetic Properties of Organic Solids State Materials*, **1991**. Fall meeting of the MRS; p. 105.

Harada, A., Lin, J. and Kamachi, M. *Nature* **1992** *356* 325.

Harada, A., Lin, J. and Kamachi, M. *Nature* **1994** *370* 126.

Harris, D.J., Bonagamba, T.J., Hong, M., Schmidt-Rohr, K. *Macromolecules* **2000** *33* 3375.

Hasegawa, R., Kobayashi, M. and Tadokoro, H. *Polymer J.* **1972a** *3* 591.

Hasegawa, R., Takahashi, Y., Chatani, Y., Tadokoro, H. *Polymer J.* **1972b** *3* 600.

He, X.W., Herz, J. and Guenet, J.M. *Macromolecules* **1988** *21* 1757.

Heine, S., Kratky, O., Porod, G., Schmitz, J.P. *Makromol. Chem.* **1961** *44* 682.

Helbert, W. PhD thesis, **1994**. Université Joseph Fourier, Grenoble, France.

Helbert, W. and Chanzy, H. *Int. J. Biol. Macromol.* **1994** *16* 207.

Henrissat, B., Marchessault, R.H., Taylor, M.G., Chanzy, H. *Polym. Commun.* **1987** *28* 113.

Hermans, J. and Hermans, J.J. *J. Phys. Chem.* **1958** *62* 1543.

Herrick, F.W. *J. Appl. Polym. Sci. Appl. Polym. Symp.* **1983** *37* 993.

Hess, K. and Trogus, C. *Ber.* **1935** *68B* 1986.

Heuser, E. and Bartunek, R. *Cellulosechemie* **1925** *6* 19.

Hickson, L. and Polson, A. *Biochim. Biophys. Acta* **1968** *165* 43.

Higgins, J.S. and Benoit, H. In: *Polymer and Neutron Scattering*, **1994**. Clarendon Press: Oxford.

Hikata, M., Sasaki, S. and Uematsu, I. *Rep. Progr. Polym. Phys. Jap.* **1977** *20* 621.

Hizukuri, S. *Agric. Biol. Chem.* **1961** *25* 45.

Hizukuri, S., Takeda, Y., Maruta, N., Juliano, B.O. *Carbohyd. Res.* **1989** *189* 227.

Honjo, G., Watanabe, M. *Nature* **1958** *181* 326.

Hoover, R. *Carbohyd. Polym.* **2001** *45* 253.

Howsmon, J.A. and Sisson, W.A. In: *Cellulose and Cellulose derivatives*, **1954**. Interscience: NY; p. 326.

Huang, L., Allen, E. and Tonelli, A.E. *Polymer* **1998** *39* 4857.

Huggins, M.L. *J. Phys. Chem.* **1942** *46* 151.

Huglin, M.B. (ed.) *Light Scattering from Polymer Solutions*, **1972**. Academic Press: London.

Hulleman, S.H.D., Helbert, W. and Chanzy, H. *Int. J. biol. Macromol.* **1996** *18* 115.

Hyde, A.J. and Taylor, R.B. *Makromol. Chem.* **1963** *62* 204.

Ianelli, P., Damman, P., Dosière, M., Moulin, J.F. *Macromolecules* **1999** *32* 2293.

Ikkala, O.T., Pietilä, L.-O., Ahjopalo, L., Österholm, H., Passiniemi, P.J. *J. Chem. Phys.* **1995** *103* 9855.

Imberty, A., Chanzy, H., Pérez, S., Buléon, A., Tran, V. *Macromolecules* **1987** *20* 2634.

Imberty, A., Chanzy, H., Pérez, S., Buléon, A., Tran, V. *J. Mol. Biol.* **1988** *201* 365.

Imberty, A. and Pérez, S. *Biopolymers* **1988** *27* 1205.

Immirzi, A., de Candia, F., Ianelli, P., Zambelli, A., Vittoria, V. *Makromol. Chem. Rapid Commun.* **1988** *9* 761.

Iovleva, M.M. and Papkov, S.P. *Polym. Sci. USSR* **1982** *24* 236.

Ishihara, N., Kuramoto, M. and Uoi, M. *Macromolecules* **1988** *21* 3356.

Ishihara, N., Seimiya, T., Kuramoto, M., Uoi, M. *Macromolecules* **1986** *19* 2465.

Itagaki, H. *Macromol. Symp.* **2001** *166* 13.

Itagaki, H. and Nakatani, Y. *Macromolecules* **1997** *30* 7793.

Itagaki, H. and Takahashi, I. *Macromolecules* **1995** *28* 5477.

Itagaki, H., Ida, R. and Mochizuki, J. *Macromol. Symp.* **2005** *222* 87.

Iuliano, M., Guerra, G., Petraccone, V., Corradini, P., Pelecchia, C. *New Polym. Mater.* **1992** *3* 133.

Izumi, Y., Kanaya, T., Shibata, K., Inoue, K. *Physica B* **1992a** *180&181* 542.

Izumi, Y., Katano, S., Funahashi, S., Furusaka, M., Arai, M. *Physica B* **1992b** *180&181* 539.

Izumi, Y., Katano, S., Funahashi, S., Furusaka, M., Arai, M. *Physica B* **1992c** *180&181* 545.

Jana, T., Chatterjee, J. and Nandi, A.K. *Langmuir* **2002** *18* 5720.

Jana, T. and Nandi, A.K. *Langmuir* **2000** *16* 3141.

Jana, T. and Nandi, A.K. *Langmuir* **2001** *17* 5768.

Janacek, J. and Ferry, J.D. *Macromolecules* **1969a** *2* 397.

Janacek, J. and Ferry, J.D. *J. Polym. Sci. A2* **1969b** *7* 1681.

Jimenez-Barbero, J., Bouffar-Roupe, C., Rochas, C., Pérez, S. *Int. J. Biol. Macromol.* **1989** *11* 265.

Johnson, D.C., Nicholson, M.D. and Haigh, F.C. *Appl. Polym. Symp.* **1976** *28* 943.

Johnson, D.L. **1970** US Patent 3,508,941.

Jones, J.L. and Marquès, C.M. *J. Phys. (Les Ullis)* **1990** *51* 1113.

Juijn, J.A., Gisolf, A. and deJong, W.A. *Kolloid Z. Z. Polym.* **1973** *251* 456.

Kainuma, K. and French, D. *Biopolymers* **1972** *11* 2241.

Karkalas, J., Ma, S., Morrison, W.R., Pethrick, R.A. *Carbohyd. Res.* **1995** *268* 233.

Katz, J.R. *Z. Physik. Chem. (Leipzig)* **1930** *150* 37.

Katz, J.R. and Derksen, J.C. *Z. Physik. Chem. (Leipzig)* **1930** *150* 100.

Katz, J.R. and Derksen, J.C. *Z. Physik. Chem. (Leipzig)* **1933** *167* 129.

Katz, J.R. and Hess, K. *Z. Physik. Chem. (Leipzig)* **1927** *122* 126.

Katz, J.R. and Rientsma, L.M. *Z. Physik. Chem. (Leipzig)* **1930** *150* 60.

Katz, J.R. and van Itallie, T.B. *Z. Physik. Chem. (Leipzig)* **1930** *150* 90.

Kawada, J. and Marchessault, R.H. *Starch/Stärke* **2004** *56* 13.

Kawai, H. *Jpn. J. Appl. Phys.* **1969** *8* 1975.

Kawanishi, K., Komatsu, M. and Inoue, T. *Polym.* **1987** *28* 980.

Kawanishi, K., Takeda, Y. and Inoue, T. *Polym. J.* **1986** *18* 411.

Kessler, Yu M., Abakumova, N.A. and Vaisman, N.A. **1981** *Zh. Fiz. Him.* 55 2682.

Kessler, Yu M., Abakumova, N.A. and Vasenin, N.V. **1982** *Trudy Inst. Mosk. Him.-Tehno. Inst. Mendeleeva* *121* 44.

Klein, M., Brûlet, A., Boué, F., Guenet, J.M. *Polymer* **1991** *32* 1943.

Klein, M., Brûlet, A. and Guenet, J.M. *Macromolecules* **1990a** *23* 540.

Klein, M. and Guenet, J.M. *Macromolecules* **1989** *22* 3716.

Klein, M., Menelle, A., Mathis, A., Guenet, J.M. *Macromolecules* **1990b** *23* 4591.

Kobayashi, M., Nakaoki, T. and Ishihara, N. *Macromolecules* **1990** *23* 78.

Koberstein, J.T., Picot, C. and Benoit, H. *Polymer* **1985** *26* 641.

Koeningsveld, R., Stockmayer, W.H. and Nies, E. *Makromol. Chem. Macromol. Symp.* **1990** *39* 1.

Kolpak, F.J. and Blackwell, J. *Macromolecules* **1975** *8* 563.

Koltisko, B., Keller, A., Litt, M., Baer, E., Hiltner, A. *Macromolecules* **1986** *19* 1207.

Kouwijzer, M. and Pérez, S. *Biopolymers* **1998** *46* 11.

Kowblansky, M. *Macromolecules* **1985** *18* 1776.

Kratochvil, P., Petrus, V., Munk, P., Bohdaneckf, M., Šolc, K. *J. Polym. Sci. Part C* **1967** *16* 1257.

Kreger, D.R. *Biochem. Biophys. Acta* **1951** *6* 406.

Kusunagi, H., Tadokoro, H. and Chatani, Y. *Macromolecules* **1976** *9* 531.

Kusuyama, H., Miyamoto, N., Chatani,Y., Tadokoro, H. *Polym. Comm.* **1983** *24* 119.

La Camera, D., Petraccone, V., Artimagnella, S., Ruiz de Ballesteros, O. *Macromolecules* **2001** *34* 7762.

Lahaye, M. and Rochas, C. **1989** *Carbohyd. Polym. 10* 289.

Lahaye, M. and Rochas, C. **1991** *Hydrobiologia 221* 137.

Langer, J.L. *Synth. Met.* **1987** *20* 35.

Le Bail, P., Buléon, A., Shiftan, D., Marchessault, R.H. *Carbohydr. Polym.* **2000** *43* 317.

Le Bail, P., Rondeau, C. and Buléon, A. *Int. J. Biol. Macromol.* **2005** *35* 1.

Lee, D.M. and Blackwell, J. *J. Polym. Sci. Polym. Phys. Ed.* **1981** *19* 459.

Lee, D.M., Blackwell, J. and Litt, M. *Biopolymers* **1983** *22* 1383.

Lee, D.M., Burnfield, K.E. and Blackwell, J. *Biopolymers* **1984** *23* 111.

Lee, J., Kim, H. and Yu, H. *Macromolecules* **1988** *21* 860.

Lehn, J.M. *Supramolecular Chemistry: Concepts and Perspectives*, **1995**. VCH-Weinheim: NY.

Lemstra, P.J., Keller, A. and Cudby, M. *J. Polym. Sci. Polym. Phys. Ed.* **1978** *16* 1507.

Li, L. and Aoki, Y. *Macromolecules* **1997** *30* 7835.

Li, L. and Aoki, Y. *Macromolecules* **1998** *31* 740.

Li, S., Cao, Y. and Xue, Z. *Synth. Met.* **1987** *20* 141.

Lieser, T. and Fichtner, F. *Liebigs Ann. Chem.* **1941** *548* 195.

Liquori, A.M., Anzuino, G., Coiro, V.M., D'Alagni, M., de Santis, P., Savino, M. *Nature* **1965** *206* 358.

Litt, M.H. and Kumar, N.G. **1970** US Patent 4,028,132.

Lòpez, D., Dahmani, M., Mijangos, C., Brûlet, A., Guenet, J.M. *Macromolecules* **1994** *27* 7415.

Lòpez, D. and Guenet, J.M. *Eur. Phys. J. B* **1999** *B12* 405.

Lòpez, D. and Guenet, J.M. *J. Phys. Chem. B* **2002** *106* 2160.

Lòpez, D., Mijangos, C., Muñoz, M.E., Santamaria, A. *Macromolecules* **1996** *29* 7108.

Lòpez, D., Reinecke, H., Hidalgo, M., Mijangos, C. *Polym. Int.* **1997** *44* 1.

Lotmar, W. and Picken, L.E.R. *Experientia* **1950** *6* 58.

Loucheux, C., Weill, G. and Benoit, H. *J. Chim. Phys.* **1958** *2* 540.

Lovinger, A.J. *J. Polym. Sci.-Phys. Ed.* **1980** *18* 793.

Lovinger, A.J. In: *Developments in Crystalline Polymers – 1*, **1981a**, D.C. Basset (ed.). Elsevier Applied Science: London; Chapter 5.

Lovinger, A.J. *Macromolecules* **1981b** *14* 322.

Lovinger, A.J. and Keith, H.D. *Macromolecules* **1979** *12* 919.

Luzzati, V. and Benoit, H. *Acta Crystallogr.* **1961** *14* 297.

MacDiarmid, A.G. and Epstein, A.J. *Synth. Met.* **1994** *65* 103.

Maeda, H., Kawada, H. and Kawai, T. *Makromol. Chem.* **1970** *131* 169.

Mahendrasingam, A., Denny, R.C., Forsyth, V.T., Greenhall, R.J., Pigram, W.J., Papiz, W.J., Fuller, W. In: *Neutron and X-rays scattering: complementary techniques*, **1990**, M.C. Fairbanks, A.N. North, R.J. Newport (eds). IOP Publishing Ltd: Bristol, UK; pp. 225–236.

Mahendrasingam, A., Forsyth, V.T., Hussain, R., Greenhall, R.J., Pigram, W.J., Fuller, W. *Science* **1986** *233* 195.

Mal, S., Maiti, P. and Nandi, A.K. *Macromolecules* **1995** *28* 2371.

Maldivi, P. *Thesis*, **1989**. Université Joseph Fourier, Grenoble.

Malik, S., Rochas, C. and Guenet, J.M. *Macromolecules* **2005a** *38* 4888.

Malik, S., Rochas, C., Schmutz, M., Guenet, J.M. *Macromolecules* **2005b** *38* 6024.

Malik, S., Rochas, C. and Guenet, J.M. *Macromolecules* **2006a** *39* 1000.

Malik, S., Roizard, D. and Guenet, J.M. *Macromolecules* **2006b** *39* 5957.

Manfredi, C., Del Nobile, M.A., Mensitieri, G., Guerra, G., Rapacciulo, M. *J. Polym. Sci. Polym. Phys. Ed.* **1997** *35* 133.

Manfredi, C., De Rosa, C., Guerra, G., Rapacciulo, M., Auriemma, F., Corradini, P. *Macromol. Chem. Phys.* **1995** *196* 2795.

Manley, R. St. John *Nature* **1964** *204* 1155.

Marchessault, R.H. and Sarko, A. *Adv. Carbohyd. Chem.* **1967** *22* 421.

McKenna, G.B. and Guenet, J.M. *J. Polym. Sci. Polym. Phys. Ed.* **1988a** *26* 267.

McKenna, G.B. and Guenet, J.M. *Polym. Commun.* **1988b** *29* 58.

Mercer, J. **1850** British Patent 13,296.

Meyer, K.H. and Misch, L. *Helv. Chim. Acta* **1937** *20* 232.

Mijangos, C., Cassagnau, P. and Michel, A. *J. Appl. Polym. Sci.* **1992** *44* 2019.

Mijangos, C., Gomez-Elvira, J.M., Martinez, G., Millán, A. *J. Appl. Polym. Sci.* **1989** *38* 1685.

Mijangos, C., Martinez, G. and Millán, A. *Makromol. Chem.* **1988** *189* 567.

Millane, R.P., Chandrasekaran, R., Arnott, S., Dea, I.C.M. *Carbohyd. Res.* **1988** *182* 1.

Miller, I.J., Falshaw, R. and Furneaux, R.H. **1994** *Carbohyd. Res.* 262 127.

Minke, R. and Blackwell, J. *J. Mol. Biol.* **1978** *120* 167.

Mittelbach, P. and Porod, G. *Acta Phys. Austriaca* **1961** *14* 405.

Monteiro, E.E.C. and Mano, E.B. *J. Polym. Sci. Polym. Phys. Ed.* **1984** *22* 533.

Morris, G.A. and Freeman, R. *J. Magn. Res.* **1978** *29* 433.

Morrison, W.R. and Karkalas, J. *Methods in Plant Biochemsitry: Starch*, Vol. 2, **1990**. Academic Press: NY; p. 323.

Moyses, S. *PhD thesis*, **1997**. Sheffield Hallam University.

Moyses, S., Sonntag, P., Spells, S.J., Laveix, O. *Polymer* **1998** *39* 3665.

Moyses, S. and Spells, S.J. *Polymer* **1998** *39* 3537.

Moyses, S. and Spells, S.J. *Polymer* **1999** *40* 3269.

Muroga, Y. *Macromolecules* **1988** *21* 2751.

Mutin, P.H. *Thesis* **1986**. ULP, Strasbourg.

Mutin, P.H. and Guenet, J.M. *Polymer* **1986** *27* 1098.

Mutin, P.H. and Guenet, J.M. *Macromolecules* **1989** *22* 843.

Myasnikova, R.M. *Vysokomol. Soyed.* **1976** *A19* 564.

Myasnikova, R.M., Titova, E.F. and Obolonkova, E.S. *Polymer* **1980** *21* 403.

Naegele, D., Yoon, D.Y. and Broadhurst, M.G. *Macromolecules* **1978** *11* 1297.

Najeh, M., Munch, J.P. and Guenet, J.M. *Macromolecules* **1992** *25* 7018.

Nakaoki, T., Katagiri, C. and Kobayashi, M. *Macromolecules* **2002** *35* 7708.

Nakaoki, T. and Kobayashi, M. *J. Mol. Struct.* **1991** *242* 315.

Natta, G., Corradini, P. and Bassi, I.W. *Nuov. Cim. Suppl.* **1960** *15* 68.

Nieduszynski, I. and Atkins, E.D.T. *Biochem. Biophys. Acta* **1970** *222* 109.

Nieduszynski, I. and Preston, R.D. *Nature* **1970** *225* 273.

Nishimura, H. and Sarko, A. *J. Appl. Sci.* **1987** *33* 867.

Okano, T. and Sarko, A. *J. Appl. Sci.* **1984** *29* 4175.

Okano, T. and Sarko, A. *J. Appl. Sci.* **1985** *30* 325.

Oster, G. and Riley, D.P. *Acta Crystallogr.* **1952** *5* 272.

Overbergh, N., Girolamo, M. and Keller, A. *J. Polym. Sci. Polym. Phys. Ed.* **1977a** *15* 1475.

Overbergh, N., Sadler, D.M. and Keller, A. *J. Polym. Sci. Polym. Phys. Ed.* **1977b** *15* 1487.

Pal, S., Maiti, P.K. and Bagchi, B. *J. Phys. Condens. Matter* **2005** *17* 4317.

Papkov, S.P. *Adv. Polym. Sci.* **1983** *59* 76.

Parrod, J., Kohler, A. and Hild, G. *Compt. Rend. Acad. Sci.* **1958** *246* 1046.

Parrod, J., Kohler, A. and Hild, G. *Makromol. Chem.* **1964** *75* 52.

Paternostre, L., Damman, P. and Dosière, M. *Polymer* **1998** *39* 4579.

Paternostre, L., Damman, P. and Dosière, M. *Macromolecules* **1999a** *32* 153.

Paternostre, L., Damman, P. and Dosière, M. *J. Polym. Sci. Polym. Phys.* **1999b** *37* 1197.

Pennings, A.J. and Kiel, A.M. *Kolloid Z. Z. Polym.* **1965** *205* 160.

Pérez, E., Vanderhart, D.L. and McKenna, G.B. *Macromolecules* **1988** *21* 2418.

Petraccone, V., La Camera, D., Pirozzi, B., Rizzo, P., De Rosa, C. *Macromolecules* **1998** *31* 5830.

Petraccone, V., La Camera, D., Caporaso, L., De Rosa, C. *Macromolecules* **2000** *33* 2610.

Petraccone, V., Esposito, G., Tarallo, O., Caporaso, L. *Macromolecules* **2005a** *38* 5668.

Petraccone, V., Tarallo, O., Venditto, V., Guerra, G. *Macromolecules* **2005b** *38* 6965.

Pfannemüller, B. *Int. J. Biol. Macromol.* **1987** *9* 105.

Plazek, D.J. and Altares, T. *J. Appl. Phys.* **1986** *60* 2694.

Point, J.J. and Coutelier, C. *J. Polym. Sci. Polym. Phys* **1985** *23* 231.

Point, J.J., Coutelier, C. and Villers, D. *J. Phys. Chem.* **1986a** *90* 3277.

Point, J.J., Jasse, B. and Dosière, M. *J. Phys. Chem.* **1986b** *90* 3273.

Point, J.J., Demaret, J.Ph. *J. Phys. Chem.* **1987** *91* 797.

Point, J.J. and Damman, P. *Macromolecules* **1991a** *24* 2019.

Point, J.J., Damman, P. and Guenet, J.M. *Polym. Comm.* **1991b** *32* 477.

Point, J.J. and Damman, P. *Macromolecules* **1992** *25* 1184.

Popov, D., Burghammer, M., Buléon, A., Montesanti, N., Putaux, J.L., Riekel, C. *Macromolecules* **2006** *39* 3704.

Porod, G. *Acta Phys. Austr.* **1948** *2* 255.

Porod, G. *Koll. Z.* **1951** *124* 83.

Porod, G. *Koll. Z.* **1952** *125* 51.

Potocki-Veronese, G., Putaux, J.L., Dupeyre, D., Albenne, C., Renaud-Simeon, M., Monsan, P., Buléon, A. *Biomacromolecules* **2005** 6 1000.

Pourahmady, N., Bak, P.I. and Kinsey, R.A. *J. Macromol. Sci.* **1992** *A29* 959.

Poux, S., Thierry, A., Fazel, N., Dahoun, A., Guenet, J.M. *Macromol. Symp.* **2001** *168* 67.

Poux, S., Thierry, A., Rochas, C., Green, M.M., Guenet, J.M. *Macromol. Symp.* **2003** *203* 265.

Prasad, A. and Mandelkern, L. *Macromolecules* **1990** *23* 5041.

Prest, W.M. Jr and Luca, D.J. *J. Appl. Phys.* **1975** *46* 4136.

Pringle, O.A., Schmidt, P.W. *J. Appl. Crystallogr.* **1971** *4* 290.

Ramesh, R., Patel, C.K. and Patel, R.D. *Makromol. Chem.* **1973** *171* 179.

Ramzi, M., Rochas, C. and Guenet, J.M. *Macromolecules* **1996** *29* 4668.

Ramzi, M., Rochas, C. and Guenet, J.M. *Macromolecules* **1998** *31* 6106.

Ramzi, M., Rochas, C. and Guenet, J.M. *Int. J. Biol. Macromol.* **2000a** *27* 163.

Ramzi, M., Mendès, E., Rochas, C., Guenet, J.M. *Polymer* **2000b** *41* 559.

Rapacciulo, M., De Rosa, C., Guerra, G., Mensitieri, G., Apicella, A., Del Nobile, M.A. *J. Mater. Sci. Lett.* **1991** *10* 1084.

Rappenecker, G. and Zugenmaier, P. *Carbohyd. Res.* **1981** *89* 11.

Rasmussen, D.H. and MacKenzie, A.P. **1968** *Nature 220* 1316.

Ray, B., ElHasri, S., Thierry, A., Marie, P., Guenet, J.M. *Macromolecules* **2002** *35* 9730.

Ray, B., ElHasri, S. and Guenet, J.M. *European Physical Journal E* **2003** *11* 315.

Rees, D.A., Steele, I.W. and Williamson, F.B. *J. Polym. Sci. Pt C* **1969** *28* 261.

Rehage, G. and Hallboth, H. *Makromol. Chem.* **1968** *119* 233.

Reinecke, H., Mijangos, C., Brûlet, A., Guenet, J.M. *Macromolecules* **1997a** *30* 959.

Reinecke, H., Fazel, N., Dosière, M., Guenet, J.M. *Macromolecules* **1997b** *30* 8360.

Reinecke, H., Mijangos, C., Lòpez, D., Guenet, J.M. *Macromolecules* **2000** *33* 2049.

Reparet, J.M., Moine, F., Arvisenet, G., Le Bail, P., Cayot, N. *J. Text. Stud.* **2006** *37* 459.

Righetti, P.G., Brost, B.C.W. and Snyder, R.S. *J. Biochem. Biophys. Methods* **1981** *4* 347.

Rizzo, P., Ruiz de Balesteros, O., De Rosa, C., Auriemma, F., La Camera, D., Petraccone, V., Lotz, B. *Polymer* **2000** *41* 3745.

Roberts, G.A.F. *Chitin Chemistry*, **1992**, George A.F. Roberts (ed.). MacMillan Press Ltd: London.

Robin, J. P., Mercier, C., Charbonniere, R., Guilbot, A. *Cereal Chem.* **1974** *51* 389.

Rochas, C. and Landry, S. In: *Gums and Stabilizers for the Food Industry*, **1988**, G.O. Phillips, D.J. Wedlock, P.A. Williams (eds). IRL Press: Oxford; p. 445.

Rochas, C., Brûlet, A. and Guenet, J.M. *Macromolecules* **1994** *27* 3830.

Roche, E. and Chanzy, H. *Int. J. Biol. Macromol.* **1981** *3* 201.

Roels, T., Deberdt, F. and Berghmans, H. *Macromolecules* **1994** *27* 6216.

Rondeau-Mouro, C., Le Bail, P. and Buléon, A. *Int. J. Biol. Macromol.* **2004** *34* 251.

Rosso, J.C., Guieu, R., Ponge, C., Carbonnel, L. *Bull. Soc. Chim.* **1973**, *9–10*, 2780.

Rousselle, M.A., Nelson, M.L., Hassenboehler, C.B., Legendre, D. *Textile Res. J.* **1976** *46* 304.

Rousselle, M.A. and Nelson, M.L. *Textile Res. J.* **1976** *46* 648.

Rudall, K.M. *Adv. Insect. Physiol.* **1963** *1* 257.

Rudall, K.M. and Kenchington, W. *Biol. Rev.* **1973** *49* 597.

Ruiz de Balesteros, O., Auriemma, F., De Rosa, C., Floridi, G., Petraccone, V. *Polymer* **1998** *39* 3523.

Rundle, R.E. and Daash, L.W. *J. Am. Chem. Soc.* **1944** *66* 130.

Sadler, D.M. and Keller, A. *Macromolecules* **1977** *10* 1128.

Sadler, D.M. and Spells, S.J. *Polymer* **1984** *25* 1219.

Saiani, A. *Thesis,* **1997**, Université Louis Pasteur, Strasbourg.

Saiani, A. and Guenet, J.M. *Macromolecules* **1997** *30* 966.

Saiani, A. and Guenet, J.M. *Macromolecules* **1999** *32* 657.

Saiani, A., Guenet, J.M. and Spěváček, J. *Macromolecules* **1998** *31* 703.

Saito, G. *Kolloidbeihefte* **1939** *49* 365.

Saito, Y., Okano, T., Putaux, J.L., Gaill, F., Chanzy, H. *Advances in Chitin Science*, Vol. II, **1997a**, A. Domard, G.A.F. Roberts and K.M. Värum (eds). J. André Publisher: Lyon.

Saito, Y., Putaux, J.L., Okano, T., Gaill, F., Chanzy, H. *Macromolecules* **1997b** *30* 3867.

Sakurada, I. and Okamura, S. *Kolloid-Z.* **1937** *81* 199.

Sarko, A. *Appl. Polym. Symp.* **1976** *28* 729.

Sarko, A. and Muggli, R. *Macromolecules* **1974** *7* 486.

Sarko, A., Southwick, J. and Hayashi, J. *Macromolecules* **1976** *9* 857.

Sarko, A. and Wu, H.-C.H. *Starch-Stärke* **1978** *30* 73.

Sarko, A. and Zugenmaier, P. In: *Fibre Diffraction Methods ACS Symp. Ser.*, **1980**, *141* 459.

Sasaki, S., Hikata, M., Shiraki, C., Uematsu, I. *Polym. J.* **1982** *14* 205.

Schacklette, L.W., Wolf, J.F., Gould, S., Baughman, R.H. *J. Chem. Phys.* **1988** *88* 3955.

Schmidt, P.W. *J. Appl. Crystallogr.* **1970** *3* 257.

Schomaker, E. and Challa, G. *Macromolecules* **1989** *22* 3337.

Segal, L. and Loeb, L. *J. Polym. Sci.* **1960** *42* 351.

Sharp, P. and Bloomfield, V.A. *Biopolymers* **1968** *6* 1201.

Shotton, M.W., Pope, L.H., Forsyth, V.T., Langan, P., Denny, R.C., Giesen, U., Dauvergne, M.T., Fuller, W. *Biophys. Chem.* **1997** *69* 85.

Shotton, M.W., Pope, L.H., Forsyth, V.T., Langan, P., Grimm, H., Rupprecht, A., Denny, R.C., Fuller, W. *Physica B* **1998** *241–243* 1166.

Sisson, W.A. *Contrib. Boyce Thompson Inst.* **1941** *12* 31.

Smith, P.L. and Pennings, A.J. *Polymer* **1974** *15* 413.

Smith, G.D., Yoon, D.Y. and Jaffe, R.L. *Macromolecules* **1994** *26* 5213.

Soni, P.L. and Agarwal, A. *Starch-Stärke* **1983** *35* 4.

Sonntag, P., Care, C.M., Spells, S.J., Halliday, I. *J. Chem. Soc. Faraday Trans.* **1995** *91* 2593.

Spells, S.J. and Sadler, D.M. *Polymer* **1984** *25* 739.

Spěváček, J. and Brus, J. *Macromol. Symp.* **1999** *138* 117.

Spěváček, J., Paternostre, L., Damman, P., Draye, A.C., Dosière, M. *Macromolecules* **1998** *31* 3612.

Spěváček, J., Saiani, A. and Guenet, J.M. *Makromol. Rapid Commun.* **1996** *17* 389.

Spěváček, J. and Schneider, B. *Makromol. Chem.* **1975** *176* 3409.

Spěváček, J. and Schneider, B. *Adv. Colloid Int. Sci.* **1987** *27* 81.

Spěváček, J. and Suchoparek, M. *Macromolecules* **1997** *30* 2178.

Spěváček, J., Suchoparek, M., Mijangos, C., Lopez, D. *Macromol. Chem. Phys.* **1998** *199* 1233.

Standt, U.D. *J. Macromol. Sci. Rev. Macromol. Chem. Phys.* **1983** *C23* 317.

Stauffer, D. *J. Chem. Soc. Farad. Trans.* **1976** *II72* 1354.

Stein, R.S. and Tobolsky, A.V. *Text. Res. J.* **1948** *18* 302.

Stevenson, D.E. and Furneaux, R.H. *Carbohyd. Res.* **1991** *210* 27.

Stipanovic, A. and Sarko, A. *Macromolecules* **1976** *9* 851.

Sugiyama, J., Rochas, C., Turquois, T., Taravel, F., Chanzy, H. *Carbohyd. Polym.* **1994** *23* 261.

Sundararajan, P.R. and Tyrer, N. *Macromolecules* **1982** *15* 1004.

Sundararajan, P.R., Tyrer, N. and Bluhm, T.L. *Macromolecules* **1982** *15* 286.

Tabb, D.L. and Konig, J.L. *Macromolecules* **1975** *8* 829.

Tadokoro, H. *Structure of Crystalline Polymers*, **1979**. Wiley and Sons: NY.

Tadokoro, H., Chatani, Y., Kobayashi, M., Yoshihara, T., Murahashi, S., Imada, K. *Rep. Prog. Polym. Phys. Jpn* **1963** *6* 303.

Tadokoro, H., Chatani, Y., Yoshihara, T., Tahara, S., Murahashi, S. *Makromol. Chem.* **1964a** *73* 109.

Tadokoro, H., Yoshihara, T., Chatani, Y., Murahashi, S. *J. Polym. Sci.* **1964b** *B2* 363.

Takahashi, Y., Kumano, T. and Nishikawa, S. *Macromolecules* **2004** *37* 6827.

Takahashi, Y. and Tadokoro, H. *Macromolecules* **1980** *12* 1317.

Takano, R., Hayashi, K. and Hara, S. *Phytochemistry* **1995** *40* 487.

Takeda, C., Takeda, Y. and Hizukuri, S. *Cereal Chemistry* **1983** *60* 313.

Tan, H.M., Hiltner, A., Moet, H., Baer, E. *Macromolecules* **1983** *16* 28.

Tarallo, O. and Petraccone, V. *Macromol. Chem. Phys.* **2004** *205* 1351.

Tarallo, O. and Petraccone, V. *Macromol. Chem. Phys.* **2005** *206* 672.

Tarallo, O., Petraccone, V., Venditto, V., Guerra, G. *Polymer* **2006** *47* 2402.

Tashiro, K., Ueno, Y., Yoshioka, A., Kobayashi, M. *Macromolecules* **2001** *34* 310.

te Nijenhuis, K.T. and Dijkstra, H. *Rheol. Acta* **1975** *14* 71.

Terech, P., Maldivi, P. and Guenet, J.M. *Europhys. Lett.* **1992** *17* 515.

Tran, V. and Buléon, A. *J. Appl. Cryst.* **1987** *21* 1887.

Trogus, C. *Cellulosechem.* **1934** *15* 104.

Trogus, C. and Hess, K. *Z. Physik. Chem. (Leipzig)* **1931** *14B* 387.

Valensin, G., Kushnir, T. and Navon, G. *J. Magn. Res.* **1982** *46* 23.

Van den Broecke, P. and Berghmans, H. *Makromol. Chem. Macromol. Symp.* **1990** *39* 59.

Vasanthan, N., Shin, I.D. and Tonelli, A.E. *Macromolecules* **1996** *29* 263.

Vikki, T., Pietilä, L.O., Österholm, H., Ahjopalo, L., Takala, A., Toivo, A., Levon, K., Passiniemi, P., Ikkala, O.T. *Macromolecules* **1996** *29* 2945.

Vikki, T., Ruokolainen, J., Ikkala, O.T., Passiniemi, P., Isotalo, H., Torkkeli, M., Serimaa, R. *Macromolecules* **1997** *30* 4064.

Vittoria, V., de Candia, F., Ianelli, P., Immirzi, A. *Makromol. Chem. Rapid Commun.* **1988** *9* 765.

Vittoria, V., Russo, R. and de Candia, F. *Makromol. Chem. Macromol. Symp.* **1990** *39* 317.

Vittoria, V., Russo, R. and de Candia, F. *Polymer* **1991a** *32* 3371.

Vittoria, V., Ruvolo Filho, A. and de Candia, F. *Polymer Bull.* **1991b** *26* 445.

Vorenkamp, E.J., Boscher, F., Challa, G. *Polymer* **1979** *20* 59.

Wagner, J.F., Dosière, M. and Guenet, J.M. *Macromol. Symp.* **2005** *222* 121.

Walters, A.T. *J. Polym. Sci.* **1954** *13* 207.

Wang, A.H.J., Quigley, G.J., Kolpak, F.J., Crawford, J.L., van Boom, J.H., Van der Marel, G., Rich, A. *Nature (London)* **1979** *282* 680.

Wang, Y.K., Savage, J.D., Yang, D., Hsu, S.L. *Macromolecules* **1992** *25* 3659.

Wang, Z.G. and Safran, S.A. *J. Chem. Phys.* **1988** *89* 5323.

Warwicker, J.O. Swelling. In: *Cellulose and Cellulose Derivatives*, Part IV, **1971**, N.M. Bikales and L. Segal (eds). J. Wiley: NY.

Warwicker, J.O. and Wright, A.C. *J. Appl. Polym. Sci.* **1967** *11* 659.

Watanabe, W.H., Ryan, C.F., Fleischer, P.C., Garrett, B.S. *J. Phys. Chem.* **1961** *65* 896.

Watase, M. and Nishinari, K. *Polym. J.* **1988** *20* 1125.

Watase, M., Nishinari, K., Clark, A.H., Ross-Murphy, S.B. *Macromolecules* **1989** *22* 1196.

Watson, J.D. and Crick, F.H.C. *Nature* **1953** *171* 737.

Weber, E. *Molecular Inclusion and Molecular Recognition Clathrates* – I. In: *Topics in Current Chemistry* **1987** *140*, E. Weber (ed.). Springer-Verlag: Berlin–Heidelberg; Chapter 1.

Weill, G. and Des Cloizeaux J. *J. Phys. (Paris)* **1979** *40* 99.

Wellinghof, S.J., Shaw, J. and Baer, E. *Macromolecules* **1979** *12* 32.

Wilkes, C.E., Folt, V. and Krimm, S. *Macromolecules* **1973** *6* 235.

Winter, W.T., Chanzy, H., Putaux, J.L., Helbert, W. *Polym. Prep.* **1998** *39* 703.

Winter, W.T. and Sarko, A. *Biopolymers* **1972** *11* 849.

Winter, W.T. and Sarko, A. *Biopolymers* **1974** *13* 1461.

Wittmann, J.C. and St John Manley, R. *J. Pol. Sci. Polym. Phys. Ed.* **1977** *15* 1089.

Wolfe, J.F., Loo, B.M. and Arnold, F.E. *Macromolecules* **1981** *14* 915.

Wu, H.C.H. and Sarko, A. *Carbohyd. Res.* **1978a** *61* 7.

Wu, H.C.H. and Sarko, A. *Carbohyd. Res.* **1978b** *61* 27.

Wunderlich, B. In: *Macromolecular Physics*, Vol. 1, *Crystal Structure, Morphology, Defects*, **1973**. Academic Press: NY.

Xia, Y., MacDiarmid, A.G. and Epstein, A.J. *Macromolecules* **1994** *27* 7212.

Yang, C.Y., Cao, Y., Smith, P., Heeger, A.J. *Synth. Met* **1993** *53* 293.

Yang, Y.C. and Geil, P.H. *J. Macromol. Sci.* **1983** *B(22)2* 463.

Yoshisaki, T. and Yamakawa, H. *Macromolecules* **1980** *13* 1518.

Yu, B., Fuji, S. and Kishihara, S. *Starch-Stärke* **1999** *51* 5.

Zarzycki, J. *Farad. Disc. Chem. Soc.* **1970** *50* 122.

Zhang, J. and Rochas, C. *Carbohyd. Polym.* **1990** *13* 257.

Index

Acetonitrile, 273, 276
Acids, 60, 68, 92, 137, 138
Adjacent re-entry, 61, 62, 226
Aerogels, 252
Agarose, 70–85, 86, 115, 116, 181, 187, 264
Aggregates, 25, 34, 41, 44, 45, 74, 75, 120, 123, 126–31, 136, 244, 256–8, 262, 273, 277, 283
Alcohol, 92, 97, 98, 134, 161
Alkalis, 60
Amines, 60, 64–6, 116
Ammonia, 64, 68
Amylopectin, 87, 92
Amylose, 87–93, 94, 95
Amylosucrase, 89
Anhydroglucose, 59, 61, 64, 66, 67
Anisotropy, 192–3
Antiferromagnetic, 205
Arthropods, 96
Atactic, 107, 109–64, 180, 255–77
Axial ratio, 139
Axial rise, 22, 81, 190

Base pairs, 101, 103
Benzene, 113, 122, 165, 212, 217, 224–7, 231–7, 242–7, 251, 266, 268, 269
Benzenethiolate, 124, 125
Benzenethiophenate, 122
Benzophenone, 159, 238–40
Benzyl alcohol, 161
Bicopper complex, 203–209
Binary, 15, 16, 70–82, 139, 155, 277, 278
Bimodal, 7, 12
Biopolymers, 12, 55, 87, 105, 284
Biphenyl, 232, 235–40
Bound solvent, 112, 276, 277
Bromobenzene, 35, 39, 112, 119, 120, 121, 170, 171, 273, 274, 275–283

Bromonaphthalene, 109, 110, 112
Brownian, 27, 28, 43, 72, 187, 277
Butyl acetate, 276
Butyl benzoate, 133
Butyrolactone, 113

Camphor-10-sulphonic acid, 138
Carbon disulphide, 257, 259, 285
Carrageenin, 70, 83, 264, 271
Caustic soda, 60, 61, 66
Cavities, 163, 190, 193, 218, 220, 251, 253
Cellulose, 57–59, 60–69
Chain, 29, 51, 60, 67, 187
Chain-folded crystals, 62, 72, 180, 181, 186, 187, 195, 196, 273
Chain-folding, 46, 47, 59, 61, 165, 200, 201, 221
Chemical potential, 3, 14, 15, 250
Chitin, 55, 96–98
Chlorobenzene, 169–72, 222, 228, 267, 273–5, 278
1-Chlorododecane, 193
Chloroform, 139, 140, 141, 217, 218, 242, 247, 248, 252
Cis-decalin, 52, 116, 181–8, 190, 193–7, 199, 201–203
Clathrate, 163, 166, 212, 222, 228, 248, 251, 265, 269
Coherent, 16, 31, 32, 74, 78, 79, 181, 206, 207
Coil–helix transition, 70
Compound, 63, 143, 164, 165, 168, 240
Conductivity, 139, 140, 142, 147, 149
Conformation, 51, 57, 63, 109, 119, 131, 165, 176, 178, 180, 210, 265, 266, 269
Congruently melting, 6, 7, 8, 9, 12, 78, 153, 155, 166, 170, 172, 174, 183, 186, 197, 228, 231, 234, 247, 248

Contrast, 33, 34, 73, 74, 78, 100, 125, 157, 228, 259
Cotton, 57, 61
Critical, 7, 14, 41, 114, 126, 127, 139, 205, 207, 216
Critical gelation concentration, 44, 279
Cross-section, 18, 19, 20, 24, 25, 28, 30, 31, 34, 42, 43, 45, 46, 48, 53, 73, 128, 166, 188, 210, 245
Crystallization, 10, 12, 30, 50, 88, 122–30, 150, 152, 166, 178, 211, 214, 216, 219, 226, 228, 236, 247
Crystallo-solvates, 76, 97, 160
Cyclodextrins, 165, 168–9
Cyclohexane, 217, 269
Cyclohexanone, 269
Cylinders, 19, 20, 21, 25–27, 48, 49, 128, 203, 278, 279

Decahydronaphthalene, 181
1-Decanol, 94
Deformation, 47, 48, 50, 51, 113, 133, 146, 150, 193, 195, 283
Deuterated, 18, 24, 31, 78, 79, 118, 176, 190, 206, 231, 244, 245, 246, 261, 262, 278, 280
Dibutyl phthalate, 132
Dibutyl oxalate, 113, 119, 128
1,2-Dichloroethane, 251
Dichloromethane, 214, 215, 220
Diesters, 121, 122, 126–35, 150, 154
Diethylene triamine, 66
Diethyl glutamate, 153
Diethyl azelate, 153
Diethyl malonate, 109, 113, 153
Diethyl oxalate, 112, 114, 129, 130, 153
Diethyl pimelate, 153
Diethyl succinate, 153
Differential scanning calorimetry (DSC), 3, 110, 196, 258
Diffraction, 17, 19–35, 99–101, 158, 165, 169, 170, 175–7, 189, 196, 198, 201, 211, 213
Dihydroxybenzene, 172
Dilute, 18, 19, 45, 59, 112, 120, 139, 214, 226, 232, 255, 273
Dimethyl formamide, 77
1,4-Dimethyl naphthalene, 224
Dimethyl sulphoxide, 92, 138
Dinonylnaphthalene sulphonic acid, 148
Dioctyl phthalate, 132
Diphenyl methane, 239, 240
Diphenyl methane, 239, 240–1

DMSO, 57, 72–80, 84–6, 92, 141
DNA, 55, 70, 99–103
DSC, 111, 112, 115, 116, 133, 134, 152, 153, 157, 166, 168, 185, 203, 219, 220, 227–9, 233–5, 258
Double helix, 24, 70, 73, 81, 99, 101, 103, 271, 272, 280–2
Double-stranded, 70, 71, 88, 90, 99, 271, 279, 282

Elasticity, 47–9, 83, 135, 136, 194, 195, 264, 283
Electric field, 74, 152
Electron diffraction, 88, 92, 95, 211
Electrostatic interactions, 78, 86, 109, 116, 121, 126, 141, 163, 284
Emeraldine base, 137, 140, 141, 142
Emptied chlathrate, 251, 252, 284
Enantiomorphous, 211
Endotherms, 5, 152, 153, 154, 181, 196, 242, 274
End-to-end distance, 41, 43, 44, 47, 128
Enthalpic, 47–9, 55, 83, 105, 135, 194, 195, 264, 283
Entropic, 47–9, 55, 83, 135, 163, 195, 283, 288
Ethylamine, 65, 67, 68
Ethyl acetoacetate, 150, 157
Ethyl benzene, 224–7, 231–2
Ethyl pelargonate, 128, 136
Ethyl valerate, 128, 136
Eutectic, 4, 5, 7, 9, 78, 156, 172, 174, 177, 273
Exotherms, 5, 181, 182, 242
Excimer, 191, 192

Fatty acids, 92
Fibres, 18, 19, 25, 26, 44, 45, 61, 64, 72, 76, 88, 92, 100, 193, 194, 195, 196, 220, 245, 264, 265, 267, 268, 285
Fibrillar, 18, 34, 41, 53, 57, 62, 70, 73, 79, 113, 125, 126, 136, 137, 147, 153, 161, 164, 180, 186, 187, 201, 207, 214, 220, 235, 238, 240, 248, 252, 284
Fibrils, 41, 83, 116, 124, 125, 128
Fickian, 216–17
Filaments, 204–208
First order, 3, 5, 50, 52, 257, 258, 261
Flexible junction, 83
Fluorescence, 191
Forced Rayleigh scattering, 259
Formamide, 77, 78
Form factor, 18, 19, 20, 24, 27, 29–32, 199

Fractal, 19, 28, 41, 43, 44, 48, 83, 126, 128, 131, 135, 136, 146, 194, 195, 283
Freely jointed, 28
Free solvent, 10, 40, 112, 194, 243, 274–6
Fringed micelle, 35
FTIR, 157, 170, 176, 178, 179, 216, 252

Gaussian, 28, 29, 38, 46, 72, 115, 187, 188, 201, 255, 256
Gelatinization, 87, 91
Gelation, 3, 5, 12, 44, 70, 72, 75, 81, 84, 110, 113, 126, 181, 184, 186, 188, 190, 191, 205, 206, 208, 216, 242, 255–7, 259–61, 279
Gelation threshold, 45, 187, 192, 195, 196, 198, 200, 205, 206, 278
Gels, 49, 50, 83–5, 113–19, 121, 127, 129, 131–6, 145–9, 180–90
Glass transition, 3, 134, 180, 186, 216, 218, 219, 228, 247, 258, 259, 273, 284
Granules, 87, 88, 90, 91
Grooves, 100
Growth, 10–12, 29, 30, 45, 88, 94, 97, 99, 123, 129–31, 136, 151, 163, 178, 205, 215, 216, 245

Half-staggered, 99
Heat capacity, 3, 14
Helices, 18–24, 70–1, 81–2, 90–8, 99–103, 212, 265–69, 271, 272, 278–82
Helix, 21, 22, 63, 99, 171, 176, 210, 229, 230, 260, 281–2
Heterotactic, 121
Hexahydroindane, 196
Hydrazine, 57, 65–8
Hydrogen bonds, 50, 55, 57, 60, 61, 65, 66, 67, 84, 86, 90, 105, 141, 145, 163, 166, 172, 173, 177
Hydrogenous, 18, 24, 31–3, 74, 78, 118, 176
Hydrolysis, 89
Hydroquinone, 169, 176

Inclusion compounds, 163, 165, 166, 168
Incoherent, 31–3, 100
Incompressibility, 17, 32
Incongruently melting, 7, 8, 9, 12, 153, 155, 186, 197, 228, 234, 243, 247
Intercalates, 55, 163, 169, 172, 212, 224, 265
Interdigitation, 117, 122, 125
Intertwining, 70, 71
Isotactic, 109, 180–209, 210, 218, 226, 242, 245, 255, 259, 271, 283

Kinetically controlled, 11, 186, 195, 196, 211
Kratky plot, 125, 199, 207, 263
Kratky representation, 118, 123, 124, 279

Labelling, 17, 31, 32, 33, 231, 235, 244, 277, 278
Latent heat, 3, 5, 13
Lattice, 58, 198, 210–15, 220, 222, 226, 230, 231, 234, 244, 251, 261, 282
Lauric acid, 94
Layer, 22, 23, 70, 71, 163, 191, 212, 282
Light scattering, 30, 33, 34, 42, 72, 126, 128, 130, 135, 137, 138, 139, 256, 257
Linalool, 94
Lipids, 94
Liquid–liquid phase separation, 5, 6, 7, 10, 11, 12, 133, 184, 186, 259, 273
Liquid–solid phase separation, 113
Liquidus, 7, 9, 12, 13, 275
Log-normal, 221
Loose helix, 70, 73, 81

Magnetic susceptbility, 208
Mass per unit area, 30, 117, 125, 201
Mass per unit length, 20, 46, 73, 74, 119, 199, 203, 245, 279
m-cresol, 139–43
Mechanical, 47, 49, 50, 111, 129, 131, 132, 133, 135, 150, 242, 263, 264
Melting enthalpy, 10, 14, 15, 47, 84, 111, 112, 143, 145, 188, 206, 234, 235, 236, 274, 275, 276
Melting point depression, 13, 14, 182
Mercerization, 59, 60
Meridional, 22, 59, 67, 71, 81, 189, 190
Mesh, 19, 34, 49, 53, 59, 79, 113, 186, 222, 238, 252, 279
Mesophase, 97, 143, 160, 203, 207, 217, 219, 266, 269, 273, 275
Mesoscopic, 79, 97, 113, 120, 283
Metallogels, 203
Metastable, 5, 10–13, 87, 172, 178, 186, 212, 229, 236, 247
Metastability, 10–13, 179
Metatectic, 7, 9, 155, 157–8
Methylene chloride, 218
Methyl formamid, 77
1-Methyl naphthalene, 192
2-Methyl resorcinol, 173–4
Microfibrils, 59, 186
Microporosity, 253
Miscibility gap, 5, 7, 186, 259

Mobility, 5, 10, 11, 39, 40, 79, 136, 169, 176, 192, 193, 215, 219, 273, 276
Moderately concentrated, 15, 214, 257, 268, 273
Modified agarose, 83–6
Modulus, 47–9, 81–5, 109–10, 129–36, 193, 194, 263, 264, 283
Molecular weight, 7, 31, 34, 88, 108, 111, 116, 130, 199, 259
Monoclinic, 57, 61, 66, 67, 88, 96, 97, 101, 151, 161, 165, 172, 222, 223, 265, 267, 268
Monoesters, 128, 129, 130, 133, 134
Monotectic transition, 6, 7, 12, 183
Morphology, 186–7, 189, 195, 198, 200, 202, 220–2, 226, 228, 235, 238–40, 244, 248, 252
Morphologies, 18, 66, 157, 164, 196, 200, 220, 222
Multiblock copolymer, 109
Multiporous, 285

Nanocomposites, 203
Naphthalene, 192, 224, 225, 232–6, 252–4
1-Naphthol, 93, 95
n-Butanol, 25
Nanoporous, 213, 214, 251, 252, 284
NaSCN, 72
Nematic, 43, 97, 133, 139, 190–1, 212, 237, 238, 261
Neutron, 17, 30, 32, 72, 73, 230, 235, 261
n-Hexane, 217
Nitric acid, 60, 68
Nitrobenzene, 121
N-methylmorpholine N-oxide, 57
N-methylpyrrolidinone, 138
NMR, 36–40, 67, 74, 83, 94, 107, 109, 121, 168, 176, 193, 214, 260, 272, 273, 276
Non-selective, 39, 276
Non-variant transformation, 13, 111, 235
Norbornadiene, 212, 213, 237, 244
n-Pentanol, 94
Nucleation, 11, 12, 13, 14, 203, 205–207, 209

Octanal, 94
Oligonucleotide, 99
Optical micrograph, 156, 174, 220
Ortho-dichlorobenzene, 169, 267
Orthorhombic, 92, 94, 100, 152, 171, 172, 175, 176, 222, 266, 269

Palmitic acid, 94
Para-dibromobenzene, 170–1

Para-dichlorobenzene, 169–72
Paraformaldehyde, 57
Para-dihalogenobenzene, 141, 165, 171, 172
para-nitrophenol, 170, 177–9
p-dodecylbenzensulphonic acid, 138, 139
Perchloric acid, 60, 68
Peritectic, 111, 197, 201
Persistence length, 52, 72, 73, 128, 135, 139, 143, 146, 187, 201, 244
Phase, 3, 8, 10, 12, 112, 196–202, 217, 219
Phase diagrams, 3–15, 52, 228–50, 258, 261, 273–6
Piezoelectric, 150
Pitch, 21, 22, 70, 71, 99, 101, 103, 190, 210, 269, 281
Plasticization, 135, 216, 219
Polarization, 86, 105, 109, 275
Polyaniline, 51, 105, 137, 140
Poly[γ benzyl L-glutamate) polynucleotides, 103
Poly[ethylene oxide], 163, 165, 166
Poly[methyl methacrylate, 164, 180, 271–83]
Poly[p-methyl styrene, 265]
Poly(p-phenylene benzobisthiazole), 160–1
Poly(p-phenylene benzobisoxaazole), 160–1
Polysaccharides, 55, 57, 70, 271
Polystyrene, 49, 74, 113, 114, 116, 119, 133, 143, 153, 164, 180–209, 210–53, 255–63, 265, 272, 283
Poly[vinyl chloride], 105, 107–35
Poly[vinylidene fluoride], 105
PVC, 38, 44, 45, 65, 80, 82, 107–36, 139, 255, 257, 259, 262, 264
Porod regime, 25, 26, 80, 114, 115, 119, 123
p-phenylenediamine, 65, 66, 67
Pseudotetragonal, 92
Purine, 101
Pyrimidine, 101
Pyrrolidin, 141

Radii, 25, 80, 123, 128, 188, 279, 280, 281
Radius, 21, 34, 245, 278, 279, 281, 282
Radius of gyration, 26, 31, 34, 41, 42, 52, 127, 128, 131, 187, 188, 226
Ramie, 66–7
Refractive index, 33, 34, 35
Regime, 18, 25, 26, 29, 80, 114, 115, 117, 119, 120, 124, 129, 200, 245, 256
Regularly folded, 29
Repeat distance, 22, 63, 90, 269
Resorcinol, 143, 145, 172, 173, 174, 175, 178
Retrogradation, 87

Rheological, 50, 85, 86, 109, 117, 122, 128, 147, 193, 204, 242, 279
Rod, 27–30, 52, 72, 117, 119, 188, 189, 199, 207, 208, 283
Rotational diffusion coefficient, 74, 75

Scale-free, 53
Scanning electron microscopy, 248
Scattering, 16–35, 72, 93, 118, 120, 121, 123, 124, 125, 187
Screening length, 72, 256
Selective, 38–9, 276, 277
Semi-crystalline, 8, 64, 90, 196, 198, 201, 216
Sheet-like, 98, 117, 124, 201, 226
Singularly melting, 7, 8, 9, 233
Smectic, 198
Soda hydrates, 61
Sol–gel transition, 81, 82, 245
Solid phase, 8, 9, 13, 113, 155, 156, 186, 234, 235, 236, 237, 239, 248
Solid–solid transformation, 3, 152
Solid solution, 7, 8, 9, 237, 239, 241
Solution-cast, 159, 202, 226, 250, 267, 268,
Sorption, 61, 64, 74, 185, 214, 216, 217, 251, 252, 260
Spherulites, 34, 50, 51, 110, 157, 172, 177, 181, 186, 195, 196, 198, 201, 214, 220, 221, 245
Spin-lattice relaxation , 37, 215, 276, 277
Spinodal, 11, 12, 72, 113, 162, 186, 188, 259, 273
Spin–spin relaxation, 37
Stable, 5, 9, 10, 11, 12, 13, 47, 60, 64, 67, 68, 73, 111, 172, 173, 175, 178, 179, 189, 219, 250
Starch, 87–92, 285
Stereocomplex, 271, 272, 278, 279, 281
Stoichiometry, 8, 77, 166, 185, 229, 240, 268, 271
Sublimation, 156, 239, 252, 254
Sucrose, 89
Sulphuric acid, 138
Supercritical, 252, 253
Supramolecular, 203, 205, 206
Surface free energy, 13, 14, 47
Surfactants, 53, 139, 147
Syndiotactic, 107–12, 210–53, 260, 262, 265–9, 270, 274, 276, 278, 279, 283, 285

Tamman's diagram, 14, 155, 158, 167, 174, 176, 182, 229, 233, 237, 239, 241, 246, 249, 274, 275
Tamman's plot, 5, 157
Ternary, 15, 31, 74, 85, 205, 206, 207
Ternary complex, 66, 74, 76–8, 81, 84, 86, 142, 143
Tetrahydronaphthalene, 248, 249
Tetrahydrofuran, 256, 268
tetrali, 242, 248, 249
Tevnia jerichonana, 97
Thermal analysis, 3, 5
Thermodynamically controlled, 187, 196
Thermograms, 171, 203
Thermoreversible, 34, 109, 180, 183, 205, 217, 232, 242
THF, 268
Thiophenate, 121, 122
Thymol, 94
Toluene, 213, 217, 220, 222, 232, 242, 244, 246, 250, 251, 273, 275, 278, 280, 283
Tosylene sulphonic acid, 142
trans-decalin, 52, 181–96, 201–208, 220, 228–31
Transitional regime, 25, 114, 115, 119
Triclinic, 176, 178
Trigonal, 166
1,3,5-Trimethyl benzene, 225
Tripropylamine, 138

Undercooling, 13, 14, 211
Urea, 165–7, 172

Valonia, 57, 58, 62, 69
van der Waals, 50, 61, 109, 142, 143, 145, 213
Variance, 3, 4, 5, 8, 41, 71, 77, 131, 188, 215, 257, 269, 283
Viscometry, 126, 139
Vitrification, 135, 136

Worm-like, 27, 43, 72, 73, 81, 187, 244, 245

X-ray diffraction, 3, 33, 61, 68, 83, 88, 92, 93, 157, 165, 170, 176, 177, 215, 216, 217, 230, 233, 237, 238, 271